W0079207

Maximum-Entropy and Bayesian Methods in Science and Engineering

Volume 1: Foundations

Fundamental Theories of Physics

An International Book Series on The Fundamental Theories of Physics: Their Clarification, Development and Application

Editor: ALWYN VAN DER MERWE
University of Denver, U.S.A.

Maximum-Entropy and Bayesian Methods in Science and Engineering

Volume 1: Foundations

edited by

Gary J. Erickson

Department of Electrical Engineering,
Seattle University, Seattle, Washington, U.S.A.

and

C. Ray Smith

Advanced Sensors Directorate
Research, Development and Engineering Center,
US Army Missile Command, Redstone Arsenal,
Alabama, U.S.A.

SPRINGER-SCIENCE+BUSINESS MEDIA, B.V.

Library of Congress Cataloging in Publication Data

ISBN 973-94-010-7871-9   ISBN 978-94-009-3049-0 (eBook)
DOI 10.1007/978-94-009-3049-0

In honour of E. T. Jaynes

Maximum-Entropy and Bayesian Methods in Inverse Problems

Edited by

C. Ray Smith and W. T. Grandy, Jr.

Department of Physics and Astronomy,
The University of Wyoming, Laramie, Wyoming, U.S.A.

This volume is the outcome of two workshops entitled "Maximum-Entropy and Bayesian Methods in Applied Statistics" and presents contributions by renowned authorities in many different fields. The purpose of these workshops was to bring together leading scientists whose research involved using the Principle of Maximum Entropy in a wide range of different applications in order to pool the experience gained and to identify common problems in need of solution. The result is stimulating and informative and provides many directions for further progress.

Audience

"Maximum-Entropy and Bayesian Methods in Inverse Problems" will be of great interest to mathematicians, physicists, geophysicists, electrical engineers, economists, and those working in communication and information theory and many other aspects of signal processing.

ISBN 978-94-010-7871-9   14

Of Related Interest

Maximum-Entropy and Bayesian Spectral Analysis and Estimation Problems

Edited by

C. Ray Smith

U.S Army Missile Command, Redstone Arsenal, Alabama, U.S.A.

and

Gary J. Erickson

*Department of Electrical Engineering,
Seattle University, Seattle, Washington, U.S.A.*

This volume contains 20 contributions by leading researchers from different fields who critically examine maximum-entropy and Bayesian methods in science, engineering, signal processing, medical physics, and other disciplines.
This is a sequel to the volume "Maximum-Entropy and Bayesian Methods in Inverse Problems', published by Reidel in 1985.

Audience

This book will be of interest to probability theorists, statisticians, electrical engineers, communication engineers, computer scientists, physicists, mathematicians, biologists, geophysicists, and those working in medical imaging.

ISBN 978-94-010-7871-9

Volume 2: Applications

CONTENTS

CONTENTS

PREFACE

This volume has its origin in the Fifth, Sixth and Seventh Workshops on "Maximum-Entropy and Bayesian Methods in Applied Statistics", held at the University of Wyoming, August 5-8, 1985, and at Seattle University, August 5-8, 1986, and August 4-7, 1987. It was anticipated that the proceedings of these workshops would be combined, so most of the papers were not collected until after the seventh workshop. Because all of the papers in this volume are on foundations, it is believed that the contents of this volume will be of lasting interest to the Bayesian community.

The workshop was organized to bring together researchers from different fields to critically examine maximum-entropy and Bayesian methods in science and engineering as well as other disciplines. Some of the papers were chosen specifically to kindle interest in new areas that may offer new tools or insight to the reader or to stimulate work on pressing problems that appear to be ideally suited to the maximum-entropy or Bayesian method. A few papers presented at the workshops are not included in these proceedings, but a number of additional papers not presented at the workshop are included. In particular, we are delighted to make available Professor E. T. Jaynes' unpublished Stanford University Microwave Laboratory Report No. 421 "How Does the Brain Do Plausible Reasoning?" (dated August 1957). This is a beautiful, detailed tutorial on the Cox-Polya-Jaynes approach to Bayesian probability theory and the maximum-entropy principle. In addition to the paper just described, we have included three more by Professor Jaynes: "The Relation of Bayesian and Maximum-Entropy Methods" (presented at the fifth workshop). "The Evolution of Carnot's Theory" (based upon a talk given at an EMBO Workshop in 1984) and "Detection of Extra-Solar System Planets" (made available for this volume). This last paper should pique the interest of anyone concerned with "superresolution". Incidentally, Professor Jaynes refers in this paper to the Ph.D. thesis of G. L. Bretthorst; much of Dr. Bretthorst's thesis appears as Chapter 5 of this volume.

These workshops and their proceedings could not have been brought to their final form without the support or help of a number of people. Professor Alwyn van der Merwe, the Editor of Fundamental Theories of Physics, and Dr. D. J. Larner of Kluwer, provided encouragement and friendship at critical times. Others who have made our work easier or more rewarding include Professor Paul D. Neudorfer of Seattle University, Mr. Robert M. Braukus, P.E., Director of Telecommunications of Puget Sound Power and Light Co., Dr. J. M. Loomis of the Radar Technology Branch of MICOM's Research, Development, and Engineering Center, and Dr. Rabinder Madan of the Office of Naval Research.

Partial support of the fifth and seventh workshops was provided by

the Office of Naval Research under Grants No. N00014.85-G-0219 and
N00014.87-G-0231.

DEDICATION

In commemoration of the thirtieth anniversary of his first papers
(published in the Physical Review) on maximum-entropy, the 1987 workshop
and these proceedings are proudly dedicated to Edwin T. Jaynes. May his
contributions continue for at least another thirty years.

How Does the Brain Do Plausible Reasoning?

E.T. JAYNES
MICROWAVE LABORATORY AND DEPARTMENT OF PHYSICS
STANFORD UNIVERSITY, STANFORD, CALIFORNIA†

ABSTRACT

We start from the observation that the human brain does plausible reasoning in a fairly definite way. It is shown that there is only a single set of rules for doing this which is consistent and in qualitative correspondence with common sense. These rules are simply the equations of probability theory, and they can be deduced without any reference to frequencies.

We conclude that the method of maximum–entropy inference and the use of Bayes' theorem are statistical techniques fully as valid as any based on the frequency interpretation of probability. Their introduction enables us to broaden the scope of statistical inference so that it includes both communication theory and thermodynamics as special cases.

The program of statistical inference is thus formulated in a new way. We regard the general problem of statistical inference as that of devising new consistent principles by which we can translate "raw" information into numerical values of probabilities, so that the Laplace–Bayes model is enabled to operate on more and more different kinds of information. That there must exist many such principles, as yet undiscovered, is shown by the simple fact that our brains do this every day.

† Present address: Wayman Crow Professor of Physics, Washington University, St. Louis. MO 63130

1

G. J. Erickson and C. R. Smith (eds.),
Maximum-Entropy and Bayesian Methods in Science and Engineering (Vol. 1), 1–24.
© 1988 by Kluwer Academic Publishers.

1. INTRODUCTION

Shannon's theorem 2, in which the formula $H(p_1 \ldots p_n) = -\sum p_i \, log \, p_i$ is deduced,[1] is a very remarkable argument. He shows that a *qualitative* requirement, plus the condition that the information measure be consistent, already determines a definite mathematical function. Actually, this is not quite true, because he chooses the condition of consistency (the composition law) in a particular way so as to make H additive. Any continuous differentiable function $f(H)$ for which $f'(H) > 0$ would also satisfy the qualitative requirements and a different, but equally consistent, composition law. Thus a qualitative requirement plus the condition of consistency determines the function H only to within an arbitrary monotonic function. The content of communication theory would, however, be exactly the same regardless of which monotonic function was chosen. Shannon's H thus involves also a convention which leads to simple rules of combination.

This interesting situation led the writer to ask whether it might be possible to deduce the entire theory of probability from a qualitative requirement and the condition that it be consistent. It turns out that this is indeed possible. In terms of the resulting theory we are enabled to see that communication theory, thermodynamics, and current practice in statistical inference, are all special cases of a single principle of reasoning.

In developing this theory we find ourselves in the fortunate position of having all the hard work already done for us. The methodology has been supplied by Shannon, the necessary mathematics has been worked out by Abel[2] and Cox[3], and the qualitative principle was given by Laplace[4]. All we have to do is fit them together.

Laplace's qualitative principle is his famous remark[4] that "Probability theory is nothing but common sense reduced to calculation." The main object of this paper is to show that this is not just a play on words, but a literal statement of fact.

One of the most familiar facts of our experience is this: that there *is* such a thing as common sense, which enables us to do plausible reasoning in a fairly consistent way[5,6]. People who have the same background of experience and the same amount of information about a proposition come to pretty much the same conclusions as to its plausibility. No jury has ever reached a verdict on the basis of pure deductive reasoning. Therefore the human brain must contain some fairly definite mechanism for plausible reasoning, undoubtedly much more complex than that required for deductive reasoning. But in order for this to be possible, *there must exist consistent rules for carrying out plausible reasoning, in terms of operations so definite that they can be programmed on the computing machine which is the human brain.* This is the "experimental fact" on which our theory is based. We know that it must be true, because we all use it every day. Our direct knowledge about this process is, however, only qualitative in much the same way as is our direct experience of temperature. For that reason it is necessary to use the methodology of Shannon.

2. LAPLACE'S MODEL OF COMMON SENSE

We now turn to development of our first mathematical model. We attempt to associate mental states with real numbers which are to be manipulated according to definite rules. Now it is clear that our attitude toward any given proposition may have a very large number of different "coordinates". We form simultaneous judgments as to whether it is probable, whether it is desirable, whether it is interesting, whether it is amusing, whether it is important, whether it is beautiful, whether it is morally right, etc. If we assume that each of these judgments might be represented by a number, a fully adequate description of a state of mind would then be represented by a vector in a space of a very large, and perhaps indefinitely large, number of dimensions.

Not all propositions require this. For example, the proposition, "The refractive index of water is 1.3", generates no emotions; consequently the state of mind which it produces has very few coordinates. On the other hand, the proposition, "Your wife just wrecked your new car," generates a state of mind with an extremely large number of coordinates. A moment's introspection will show that, quite generally, the situations of everyday life are those involving the greatest number of coordinates. It is just for this reason that the most familiar examples of mental activity are the most difficult ones to reproduce by a model. We might speculate that this is the reason why natural science and mathematics are the most successful of human activities; they deal with propositions which produce the simplest of all mental states. Such states would be the ones least perturbed by a given amount of imperfection in the human brain.

The simplest possible model is one–dimensional. We allow ourselves only a single number to represent a state of mind, and wish to discover how much of mental activity we can reproduce subject to that limitation. For the time being we call these numbers *plausibilities*, reserving the term "probability" for a particular quantity to be introduced later.

The way in which states of mind are to be reduced to numbers is at this stage very indefinite. For the time being we say only that greater plausibility must always correspond to a greater number, and we assume a continuity property which can be stated only imprecisely: infinitesimally greater plausibility should correspond only to an infinitesimally greater number.

We denote various propositions by letters $A, B, C, \ldots$. By the symbolic product AB we mean the proposition "Both A and B are true." The expression $(A + B)$ is to be read, "At least one of the propositions A, B is true." The plausibility of any proposition A will in general depend on whether we accept sme other proposition B as true. We indicate this by the symbol

$$(A|B) = \text{conditional plausibility of } A, \text{ given } B.$$

Thus, for example,

$$(AB|C) = \text{plausibility of } (A \text{ and } B), \text{ given } C.$$
$$(A + B|CD) = \text{plausibility that at least one of the propositions } A, B \text{ is true,}$$
$$\text{given that both } C \text{ and } D \text{ are true,}$$
$$(A|C) > (B|C) \text{ means that, on data } C, A \text{ is more plausible than } B.$$

In order to find rules for manipulation of these symbols, we are guided by two requirements:

1) *The rules must correspond qualitatively to common sense.* (2-1)
2) *The rules must be consistent.* This is used in two ways:

$$\left\{ \begin{array}{l} \textit{If a result can be arrived at in more than one way,} \\ \textit{we must obtain the same result for every possible} \\ \textit{sequence of operations on our symbols.} \end{array} \right\} \quad (2\text{-}2)$$

$$\left\{ \begin{array}{l} \textit{The rules must include deductive logic as a special case.} \\ \textit{In the limit where propositions become certain} \\ \textit{or impossible in any way, every equation must reduce} \\ \textit{to a valid example of deductive reasoning.} \end{array} \right\} \quad (2\text{-}3)$$

By a *successful model* we mean any set of rules satisfying these conditions. If we find that we have any freedom of choice left after imposing them, we can exercise that freedom to adopt conventions so as to make the rules as simple as possible. If we find that these requirements are so restrictive that there is in effect only one possible model satisfying them, are we entitled to claim that we have discovered the mechanism by which the brain does "one–dimensional" plausible reasoning? Except for the proviso that the human mind is imperfect, it seems that to deny that claim would be to assert that the human mind operates in a deliberately inconsistent way.

We now seek a consistent rule for obtaining the plausibility of AB from the plausibilities of A and B separately. In particular, let us find the plausibility $(AB|C)$. Now in order for AB to be true on data C, it is first of all necessary that B be true; thus the plausibility $(B|C)$ must be involved. If B is true, it is further necessary that A be true; thus $(A|BC)$ is needed. If, however, B is false, then AB is false independently of any statement about A. Therefore $(A|C)$ is not needed; it tells us nothing about AB that we did not already have in $(A|BC)$. Similarly, $(A|B)$ and $(B|A)$ are not needed; whatever plausibility A or B might have in the absence of data C, could not be relevant to judgments of a case where we know from the start that C *is* true.

We could, of course, interchange A and B in the above paragraph, so that knowledge of $(A|C)$ and $(B|AC)$ would also suffice. The fact that we must obtain the same value for $(AB|C)$ no matter which procedure we choose is one of our conditions of consistency.

Thus, we seek some function $F(x, y)$ such that

$$(AB|C) = F[(A|BC), (B|C)]. \tag{2-4}$$

It is easy to exhibit special cases which show that no relation of the form $(AB|C) = F[(A|C), (B|C)]$, or of the form $(AB|C) = F[(A|C), (A|B), (B|C)]$; could satisfy conditions (2-1), (2-2), (2-3).

Condition (2-1) imposes the following limitations on the function $F(x, y)$. An increase in either of the plausibilities $(A|BC)$ or $(B|C)$ must never produce a decrease in $(AB|C)$. Furthermore, $F(x, y)$ must be a continuous function. otherwise we could produce a situation where an arbitrarily small increase in $(A|BC)$ or $(B|C)$ still results in the same large increase in $(AB|C)$. Finally, an increase in either of the quantities $(A|BC)$ or $(B|C)$ must always produce *some* increase in $(AB|C)$, unless the other one happened to represent impossibility. Thus condition (2-1) requires that

$$\left\{ \begin{array}{l} F(x, y) \text{ must be a continuous function, with } \left(\dfrac{\partial F}{\partial x}\right) \geq 0 \\[2mm] and \left(\dfrac{\partial F}{\partial y}\right) \geq 0. \text{ The equality sign can apply only when} \\[2mm] (AB|C) \text{ represents impossibility.} \end{array} \right\} \tag{2-5}$$

The condition of consistency (2-2) places further limitations on the possible form of the function $F(x, y)$. For we can calculate $(ABD|C)$ from (2-4) in two different ways. If we first group AB together as a single proposition, two applications of (2-4) give us

$$(ABD|C) = F[(AB|DC), (D|C)] = F\{F[(A|BDC), (B|DC)], (D\ C)\}.$$

But if we first regard BD as a single proposition, (2-4) leads to

$$(ABD|C) = F[(A|BDC), (BD|C)] = F\{(A|BDC), F[(B|DC), (D|C)]\}$$

Thus, if (2-4) is to be consistent, $F(x, y)$ must satisfy the functional equation

$$F[F(x, y), z] = F[x, F(y, z)]. \tag{2-6}$$

Conversely, it is easily shown by induction that if (2-6) is satisfied, then (2-4) is automatically consistent for all possible ways of finding any number of joint plausibilities, such as $(ABCDEF|G)$. This functional equation turns out to be one which was studied by N.H. Abel.[2] Its solution, given also by Cox,[3] is

$$p[F(x, y)] = p(x)\ p(y), \tag{2-7}$$

where $p(x)$ is an arbitrary function. By (2-5) it must be a continuous monotonic function. Therefore our rule necessarily has the form

$$p[(AB|C)] = p[(A|BC)]\ p[(B|C)],$$

which we will also write, for brevity, as[7]

$$p(AB|C) = p(A|BC)\ p(B|C). \qquad (2\text{-}8)$$

The condition (2-3) above places further restrictions on the function $p(x)$. Assume first that A is certain, given C. Then $(AB|C) = (B|C)$, and $(A|BC) = (A|C) = (A|A)$. Equation (2-8) then reduces to

$$p(B|C) = p(A|A)\ p(B|C)$$

and this must hold for all $(B|C)$. Therefore,

$$Certainty\ must\ be\ represented\ by\ p = 1. \qquad (2\text{-}9)$$

If for some particular degree of plausibility $(A|BC)$, the function $p(A|BC)$ becomes zero or infinite, then (2-8) says that $(B|C)$ becomes irrelevant to $(AB|C)$. This contradicts common sense unless $(A|BC)$ corresponds to impossibility. Therefore

$$p\ cannot\ become\ zero\ or\ infinite$$

$$for\ any\ degree\ of\ plausibility\ other\ than\ impossibility. \qquad (2\text{-}10)$$

Now assume that A is impossible, given C. Then $(AB|C) = (A|BC) = (A|C)$, and (2-3) reduces to

$$p(A|C) = p(A|C)\ p(B|C)$$

which must hold for all $(B|C)$. There are three choices for $p(A|C)$ which satisfy this; $p(A|C) = 0$, $or + \infty$, $or - \infty$. But by (2-9) and (2-10) the choice $-\infty$ must be excluded, for any continuous monotonic function which has the values $+1$ and $-\infty$ at two given points necessarily passes through zero at some point between them. Therefore

$$Impossibility\ must\ be\ represented\ by\ p = 0,\ or\ p = \infty. \qquad (2\text{-}11)$$

Evidently the plausibility that A is false is determined by the plausibility that A is true in some reciprocal fashion. We denote the denial of any proposition by the corresponding small letter; i.e.

$$a = \text{``A is false''}$$

$$b = \text{``B is false''}$$

We could equally well say that $A = $ "a is false," etc. Clearly, $(A + a)$ is always true, and Aa is always false.

Since we already have some rules for manipulation of the quantities $p(A|B)$, it will be convenient to work with $p(A|B)$ rather than $(A|B)$. For brevity in the following derivation we use the notation

$$[A|B] = p(A|B).$$

Now there must be some functional relationship of the form

$$[a|B] = S[A|B] \tag{2-12}$$

where by (2-1), $S(x)$ must be a monotonic, decreasing function. Since the propositions a and A are reciprocally related, we must have also

$$[A|B] = S[a|B]. \tag{2-13}$$

Therefore the function $S(x)$ must satisfy the functional equation

$$S[S(x)] = x. \tag{2-14}$$

To find another condition which $S(x)$ must satisfy, apply (2-8) and (2-12) alternately as follows:

$$[AB|C] = [A|BC][B|C] = S[a|BC][B|C] = [B|C] S \left\{ \frac{[aB|C]}{[B|C]} \right\}. \tag{2-15}$$

The original expression $[AB|C]$ is symmetric in A and B. So also, therefore, is the final expression; thus

$$[AB|C] = [A|C] S \left\{ \frac{[bA|C]}{[A|C]} \right\}. \tag{2-16}$$

The expressions (2-15) and (2-16) must be equal whatever A, B, C, may be. In particular, they must be equal when $b = AD$. But in this case,

$$[bA|C] = [b|C] = S[B|C],$$
$$[aB|C] = [a|C] = S[A|C].$$

Substituting these into (2-15) and (2=16), we see that $S(x)$ must also satisfy the functional equation

$$x S \left[\frac{S(y)}{x} \right] = y S \left[\frac{S(x)}{y} \right]. \tag{2-17}$$

R. T. Cox[3] has shown that the only continuous differentiable function satisfying both (2-14) and (2-17) is

$$S(x) = (1 - x^m)^{1/m} \tag{2-18}$$

where m is any non–zero constant. Therefore the reciprocal relation between $[a|B]$ and $[A|B]$ necessarily has the form

$$[A|B]^m + [a|B]^m = 1. \tag{2-19}$$

Suppose we represent impossibility by $p = 0$. Then, from (2-19), m must be chosen positive. However, use of different values for m does not represent any freedom of choice that we did not already have in the arbitrariness of the function

$p(x)$. The only condition on $p(x)$ is that it be a continuous monotonic function which increases from 0 to 1 as we go from impossibility to certainty. If the function $p_1(x)$ satisfies this condition, so also does the function

$$p_2(x) = [p_1(x)]^m .$$

Therefore if we write (2-19) in the form

$$p(A|B) + p(a|B) = 1 \qquad (2\text{-}20)$$

in which $p(x)$ is understood to be an arbitrary monotonic function, Eq. (2-20) is just as general as is (2-19).

Suppose, on the other hand, that we represent impossibility by $p = \infty$. Then we must choose m negative. Once again, to say that we can use different values of m does not say anything that is not already said in the statement that $p(x)$ is an arbitrary monotonic function which increases from 1 to ∞ as we go from certainty to impossibility. The equation

$$\frac{1}{p(A|B)} + \frac{1}{p(a|B)} = 1 \qquad (2\text{-}21)$$

is also just as general as (2-19).

An entire consistent theory of plausible reasoning can be based on (2-21) as well as on (2-20). They are not, however, different theories, for if $p_1(x)$ satisfies (2-21), the equally good function

$$p_2(x) = \frac{1}{p_1(x)}$$

satisfies (2-20), and says exactly the same thing. If we agree to use only functions of type (2-20), we are not excluding any possibility of representation, but only removing a certain redundancy in the mathematics.

From (2-20) we can derive the last of our fundamental equations. We seek an expression for the plausibility of $(A + B)$, the statement that at least one of the propositions A, B is true. Noting that if $D = A + B$, then $d = ab$, we can apply (2-20) and (2-8) in alternation to get

$$p(A + B|C) = 1 - p(ab|C) = 1 - p(a|bC)\, p(b|C)$$
$$= 1 - [1 - p(A|bC)]\, p(b|C) = p(B|C) + p(Ab|C)$$
$$= p(B|C) + p(A|C)\, [1 - p(B|AC)]$$

or,

$$p(A + B|C) = p(A|C) + p(B|C) - p(AB|C). \qquad (2\text{-}22)$$

Equations (2-8) and (2-22) are the fundamental equations of the theory of probability. From them all other relations follow.

We have found that the most general consistent rules for plausible reasoning can be expressed in the form of the product and sum rules (2-8) and (2-22), in which $p(x)$ is an arbitrary continuous monotonic function ranging from 0 to 1. It might appear that different choices of the function $p(x)$ will lead to models with different content, so that we have found in effect an infinite number of different possible consistent rules for plausible reasoning. This, however, is not the case, for regardless of which function $p(x)$ we choose, when we start to use the theory we find that it is always p, not x, that has a definitely ascertainable numerical value. To demonstrate this in the simplest case, consider n propositions $A_1, A_2, \ldots, A_n$ which are mutually exclusive; i.e., $p(A_i A_j | C) = p(A_i | C) \delta_{ij}$. Then repeated application of (2-22) gives the usual sum rule

$$p(A_1 + \ldots + A_n | C) = \sum_{k=1}^{n} p(A_k | C). \qquad (2\text{-}23)$$

If now the A_k are all equally likely on data C (this means only that data C gives us no reason to expect that one of them is more valid than the others), and one of them must be true on data C, the $p(A_k | C)$ are all equal and their sum is unity. Therefore we necessarily have

$$p(A_k | C) = \frac{1}{n}. \qquad (2\text{-}24)$$

This is Laplace's "Principle of Insufficient Reason." No matter what function $p(x)$ we choose, there is no escape from the result (2-24). Therefore, rather than saying that p is an arbitrary monotonic function of $(A|C)$, it is more to the point to say that $(A|C)$ is an arbitrary monotonic function of p, in the interval $0 \leq p \leq 1$. It is the connection of the numbers $(A|C)$ with intuitive states of mind that never gets tied down in any definite way. In changing the function $p(x)$, or better $x(p)$, we are not changing our model, but just displaying the fact that our intuitive sensations provide us only with the relation "greater than," not any definite numbers. Throughout these changes, the numerical values of and relations between, the quantities p remain unchanged.

All this is in very close analogy with the concept of temperature, which also originates only as a qualitative sensation. Once it has been discovered that, out of all the monotonic functions represented by the readings of different kinds of thermometers, one particular definition of temperature (the Kelvin definition) renders the equations of thermodynamics especially simple, the obvious thing to do is to recalibrate the scales of the various thermometers so that they agree with the Kelvin temperature. The Kelvin temperature is no more "correct" than any other; it is simply more convenient.

Similarly, the obvious thing for us to do at this point is to adopt the *convention* $p(x) = x$, so that the distinction between a plausibility and the quantity p (which we henceforth call the *probability*) disappears. This means only that we have found a way of calibrating our "plausibility–meters" so that the consistent rules of reasoning take on a simple form. The content of the theory would, however, be exactly the same no matter what function p(x) was chosen. Thus, *there is only one consistent model of common sense.*

From now on, we write our fundamental rules of calculation in the form

$$(AB|C) = (A|BC)(B|C) = (B|AC)(A|C) \tag{2-25}$$

$$(A + B|C) = (A|C) + (B|C) - (AB|C). \tag{2-26}$$

Laplace's model of common sense consists of these rules, with numerical values determined by the principle of insufficient reason.

Out of all the propositions which we encounter in this theory, there is one which must be discussed separately. The proposition X stands for all of our past experience. *There can be no such thing as an "absolute" or "correct" probability; all probabilities are conditional on X at least, and X is not only different for different people, but it is continually changing for any one person.* If X happens to be irrelevant to a certain question, then this observation is unnecessary but harmless. We often suppress X for brevity, with the understanding that even when it does not appear explicitly, it is still "built into" all bracket expressions: $(A|B) = (A|BX)$. Any probabilities conditional on X alone are called *a–priori* probabilities. In an a–priori probability we will always insert X explicitly: $(A|X)$.

It is of the greatest importance to avoid any impression that X is some sort of hidden major premise representing a universally valid proposition about nature; it is simply whatever initial information we have at our disposal for attacking the problem. Alternatively, we can equally well regard X as a set of hypotheses whose consequences we wish to investigate, so that all equations may be read, "If X were true, then $\cdots$" It makes no difference in the formal theory.

3. DISCUSSION

It is well known that criticism of the theory of Laplace, and pointing out of its obvious absurdity, has been a favorite indoor sport of writers on probability and statistics for decades. In view of the fact that we have just shown it to be the only way of doing plausible reasoning which is consistent and in agreement with common sense, it becomes necessary to consider the objections to Laplace's theory and if possible to answer them.

Broadly speaking, there are three points which have been raised in the literature. The first is that any quantity which is only subjective, i.e. which represents a "degree of reasonable belief," in Jeffreys' terminology,[8] cannot be measured numerically, and thus cannot be the object of a mathematical theory. Secondly, there is a widespread impression that even if this could be accomplished, a quantity which is different for different observers is not "real," and cannot be relevant to application.[9] Thirdly, there is a long history of pathology associated with this view; it is tempting and easy to misuse it.

The latter is of course not a valid objection to any theory, and we need only answer the first two. The arguments of Sec. 2 almost answer the first, but there remains the question of finding numerical values of probabilities in cases where there is no apparent way of reducing the situation to one of "equally possible" cases. We must hasten to point out that the notion of "equally possible" has, at this stage, nothing whatsoever to do with frequencies. The notion of frequency has not yet appeared in the theory. Now the question of how one finds *numerical values* of probabilities is evidently an entirely different problem than that of finding a consistent *definition* of probability, and consistent rules for calculation. In physics, after the Kelvin temperature is defined, there remains the difficult problem of devising experiments to establish its numerical value. Similarly, after our model has been set up, the problem of reducing "raw" information to a statement of probability numerical values remains.

Most of the objections to Laplace's theory which one finds in the literature[11] consist of applying it to some simple problem, and pointing out that the result flatly contradicts common sense. However, study of these examples will show that *in every case where the theory leads to results which contradict common sense, the person applying the theory has additional information of some sort, relevant to the question being asked, but not actually incorporated into the equations.* Then his common sense utilizes this information unconsciously and of necessity comes to a different conclusion than that provided by the theory.

Here is one of Polya's examples.[11] A boy is ten years old today. According to Laplace's law of succession, he has the probability $\frac{11}{12}$ of living one more year. His grandfather is 70. According to the same law, he has the probability $\frac{71}{72}$ of living one more year. Obviously, the result contradicts common sense. Laplace's law of succession, however, applies only to the case where we have absolutely no prior information about the problem.[13] In this example it is even more obvious that we *do have a great deal* of additional information relevant to this question, which our common sense used but we did not allow Laplace's theory to use.

Laplace's theory gives the result of consistent plausible reasoning on the basis of the information *which was put into it.* The additional information is often of a vague nature, but nevertheless highly relevant, and it is just the difficulty of translating it into numerical values which causes all the trouble. This shows that the human brain must have extremely powerful means, the nature of which we have not yet imagined, for converting raw information into probabilities.

We can see from this why Laplace's theory was incomplete and why it will always remain incomplete. It is simply that there is no end to the variety of kinds of partial information with which we might be confronted, and therefore no end to the problem of finding consistent ways of translating that information into probability statements. Here again there is a close analogy with physics. Whenever research involving temperature extends into some new field, science is dependent on the ingenuity of experimenters in devising new procedures which will give the Kelvin temperature in terms of observed quantities. Physicists must continually invent new kinds of thermometers; and users of probability theory must continually invent new kinds of "plausimeters." Laplace's theory is incomplete in the same sense, and for the same reason, that physics is incomplete; but Laplace's basic model occupies the same fundamental position in statistics as do the laws of thermodynamics in physics.

The principle of insufficient reason is only one of many techniques which one needs in current applications of probability theory, and it needs to be generalized before it is applicable to a very wide range of problems.[14] In the following sections we will show two principles available for doing this. The first has been made possible by information theory, and the second comes from a relation between probabilities and frequencies.

Consider now the second objection, that a probability which is only subjective and different for different people cannot be relevant to applications. It seems to the writer that this is the exact opposite of the truth; *it is only a subjective probability which could possibly be relevant to applications.* What is the purpose of any application of probability theory? Simply to help us in forming reasonable judgments in situations where we do not have complete information. Whether some other person may have complete information is quite irrelevant to *our* problem. We must do the best we can with the information we have, and it is only when this is incomplete that we have any need for probability theory. The only "objective" probabilities are those which describe frequencies observed in experiments already completed. Before they can serve any purpose in applications they must be converted into subjective judgments about other situations where we do *not* know the answer.

If a communication engineer says, "The statistical properties of the message and noise are known," he means only that he has some knowledge about the *past* behavior of some particular set of messages and some particular sample of noise. When he infers that some of these properties will hold also in the future and designs a communication system accordingly, he is making a subjective judgment of exactly the type accounted for by Laplace's theory, and *the sole purpose of the statistical analysis of past events was to obtain that subjective judgment.*

Two engineers who have different amounts of statistical information about messages will assign different n-gram probabilities and design different coding systems. Each represents rational design on the basis of the available information, and it is quite meaningless to ask which is "correct." Of course, the man who has more advance knowledge about what a system is to do will generally be able to utilize that knowledge to produce a more efficient design, because he does not have to provide for so many possibilities. This is in no way paradoxical, but just simple common sense.

Similarly, if a medical researcher says, "This new medicine is effective in 85 per cent of the cases," he means only that this is the frequency observed in *past* experiments. If he infers that it will hold approximately in the future, he is making a subjective judgment which might be (and often is) entirely erroneous. Nevertheless, it was the most reasonable judgment he could have made on the basis of the information available. The judgment, and also its level of significance, are accounted for by Laplace's theory. Its conclusions are, for all practical purposes, identical with those provided by the method of confidence intervals,[15] and it is our contention that the validity of the latter method depends on this agreement.

4. THE PRINCIPLE OF INSUFFICIENT REASON

Two conditions are necessary before we can assign probabilities by means of the principle of insufficient reason:

$$\left\{ \begin{array}{l} We\ must\ be\ able\ to\ analyze\ the\ situation\ into\ an \\ enumeration\ of\ the\ different\ possibilities\ which \\ we\ recognize\ as\ mutually\ exclusive\ and\ exhaustive. \end{array} \right\} \quad (4\text{-}1)$$

$$\left\{ \begin{array}{l} Having\ done\ this,\ we\ must\ then\ find\ that\ the \\ available\ information\ gives\ us\ no\ reason\ to\ prefer \\ any\ possibility\ to\ any\ other. \end{array} \right\} \quad (4\text{-}2)$$

In practice these conditions are hardly ever met unless there is some evident element of symmetry in the problem, as is usually the case in games of chance. Note, however, that there are two different ways in which condition (4-2) may be satisfied. It may be the consequence of complete ignorance, or it may be the consequence of positive knowledge.

Suppose a person, known to be very dishonest, is going to toss a die. Observer A is allowed to examine the die, and he has at his disposal all the facilities of the National Bureau of Standards. He performs thousands of experiments with scales, calipers, microscopes, magnetometers, x–rays, neutron beams, etc., and finally is convinced that the die *is* perfectly symmetrical. Observer B is not told this; he knows only that a die is being tossed by a shady character. He suspects that it is biased, but has no idea in which direction. Condition (4-2) is satisfied for both, and they will both assign probability $\frac{1}{6}$ to each face. The same probability assignment may describe either knowledge or ignorance. This seems paradoxical: why doesn't A's extra knowledge make any difference?

Well, it *does* make a difference, and a very important one, but the difference requires time to "develop." Suppose that the first toss gives a "3." To observer B this constitutes evidence that the die is biased to favor 3, and so on the second throw B will assign different probabilities which take this into account. Observer A, however, will continue to assign probability $\frac{1}{6}$ to each face, because to him the evidence of symmetry carries overwhelmingly greater weight than does the evidence of one throw.

It is now fairly clear what will happen. To observer B, every throw of the die represents new evidence about its bias, which causes him to change his probability assignments for the next throw. Under certain circumstances, his assignments are given by a generalization of Laplace's law of succession. To observer A, the evidence of symmetry continues to carry greater weight than does the evidence of the random experiment, and he persists in assigning probability $\frac{1}{6}$. Each observer has done consistent plausible reasoning on the basis of the information available to him, and Laplace's theory accounts for the behavior of each (Sec. 6).

This difference in behavior is not, however, accounted for by any theory based on a frequency definition of probability, because when you define a probability simply as a frequency you deprive yourself of any way of saying that you *have* evidence unless it is in the form of an observed frequency. Everything which the National Bureau of Standards can tell us must be ignored, because it has no frequency interpretation.

5. THE ENTROPY PRINCIPLE

A biased die, colored black with white spots, has been tossed many times onto a black table, and we have recorded the experiment with a camera, obtaining a multiple exposure of uniform density. From the blackening of the film we cannot determine the relative frequencies of the different faces, but only the *average* number of spots which were on top. This average is not 3.5, as we might expect from an honest die, but 4.5. On the basis of this information, what are the probabilities for the different faces?

Automobiles of make i have weight W_i and length L_i. We observe a cluster of 1000 cars packed bumper to bumper, occupying a total length of 3 miles. As these cars pass an intersection they go over a machine which weighs each one and totals the result, not retaining the record of the individual weights. Therefore we have only the total length and total weight of the 1000 cars. What can we infer about the number of cars of each make in the cluster?

During an earthquake, 100 windows were broken into 1000 pieces. What is the probability for a window to be broken into exactly m pieces?

These are examples of problems where condition (4–1) is satisfied but not condition (4–2). They can be formulated in a general way as follows. The quantity x can assume the discrete values $x_1 \ldots x_n$. There are k functions $f_1(x), \ldots, f_k(x)$ for which we know the average values

$$f_r = \sum_{i=1}^{n} p_i f_r(x_i), \qquad 1 \leq r \leq k. \tag{5-1}$$

The problem is to find the p_i. If $k < (n-1)$, there are not enough conditions to determine the p_i in the sense of a mathematical solution of (5–1) and $\sum p_i = 1$. We cannot use the principle of insufficient reason because we have too much information; there *are* reasons for preferring some possibilities to others. There are many probability assignments which would all agree with the available information. Which is the most reasonable one to adopt?

Consider the third example above, and restate it as: the average window is broken into 10 pieces. If we were to conclude that *each* window is broken into 10 pieces, this would be in complete agreement with all the available information. However, our common sense tells us that it would not be a *reasonable* probability assignment; we would be assuming far more than was given in the statement of the problem. It is more reasonable to assign probability $p_m = \frac{1}{5}$ for a window to be broken into m pieces, where $m = 8, 9, 10, 11, 12$. But this still assumes more than was warranted by the given information. It says, for example, that it is impossible for a window to be broken into 13 pieces. Evidently we regard a broad distribution as more reasonable than a sharply peaked one, and there is no value of m for which we would be justified in assigning $p_m = 0$.

To make a long story short, we want the probability assignment which assumes *nothing* beyond what was given in the statement of the problem. Shannon's theorem 2 tells us that the consistent measure of the "amount of uncertainty" in a probability distribution is its entropy, and therefore we must choose the distribution which has maximum entropy subject to the constraints (5-1). Any other distribution would represent an arbitrary assumption of some kind of information which was not given to us. The maximum–entropy distribution is "maximally noncommittal" with respect to missing information.

The solution follows immediately from the method of Lagrangian multipliers, by arguments which are very well known in a different context. The results are expressed compactly if we define the *partition function*:

$$Z\left(\lambda_1 \ldots \lambda_k\right) = \sum_{i=1}^{n} exp\left[-\lambda_1 f_1\left(x_i\right) - \ldots - \lambda_k f_k\left(x_i\right)\right]. \tag{5-2}$$

Then the maximum–entropy distribution is

$$p_i = exp\left[-\lambda_0 - \lambda_1 f_1\left(x_i\right) - \ldots - \lambda_k f_k\left(x_i\right)\right] \tag{5-3}$$

with the λ_r determined by

$$\lambda_0 = \log Z \tag{5-4}$$

and

$$\langle f_r\left(x\right)\rangle = -\frac{\partial}{\partial \lambda_r}\log Z, \qquad 1 \leq r \leq k. \tag{5-5}$$

At first glance it seems idle and trivial that we should have to do all this in order to learn how to say nothing. The important point, however, is that we have here found a consistent way of saying nothing in a new language; the language of probability theory. The triviality fades away entirely when we notice that the problem of inferring the macroscopic properties of matter from the laws of atomic physics is of exactly the type we are considering. *All of thermodynamics, including the prediction of every experimentally reproducible feature of irreversible processes, is contained in the above solution.*[16,17,18]

This is so easy to demonstrate that we will sketch the argument here. In any macroscopic experiment the exact microscopic state of a system is never under control or observation; there will be perhaps

$$10^{10^{20}} = \left(10^{10^{10}}\right)^{10^{10}}$$

different quantum states compatible with a given set of experimental conditions. Although the microscopic state is changing rapidly, the time required for any reasonably complete "sampling" of so many states is still rather long; perhaps $10^{10^{10}}$ years. When we repeat the experiment we will surely not repeat the microscopic state. Therefore, any property which is experimentally reproducible must be characteristic of *each* of the great majority of the class C_e of microscopic states allowed by the experimental conditions. This is not necessarily the same as the *subjective* class C_s consisting of all reasonably probable states in the maximum–entropy distribution.[19] Clearly, the only properties which we will be able to predict definitely from the maximum–entropy distribution will be those characteristic of the great majority of the states in class C_s.

Now if it is found that the class P_s of properties predictable by maximum–entropy inference is identical with the class P_e of experimentally reproducible properties, the theory is entirely successful. This would by no means imply that the class C_s is identical with the class C_e. If, however, the class P_s is found to differ in any way from the class P_e, we would be forced to conclude that $C_s \neq C_e$. But this could be true only if there exist new physical states, or new constraints on the possible physical states, which we did not take into account in our initial numeration.

Therefore, strictly speaking, we should not assert that maximum–entropy inference *must* lead to correct predictions. But we can assert something even more important: *if the class of predictable properties is found to differ in any way from the class of experimentally reproducible properties, that fact would in itself demonstrate the existence of new laws of physics.* Assuming that this occurs and the new laws are eventually worked out, then maximum–entropy inference based on the new laws will again have this property.

From this we see that maximum–entropy inference is precisely the appropriate tool for reasoning from the microscopic to the macroscopic. Its characteristic property is that it does not allow us to form any conclusions which are not indicated by the available evidence. Any other distribution *would* permit one to draw conclusions not warranted by the evidence.

Historically, maximum–entropy inference was discovered, in its mathematical aspects, by Boltman about 1870, and greatly advanced by Gibbs around 1900. The result is what the physicist calls statistical mechanics. However, the *interpretation* of the mathematical rules has always been a subject of great confusion, because of the illusion that probabilities must be given a frequency interpretation. This made it appear that the rules could be justified only by demonstrating a certain physical property called ergodicity, or in modern terms, metric transitivity. All attempts to demonstrate this have, however, failed. Until the discovery of Shannon's theorem 2, it was not possible to understand just what we were doing in statistical mechanics, or to have any confidence in it for the prediction of irreversible processes. However, we can now see that statistical mechanics is a much more powerful tool than physicists had realized.

6. PROBABILITY AND FREQUENCY

Although the word "frequency" has appeared a few times above, we have not so far made any use of it in developing the basic theory or in demonstrating its application to thermodynamics. This has been done deliberately in order to emphasize the fact that the notions of probability and frequency are entirely distinct. Many of the most important applications of probability theory can be justified and carried to completion without ever introducing the notion of frequency. However, in cases where a random experiment provides most or all of the available information, there should exist some relationship between the observed frequency of the event and the probability which we assign to it. Similarly, if an event can be regarded as a possible result of a random experiment, there may in some cases be a relation between the probability which we assign to it, and the relative frequency with which we expect it to occur. Such relations must, of course, be deduced from the theory and not postulated.

To demonstrate the latter relation, we introduce the propositions

$$A_p = \text{``The probability of } A \text{ in each case is } p\text{.''} \tag{6-1}$$

$$N_n = \text{``In } N \text{ trials, } A \text{ was (or will be) true } n \text{ times.''} \tag{6-2}$$

The probability $(N_n|A_p)$, obtained immediately from the sum and product rules (2-25), (2-26), is the binomial distribution

$$(N_n|A_p) = \binom{N}{n} p^n (1-p)^{N-n} \tag{6-3}$$

As a function of n, this attains a maximum value when n is within one unit of Np, so that the most probable frequency is substantially equal to the probability.

Note that the phrase "in each case," in (6-1) is essential. To demonstrate this, we look more closely at the derivation of (6-3) from our basic rules. Define the proposition

$$B_n = \text{``}A \text{ is true in the } n\text{'th trial.''} \tag{6-4}$$

Now according to (2-25) we have

$$(B_2 B_1 | A_p) = (B_2 | B_1 A_p)(B_1 | A_p)$$

which reduces to

$$(B_2 | A_p)(B_1 | A_p) = p^2$$

only if $(B_2|B_1 A_p) = (B_2|A_p)$; i.e. the probability of A at the second trial which is involved in (6-3) is that based on A_p *and* knowledge of the result of the first trial. It is equal to p, as assumed in (6-3), only if knowing the result of the first trial would have given us no reason to change the assignment. This in spite of the fact that in (6-3) we are predicting a frequency entirely on the basis of A_p, since only A_p appears to the right of the vertical stroke. Even though we are not given the results of any trial, the expected frequency still depends on whether such knowledge would have been relevant.

This again corresponds to common sense. To take the most extreme case, suppose we are tossing a coin and A stands for "heads." Let it be a very dishonest coin, and define the proposition

$$C_p = \text{``The coin has either two heads or two tails,}$$

$$\text{and the probability of the former is } p\text{.''} \tag{6-5}$$

Now on the basis of this evidence alone, it is still true that the probability of "heads" in each particular throw is p. But no one expects the relative *frequency* of heads to be p: We now have $(B_2|B_1 C_p) = 1$, so that

$$(B_2 B_1|C_p) = (B_2|B_1 C_p)(B_1|C_p) = p$$

and by repeated applications of (2-25), we find that the only sequences of N throws which do not have probability zero, correspond to

$$(B_N \ldots B_2 B_1|C_p) = p$$
$$(b_N \ldots b_2 b_1|C_p) = 1 - p$$

so that in place of (6-3) we have

$$(N_n|C_p) = p\delta(n, N) + (1 - p)\delta(n, 0), \tag{6-6}$$

which is exactly what our common sense told us without any calculation.

This shows that *before we can infer any definite frequency from a probability assignment, the evidence on which that probability assignment is based must be very good evidence indeed.* It corresponds to that possessed by the man from the Bureau of Standards in the dice game of Section 3. In order for (6-3) to hold, the evidence on which A_p is based must carry overwhelmingly more weight than does the evidence of N throws. For this reason, the probabilities obtained from maximum–entropy inference have no reasonable frequency interpretation, and we can see why statistical mechanics was so confusing as long as we tried to interpret it this way.[18] Now introduce the proposition,

$$D_f = \text{"In an infinitely long sequence of trials,}$$
$$\text{the relative frequency of } A \text{ approaches } f.\text{"} \tag{6-7}$$

In the limit as $N \to \infty$, the binomial distribution becomes infinitely sharp, and so we obtain the Dirac delta- function[20]

$$(D_f|A_p) = \delta(f - p). \tag{6-8}$$

Equation (6-8) is loaded with logical booby–traps, which we must hasten to point out. Note first that it by no means says that the relative frequency $f = p$ *must* occur. It says only that, on the basis of the information which led to the assignment A_p, this is the only relative frequency which it is reasonable to expect; the available evidence gives no support at all to any other value. The probability (6-8) is still only a subjective quantity.

Equation (6-8) represents a limiting case which can never be justified in practice, because in order for (6-3) to continue to hold as $N \to \infty$, the evidence on which A_p is based must carry overwhelmingly more weight than do the results of an infinite number of trials. Not even the Bureau of Standards can provide us with evidence this good.

But there is still a paradox here. Suppose that the evidence A_p *was* perfectly reliable. It would still represent only partial information about the random experiment. According to (6-8), the probability that the limiting frequency lies in the interval $(p - \epsilon) < f < (p + \epsilon)$ is

$$\int_{p-\epsilon}^{p+\epsilon} (D_f|A_p)\, df = 1;$$ (6-9)

i.e., f was certain, on data A_p, to lie in this interval. How could we have been certain of *anything* on the basis of only partial information? How could we have been certain that a limiting frequency even exists?

Well, Eq (6-8) is actually a logical contradiction, but a useful one. We have asked the theory a foolish question, and it has given us a foolish answer. Equation (6-8) refers only to an infinite number of trials. If N is finite, there is no n in $0 \leq n \leq N$ for which $(N_n|A_p) = 0$. We are *not* certain of the result of any *possible* experiment. It is only when the experiment is *impossible* that we can be certain of the result! Any attempt to define a probability as the limit of a frequency is evidently subject to the same logical difficulty, but in a much more acute form, because there is no way at all of avoiding it.

In spite of this, (6-8) is useful if we understand how to use it. If N is large and the supporting evidence A_p fairly good, it may be a perfectly valid approximation to (6-3) for some purposes, and it will then lead to simpler formulas than would (6-3).

Equation (6-8) can also be used in a different way. If we *had* evidence about limiting frequencies, that evidence would be equivalent to a perfectly reliable assignment A_p . Thus, if E is any proposition, and A_p is perfectly reliable so that (6-8) holds, we would have

$$(E|D_f) = (E|A_p), \qquad f = p.$$

In particular,

$$(N_n|D_f) = \binom{N}{n} f^n (1 - f)^{N-n}$$ (6-10)

which is the form used in the frequency theory.

The inverse problem, of inferring a probability from an observed frequency, is much more difficult. The quantity which we have here to evaluate is $(B_{N+1}|N_n X)$, where we denote, as in Sec. 2, the prior evidence by X. It does not seem possible to carry out this calculation once and for all in the most general case, because the prior evidence might provide intricate relations between the probabilities at different trials, in an infinite number of different ways. The order in which "A true" and $a = $ "A false" occurred would in general be relevant to the probability of B_{N+1}, but the above notation implies that we are not going to consider that evidence.

The only case which the frequency school of thought can treat is the one where we ignore completely all the prior evidence; the frequency school regards a–priori probabilities as nonsense. This simplifies our problem, because it is only that case that we need to exhibit here in order to establish the relation between the frequency theory and Laplace's theory. In other words, the prior evidence X is now to tell us nothing whatsoever. We have, from (2-25) and (2-26),

$$(B_{N+1}|N_n) = \int_0^1 (B_{N+1}D_f|N_n)\, df = \int_0^1 (B_{N+1}|D_f N_n)(D_f|N_n)\, df. \qquad (6\text{-}11)$$

Also, by (2-25),

$$(D_f|N_n) = (D_f|X)\frac{(N_n|D_f)}{(N_n|X)}. \qquad (6\text{-}12)$$

The a–priori probabilities $(D_f|X)$ and $(N_n|X)$ must now say nothing about the values of f or n. The consistent way of saying this is, from the principle of maximum entropy,

$$(D_f|X) = 1; \qquad (N_n|X) = \frac{1}{N+1} \qquad 0 \le n \le N.$$

Furthermore, the evidence D_f carries overwhelmingly more weight than does N_n , so that

$$(B_{N+1}|D_f N_n) = (B_N + 1|D_f) = f.$$

Substituting these results and (6-10) into (6-11), we have

$$(B_{N+1}|N_n) = (N+1)\binom{N}{n}\int_0^1 f^{n+1}(1-f)^{N-n}\, df = \frac{n+1}{N+2}, \qquad (6\text{-}13)$$

which is Laplace's law of succession. If N is sufficiently large, the probability which we assign to A at the next trial is substantially equal to its observed frequency in the previous trials.

From these results we conclude that the general relation between the two theories is the following. Whenever all of the available evidence consists of observed frequencies, the conclusions obtained from the frequency theory approach those given by Laplace's theory asymptotically as the number of observations increases. If we have additional evidence not expressible in terms of frequencies, the conclusions of the theories may differ widely, and it is Laplace's theory which will agree with common sense.

As a simple example of this, suppose that two observers listen to a geiger counter, known by both to have an efficiency of 10 per cent. 0_1 has no knowledge about the source of the particles being counted. 0_2 knows that the source is a radioactive sample of long lifetime, in a fixed position. He does not know anything about its strength except, of course, that it is not infinite. During the first minute, 10 counts are registered. 0_1 infers, by maximum–likelihood, that about 100 particles actually passed through the counter, and 0_2 agrees. During the second minute, 16 counts are registered. 0_1 infers that about 160 particles were present, and he does not change his estimate for the first minute. 0_2 , using Bayes' theorem, concludes that the most probable value is only 137, and he revises his estimate for the first minute to 123. Each has done consistent plausible reasoning, but *prior evidence which has no frequency interpretation can completely change the conclusions which we draw from random data, and their degree of reliability.*

7. "SUBJECTIVE" COMMUNICATION THEORY

Laplace's theory is of such wide scope that in principle it includes every example of plausible reasoning, and thus *a fortiori*, communication theory. In particular, much of communication theory can be regarded as an application of maximum–entropy inference. This viewpoint may or may not lead to new mathematical results unlikely to be found without it. However, the conditions for validity of some known results can be extended. Also, it clarifies a constantly recurring question: what parts of communication theory describe measurable properties of messages, and what parts describe only the state of knowledge of some observer?

The current tendency is to state and prove theorems using the frequency terminology. Mathematical properties needed for the proof must then be regarded as objective properties of the messages or noise, and this makes it appear that the theorem is valid only if these properties can be demonstrated as "true." For example, Shannon's proofs of theorems often "assume the source to be ergodic so that the strong law of large numbers can be applied." But how are we to decide whether a source is "really" ergodic? What measurements could we perform on it? Ergodicity has a precise frequency interpretation only for behavior over infinite periods of time. From an operational viewpoint it is therefore meaningless. How, then, can we ever trust the result of the theorem?

If we look at the problem in Laplace's way this difficulty disappears. When we say, "The source is ergodic," we are not describing the source, but rather our state of knowledge about the source. We mean only that nothing in the available evidence leads us to expect that it has a sub–class of states in which it can get stuck. *As far as we know*, there is always a possible route by which it can get from any state to any other.

Whether or not this is actually true is irrelevant for the use we make of the theorem. Our job, again, is only to do the best we can with the information we have, and it would be quite unjustified to assume an invariant sub–class of states unless we have evidence to support this. It could, for example, lead to design of a communication system which turns out to be incapable of handling the actual messages. Ergodicity of this subjective kind is a consequence only of our being conservative and avoiding unwarranted assumptions; the resulting probabilities are the ones which maximize the entropy subject to whatever we *do* know. Exactly the same argument applies to ergodicity in statistical mechanics.

Many of the fundamental theorems of communication theory can be reinterpreted in this way, and we then see that they are valid and useful in far more general conditions than one would suppose from the frequency definition of probability.

Consider an observer 0_n who knows in advance the n-gram frequencies which a source is going to generate, but has no other knowledge about it, what communication system represents rational design on the basis of this much knowledge, what is the best way of encoding into binary digits for the noiseless case, and what channel capacity does 0_n require? In principle, the answer is always the same; we need to find the probabilities $p(M)$ which 0_n assigns to each of the conceivable messages, and use the method of Fano and Shannon.[21]

We wish to emphasize that it makes no sense whatever to say that there exists a "correct" distribution $p(M)$ for this problem; $p(M)$ is an entirely subjective quantity. This becomes especially clear if we suppose that only a single message is ever going to be sent over the communication system, but we wish to transmit it as quickly as possible. Thus there is no conceivable procedure by which $p(M)$ could be measured. This would in no way affect the problem of engineering design which we are considering.

In choosing a distribution $p(M)$, it would by possible to assume a particular message structure beyond n symbols. But from the standpoint of 0_n this could not be justified, for *as far as he knows,* an encoding system based on any such structure is as likely to hurt as to help. From 0_n's standpoint, rational conservative design consists in carefully *avoiding* any such assumption. This means, in short, that 0_n should choose the distribution $p(M)$ by maximum–entropy inference based on the known n–gram frequencies.[22] For 0_1 and 0_2 the solution is well known in a different context; the physicist calls them the linear Ising chain with no interactions, and with nearest-neighbor interactions respectively.[23]

Laplace's point of view is helpful also in the problem of detecting a radar signal in noise. Anyone who studies this problem comes to the conclusion that there is no way of evading the notion of a–priori probabilities of different signals. They are an essential part of the problem, because any prior knowledge we have about the signal is extremely relevant to the proper engineering design. The question of how one finds their "true" numerical values then becomes quite embarrassing. They can be given a frequency interpretation only by devices so arbitrary and forced that they could have no relevance to the problem.

We can now see the answer to this. In the first place, *no one needs to apologize for, or do any cautious egg-walking around, the use of Bayes' theorem and a–priori probabilities.* This is in fact the only consistent way of handling the problem. We have at present no known procedure for translating our prior knowledge about signals into numerical values of probabilities. At least not on paper. But we still have our brains, and until new principles are discovered, we will have to use them. We must take into account everything we know about the signal, and then *guess* the a–priori probabilities.

8. CONCLUSION

We have tried to show above how a re–interpretation of the probability concept can clarify and extend the power of statistical methods for current applications in science and engineering. Laplace's view of probability theory as the symbolic logic of plausible reasoning enables us to follow the process which our brains must be using, in every case where numerical values of probabilities can be found. It enables us to do this in far greater detail than is possible on the frequency theory, and to take into account additional evidence which cannot even be stated in terms of frequencies.

The analysis of Sec. 2 above is, of course, far from rigorous in the modern sense of the term. However, I believe that all the necessary epsilons and deltas can be supplied by anyone sophisticated enough to feel the need for them. There is always a danger that too much generality will obscure the important points of an argument. Finally, it is interesting to note the increasing importance of the theory of functional equations in this field, shown also by Bellman and Kalaba.[24]

REFERENCES

1 C.E. Shannon, "A Mathematical Theory of Communication," Bell. Syst. Tech. Jour. Vol. 27, pp. 379-423, 623-655; July, October, 1948. Also in C. E. Shannon and W. Weaver, "The Mathematical Theory of Communication," University of Illinois Press, Urbana, 1949.

2 N. H. Abel, *Crelle's Jour.*, Bd. 1 (1826).

3 R. T. Cox, "Probability, Frequency, and Reasonable Expectation," *American Journal of Physics.* Vol. 14, p. 1 (1946). This is a very important, but unfortunately little–known, paper which comes quite close to solving the problem of Sec. 2.

4 "La théorie des probabilités n'est que le ben sens reduit au calcul." This occurs in the Introduction to P.S. Laplace, "Exposition de la théorie des chances at des probabilités," Paris, 1843. The same statement, with slightly different wording, is found in the Truscott–Emory translation of P.S. Laplace, "A Philosophical Essay on Probabilities," Dover Publications, N. Y. (1951), p. 196.

5 G. Polya, "Mathematics and Plausible Reasoning," Volumes I and II, Princeton University Press, 1954.

6 G. Polya, "How to Solve It," Princeton University Press, 1945; Second paperbound edition by Doubleday Anchor Books, Inc., Garden City, N.Y., 1957.

7 This notation is perhaps confusing. It can be made clearer if we suppose that the symbol for a plausibility is not $(A|B)$, but just $A|B$, the parentheses being unnecessary. However, when one writes down more involved equations, the absence of parentheses can cause even greater confusion.[3] The notation adopted here, while not entirely consistent, appears to the writer as the lesser of two evils.

8 H. Jeffreys, "Theory of Probability," Oxford University Press, 1939.

9 This is not a direct quotation from any particular author, but a statement of what is implied by many authors. For example, see Ref. 10, pp. 150-151, or Ref. 12, pp 4-6.

10 H. Cramér, "Mathematical Methods of Statistics," Princeton University Press, 1946.

11 Reference 5, Vol. II, p. 136. For other examples, see Ref. 8, pp. 107-110, and Ref. 12, p. 64.

12 W. Feller, "An Introduction to Probability Theory and its Applications," John Wiley and Sons, Inc., N.Y., 1950. Any reader familiar with this book will see at once that the present paper is largely a reaction against and search for an alternative to, the philosophical views expressed therein. I believe this is necessary if probability theory is to meet all the needs of science and engineering. But no one can challenge Feller's beautiful mathematical results, the validity of which does not depend on how we choose to interpret them. They are as useful in Laplace's theory as in the frequency theory.

13 This is far from being a precise statement. The derivation of Eq. (6-13) shows in more detail what is required for the law of succession to apply.

14 However, it served Laplace very well indeed. The following procedure led him to some of the most important discoveries in celestial mechanics. Noting a discrepancy between observation and existing theory, he would break down the situation into alternatives which seemed intuitively "equally possible." He would then compare the probability that a discrepancy of this size is due to a systematic effect, with the probability that it is due to errors of observation. Whenever the ratio was sufficiently high, he would decide that this is a problem worth working on, and attack it. He was, in fact, using Wald's decision theory, in exactly the way developed recently by Middleton, van Meter, and others for the detection of signals in noise.

15 Ref. 10, pp. 507-524.

16 E. T. Jaynes, "Information Theory and Statistical Mechanics," *Physical Review*, Vol. 106, pp. 620-630; May 15, 1957. At the time of writing this, I was under the impression that the frequency theory and Laplace's theory are parallel, co-equal theories using the same mathematical rules. However, the arguments of the present paper show that the frequency theory is only a special case of Laplace's theory.

17 E.T. Jaynes, "Information Theory and Statistical Mechanics II," Submitted to the *Physical Review*.

18 E. T. Jaynes, "Poincaré Recurrence Times and Statistical Mechanics," Submitted to the *Physical Review*.

19 This can be stated in a more precise epsilon–delta language, but the reader will anticipate that the conclusions are largely independent of what we mean by "reasonably probable," for the same reason as in Shannon's theorem 4.

20 $(D_f|A_p)$ is a probability density, $(D_f|A_p)\,df$ being a probability. Since, however, the differentials cancel out of equations and the distinction is already determined by whether the variable is continuous or discrete, there is no need to invent a new notation. On the other hand, it is essential in this theory that we *do* distinguish in notation between a probability and a frequency.

THE RELATION OF BAYESIAN AND MAXIMUM ENTROPY METHODS

E. T. Jaynes
Arthur Holly Compton Laboratory of Physics
Washington University, St. Louis, Missouri 63130, U.S.A.

Abstract. Further progress in scientific inference must, in our view, come from some kind of unification of our present principles. As a prerequisite for this, we note briefly the great conceptual differences, and the equally great mathematical similarities, of Bayesian and Maximum Entropy methods.

We are all pleased at the progress that has been made, in many different fields, as a result of recent recognition of the power of Bayesian inference and the Maximum Entropy Principle (MAXENT). But this is not to say that further clarifications and technical developments aren't needed. It is a truism that every new level of understanding reached only reveals to us new questions of which we were unaware before.

Therefore, in spite of present successes, this is no time for relaxing our efforts to develop still better pragmatic algorithms and a more unified theoretical structure. Indeed, because of the pressure of new applications opened up by these very successes, the field of scientific inference has never been in greater need of new creative thought. But before we can hope to make much further progress, some clarification of our present principles is needed.

We have at present two principles, Bayes' theorem and MAXENT, that are held to have some fundamental status in the new domains. The practitioners of the art sometimes use one, sometimes the other; but beginners and critics alike seem puzzled by how we choose between them. How are these principles related to each other? Are they mutually consistent or in conflict? What is the proper place of each in our toolbox? Since nearly every conceivable opinion on these matters has been expressed already, one is hard put to say anything really new; but perhaps we may sift things out a bit.

At the most fundamental level as now perceived, by "applying Bayes' theorem" we mean *calculating* the probability

$$p(H \mid DI) = p(H \mid I)p(D \mid HI)/p(D \mid I) \tag{1}$$

where, in our applications, H stands for some hypothesis whose truth we want to judge, D for a set of data, and I for whatever "prior information" we have in addition to the data. The prior probability $p(H \mid I)$ of H gets updated to the posterior

25

probability p(H|DI) as a result of acquiring the data D. This includes parameter estimation, since H might be a statement about some property of a parameter θ.

By "applying MAXENT" we mean *assigning* a distribution $(p_1 \cdots p_n)$ on some "hypothesis space" by the criterion that it shall maximize the information entropy

$$S_I = -\sum p_i \log p_i \tag{2}$$

subject to constraints that express properties we wish the distribution to have, but are not sufficient to determine it. Entropy is used as the criterion for resolving the ambiguity remaining when we have stated all the conditions we are aware of.

On the face of it, it is hard to imagine two procedures more different, mathematically or logically. Bayes' theorem expresses nothing more than that Aristotelian logic is commutative. The propositions

$$HD = \text{"H and D are both true"}$$
$$DH = \text{"D and H are both true"}$$

say the same thing, so they must have the same truth value and the same probability whatever our information I. Then in the product rule of probability we may interchange D and H:

$$p(DH|I) = p(D|HI)p(H|I) = p(H|DI)p(D|I) \tag{3}$$

which is Bayes' theorem. Obviously, then, anyone who reasons in a way that conflicts with Bayes' theorem is violating a rather elementary principle of logic.

Fundamentally, a single application of Bayes' theorem gives us only a probability; not a probability distribution. Indeed, Bayes' theorem makes no reference to any sample space or hypothesis space; (H, D, I) may stand for any propositions with well-defined meanings. Just for that reason, Bayes' theorem cannot determine the numerical value of any probability directly from our information; to apply it one must first use some other principle to translate our information into numerical values for p(H|I), p(D|HI), p(D|I).

In scientific inference, therefore, before we can apply Bayes' theorem our problem must be developed beyond the "exploratory phase", to the point where it has enough structure to determine p(D|HI).

In contrast, MAXENT requires that we specify in advance a definite hypothesis space $H_1 \cdots H_n$ which sets down the possibilities to be considered. It gives us necessarily a probability distribution, not just a probability; it does not make sense to ask for the MAXENT probability of an isolated proposition H, that is not embedded in some hypothesis space of alternative propositions. But MAXENT does not require for input the numerical values of any probabilities on that space; rather it assigns those numerical values for us, directly out of our information, as expressed by our choice of hypothesis space and constraints. Therefore MAXENT can be applied in -- and is indeed most useful in -- the exploratory phase of a problem.

In these functional respects, MAXENT does for us almost the opposite of what Bayes' theorem does. How, then is it possible that two principles so different could be confused? This comes from two circumstances.

In the first place they have, after all, one feature in common; the updating of a state of knowledge. In MAXENT, for example, one may consider a problem with constraints X and Y, and find the solution $p_i(X,Y)$. Then a third constraint Z is added, and we re-maximize the entropy subject to all three constraints, leading to an updated solution $p_i(X,Y,Z)$. There is indeed a superficial resemblance to Bayes' theorem; and for some it requires only a sloppy notation and terminology -- calling these two MAXENT distributions "prior probabilities" and "posterior probabilities" -- to confuse them thoroughly.

Secondly, there is a technical circumstance which has caused trouble throughout the history of probability theory; different problems may lead to the same computational procedure. In some cases application of Bayes' theorem in one hypothesis space, and MAXENT in another, leads us to nearly identical calculations.

For example, starting with Darwin & Fowler in the 1920's, many have noted that the MAXENT procedure on the space S of a single trial, and the Bayes' theorem procedure on the extension space S^n of n trials are asymptotically equivalent as n becomes very large; the latter circumstance may be taken as the basis of the combinatorial rationale for MAXENT, which differs from the more fundamental probabilistic one noted above. Some other examples of this Bayes-MAXENT mathematical correspondence are given in Jaynes (1968). In a sense, this only illustrates their mutual consistency; but it can also be a rich source of confusion.

The recent literature has many attempts to clarify the relation of these principles. Williams (1980) sees Bayes' theorem as a special case of MAXENT, while van Campenhout & Cover (1981) see MAXENT as a special case of Bayes' theorem. In our view, both are correct as far as they go; but they consider only special cases. Zellner (1987) generalizes Williams' result.

Thus Williams considers the case where we have a set of possibilities $(H_1 \cdots H_n)$, and some new information E confines us to a subset of them. Such primitive information can be digested by either Bayes' theorem or MAXENT, leading of course to the same result; but Bayes' theorem is designed to cover far more general situations. Likewise, van Campenhout & Cover consider only the Darwin-Fowler scenario; MAXENT is designed to cover more general situations, where it does not make sense to speak of "trials".

Attempts to evade Bayes' theorem have been underway unceasingly since the rise of the "sampling theory" school of thought in the early 1900's. But we have already surveyed the results (Jaynes, 1983); whenever sampling theory methods have led us to different conclusions, closer examination has always shown the Bayesian results to be superior. Likewise, to the best of our knowledge, all attempts to extend Bayes' theorem [such as that of Jeffrey (1983)] have proved on closer examination to be satisfactory only in the cases where they agree with Bayes' theorem. Further strong evidence is given by Bretthorst (1987).

That MAXENT is in a similar position is indicated by the fact that the MAXENT procedure is in constant use, either as an analytical tool or as a computational algorithm, in a variety of very different problems; and no alternative has been found. Even those who reject the MAXENT principle are often led, by long and different reasoning, to the actual MAXENT algorithm and result.

To the best of our knowledge, all attempts to evade MAXENT in problems where we consider it appropriate, or to extend it to new problems, have been no more successful than the attempts to evade or extend Bayes' theorem. One does so only at the cost of getting results that can be shown to be defective or incomplete, in that they either fail to use all the relevant information or assume false information. We hope to discuss the evidence for this conclusion in much greater detail elsewhere.

There is indeed something fundamental about these principles, although we think that more unified ways of presenting them are still needed and will be found in the future. But we stress that our present principles and practice are fairly good; they have many demonstrable optimality properties and impressive pragmatic success. So unless we can recognize, and clearly understand the reason for, some specific defect in our present principles, we are hardly in a position to improve on them; as so much past experience has shown, we are far more likely to lose some of the good performance features already accomplished.

An old adage among moralists is that "Virtue cannot be taught; only demonstrated"; and we must admit that today more worked-out examples of their analytical and numerical details, in a wider variety of real problems, are much needed to demonstrate how to apply them and what kind of results are to be expected. But with more and more books and Bayesian/MAXENT computer programs being written, this need should be filled soon.

REFERENCES

Bretthorst, L. (1987), Ph.D. Thesis, Department of Physics, Washington University, St. Louis, Missouri.

Jaynes, E. T. (1968), "Prior Probabilities", IEEE Trans. Systems Sci. Cybern. SSC-4, 227-241 (1968). Reprinted in V. M. Rao Tummala and R. C. Henshaw, Eds., Concepts and Applications of Modern Decision Methods (Michigan State University Business Studies Series, 1976), and in Jaynes (1983).

Jaynes, E. T. (1983), Papers on Probability, Statistics and Statistical Physics", R. D. Rosenkrantz, Editor, D. Reidel Pub. Co., Dordrecht-Holland. Reprints of 14 papers dated 1957-1980.

Jeffrey, R. C. (1983), The Logic of Decision, 2nd Edition, Univ. of Chicago Press.

van Campenhout, J. & Cover, T. M. (1981), IEEE Trans. Inform. Theory IT-27, 483.

Williams, P.M. (1980), "Bayesian Conditionalisation and the Principle of Minimum Information", Brit. J. Phil. Sci. 31, 131-144. The same mathematical fact was noted by R. D. Rosenkrantz, Inference, Method and Decision, D. Reidel, Dordrecht-Holland (1977); 55-57.

Zellner, A. (1987), "Optimal Information Processing and Bayes' Theorem" The American Statistician (to be published).

AN ENGINEER LOOKS AT BAYES

Myron Tribus
Acting Director
American Quality and Productivity Institute

Seventh Annual Workshop
Maximum Entropy and Bayesian Methods
Seattle University
August 1987

INTRODUCTION

These notes are intended to accompany a tutorial session on the fundamental ideas behind the use of Bayesian methods and the principle of Maximum Entropy. The material on which these notes is based is scattered throughout the literature and seldom brought together in a coherent whole. In this presentation many difficult steps in mathematics are skipped so as to bring out the flow of ideas. References at the end provide the missing detail.

PROBABILITY IN SCIENCE AND ENGINEERING

I am an engineer whose education included only one course in statistics, that one taken after WWII at the graduate level. It turned me off. I found very little of value in it. It was dull and dry.

That graduate course was not my first encounter with statistical reasoning. The first encounter came much earlier, as an undergraduate chemistry major, when Professor Giauque, a Nobelist at Berkeley, introduced us to statistical mechanics. For me it was like a small boy suddenly discovering a book on the facts of life. Statistical reasoning removed so many mysteries from the field of thermodynamics that I was hooked on it.

After World War II when I became an instructor, I tried to introduce statistical reasoning in the undergraduate course in engineering thermodynamics. It was then that I began to learn about the difficulties with the teaching of statistics to engineers. There were many points in the logical development of statistical mechanics, which did not hold together properly and although I struggled for ten years, I could not provide a convincing logical development. All the time I kept trying to

G. J. Erickson and C. R. Smith (eds.),
Maximum-Entropy and Bayesian Methods in Science and Engineering (Vol. 1), 31–52.
© 1988 by Kluwer Academic Publishers.

bring together what I found in the books on statistics and what was to be found in books on statistical mechanics. It was impossible because of flaws in the logical foundations of each subject as classically presented.

The end of World War II was a period of intense activity in science and engineering. Wiener's work on Cybernetics was published. John Von Neumann's theory of games appeared. Operations Research flourished as a new discipline. Shannon's theory of communication suddenly found many applications beyond communication. These contributions, Cybernetics, the Theory of Games, the Theory of Communication, had something in common which I suppose few of us noticed at the time.

1. They dealt with man-made systems. They were not aimed at discovering "laws of nature".

2. They were "inventions". They were not "discoveries". They were designed.

Shannon's work in information theory was particularly inspiring to many of us in thermodynamics and statistical mechanics, for Shannon used a function which he called "entropy" to represent the uncertainty in the mind of someone about to receive a message. Shannon's entropy is defined by:

$$S = -k \sum_i p_i \ln p_i$$

This same function appears in statistical mechanics and, on the advice of John Von Neumann, Claude Shannon called it "entropy". I talked with Dr. Shannon once about this, asking him why he had called his function by a name that was already in use in another field. I said it was bound to cause some confusion between the theory of information and the thermodynamics. He said that Von Neumann had told him: "No one really understands entropy. Therefore, if you know what you mean by it and you use it when you are in an argument, you will win every time".

Now just because a function appears in two different fields it does not mean that the fields are analogous. For example sin(x) appears in both surveying and in electrical engineering and no one thinks that therefore the two fields are logically the same. We must look deeper to find the connection.

Anyone looking at the definition of entropy can see that there are only 8 symbols in it. S is the quantity to be defined. The symbols "=", "-", "Σ", and "ln" have well understood meanings. "k" requires no explanation, for it is just a constant which determines the scale for S.   That leaves only the symbols "p" and "i" to be defined in order to remove all ambiguity (or at least to remove the ambiguity to somewhere else).

The search for an understanding of the link between information theory and thermodynamics, therefore, resolves into a search for a better understanding of what the word "probability" means.

As all of you probably know, the search for a satisfactory meaning for the word "probability" has occupied the best minds for over two centuries. Hume, Laplace, Bernoulli and many, many others have wrestled with the challenge.   Then in 1959 one of my graduate students showed me a remarkable paper that in one fell swoop removed my perplexity.   The paper was written by Edwin T. Jaynes, a physicist now at Washington University in St. Louis, and published in Physical Reviews, "Information Theory and Statistical Mechanics", (106, 620, 1957 and 108, 171, 1957).   This paper, itself, was based on another very little known work by Richard T. Cox, another physicist at Johns Hopkins University,   published in the American Journal of Physics in 1946.   Together these papers provided a basis for the connections among statistical mechanics, classical macroscopic thermodynamics and information theory.   When I understood what these papers were saying, I could not sleep for a week.   For the first time it became possible to develop the ideas of thermodynamics and statistical mechanics from a single logical base in information and quantum theories and to make the laws of thermodynamics consequences and not premises. Moreover, the reconciliation between what statisticians were saying and what was being said in statistical mechanics could now be attempted. Being an outsider, I did not comprehend, at first, that this reconciliation would require a recasting of the foundations of statistics and probability theory.

This development of a new view in thermodynamics brought me into considerable controversy as defenders of the old ways to think about thermodynamics on the one hand and statistics on the other attacked with great vigor and energy.   I began to realize that thermodynamics was considered by many to be a branch of theology.   Statisticians seemed to feel that a discussion of the foundations of their subject was an attack upon their sanity.

THE BASIS FOR A NEW RATIONALE..REFORMULATING THE PROBLEM

Whether the subject be thermodynamics, quality, reliability, communication theory, census taking or even management, there exists a common problem which may be formulated this way:

> You know <u>something</u>, but you do not know <u>everything</u>.
> How should you describe your incomplete knowledge
> to someone else so that you tell neither more nor less
> than you truly know?   How should that person make
> use of the incomplete information?
>
> What is the proper <u>code</u> for the transmission of
> incomplete information?   How should a person who
> receives such coded information from more than one
> source combine the information and how should
> inferences be drawn from such coded information?

When the question is put in this manner it may be recognized as a <u>design problem</u>. This is something an engineer can sink his teeth into.   What is required is a <u>code</u> which handles incomplete information from several sources and a <u>logical machine</u> which processes it to produce inferences.

As with all design problems, it is important at the outset to say what shall be the criteria by which the design is to be judged.   Failure to decide ahead of time how the results will be valued often causes great disappointment with the product.   (This is just a paraphrase of Deming's admonition to keep the customer in mind.)

In this presentation I am borrowing heavily from the works of Edwin T. Jaynes and I would urge you to read his many publications.   One of the tricks I have seen him use in discussing this theoretical development is to introduce a logical machine which we shall call "Robby the Robot".   Robby is a creature we are going to design. It is a machine which functions on incomplete information and even when we do not tell it everything it should know about a situation, it nevertheless gives us the "best" conclusions it can.   Unlike current computers, it does not go into an endless loop or just sit there when given an incompletely defined problem.   By introducing this machine we get around discussions of how people think or how you or I ought to think.

If the machine is to do the "best" it can, then it is up to us, the customers , to define what we mean by "best".   Our criteria will then provide the basis for deciding if the design is of high quality.
Our task is simply to design:

1. The hardware--the logical processor.

2. The software--the language with with we communicate with the machine.

3. The teachware--the instruction manual for the user.

The troublesome word here is "best".   To provide a better understanding of what we mean by "best" it is helpful to define the opposite.   To begin therefore, we consider a set of characteristics which will unerringly define a flawed machine.   Here are some criteria which permit us to recognize a machine as so badly flawed we would disqualify it at once:

INCONSISTENCY.   If Robby gives self contradictory results, we shall say it is inconsistent and refuse to use it.   For example if it solves the same problem posed two different ways and comes back with two different answers, which we cannot reconcile, we shall refuse to rely upon its answers.   Robby should supply answers, not quandries.

AMBIGUITY .   If it is not possible to decode the answers and decide what they mean, we shall reject Robby.   We want a reasoning machine, not a mystic.

SPECIAL PURPOSE.   If Robby the Robot has to be redesigned for every problem, we shall be uninterested in it.   We do not want an ad-hoc product.

DECEIT.   If Robby's methods are impossible for us to reconcile with any logical process or if some steps of the process are inscrutable, we shall reject the answers.   Robby must not insist that we accept the results on faith.   He should not say, merely, "Trust me".

These criteria are disqualifiers.   By taking their opposites, therefore, we may define four "desiderata" which, while not guaranteeing we shall find Robby an acceptable machine, will guarantee that we have not **deliberately** made Robby unacceptable.

This reasoning, therefore, leads to the following 4 criteria:

1. <u>UNAMBIGUITY</u>.   Robby should not yield uninterpretable answers.

2. <u>UNIVERSALITY.</u>   Robby should not be a "special purpose" machine. It should not be redesigned for every problem. If Robby is given complete information it should provide the same answers as deterministic machines given the same information.

3. <u>CONSISTENCY</u>. Robby should give the same answer whenever given the same information. Robby should <u>not</u> give different answers whenever the same problem is presented in different ways. If the problem may be solved in more than one way, all ways should lead to the same answer.

4. <u>CANDOR.</u>   This constraint is laid upon both the user and the machine. User's are not supposed to ask question Q_1 and then complain because Robby did not answer question Q_2. All machine operations should be   understandable.

Here is a picture of how far we have come in the design process:

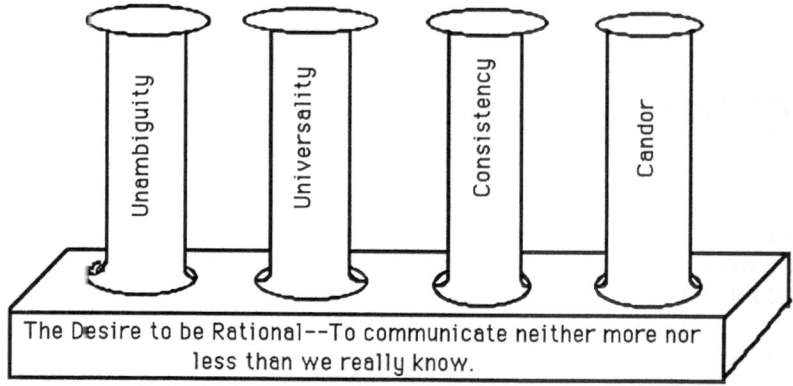

We have erected four pillars upon which to construct the theory, but we are not sure they will support what we eventually choose to erect. For one thing, we are aware of Goedel's theorem, so we know that there is no way we can <u>guarantee</u> that these desiderata will be met. The best we as designers can do is to be sure that we do not <u>deliberately</u> introduce flawed components into the design. We know how to recognize some things which will guarantee failure. We do not know what to do to guarantee success. If we knowingly select a design which violates the desiderata, we shall know that the structure is flawed.

UNAMBIGUITY

To satisfy the "<u>unambiguity</u>" requirement, we shall require that the user and Robby be able to distinguish what is said from what is not said. This suggests the use of a notation based on Boolean algebra. We set up a name or symbol for whatever we are talking about and at the same time set up a symbol for what we are <u>not talking about.</u> In the parlance of Boolean algebra, it means setting up a symbol for the "contradictory". In our machine it means that for every proposition we have to set up at least <u>two</u> memory locations for every proposition we ask Robby to consider.

The user requires a notation for communication with Robby. (Computer designers call this the I/O part of the computer).

We adopt this notation from among the many we could use. Let an upper case bold (i.e., A) letter represent a proposition and the lower case bold (i.e., a) letter represent the contradictory. Current machines only use one register per proposition. This means they do not provide a clear indication as to what is the denial of a statement. When there are several statements forming a mutually exclusive and exhaustive set, we treat them this way. We let the "+" sign represent "or" and write (if A_1, A_2 and A_3 constitute an exhaustive set):

A_1 = first propostion

a_1 = denial of first propostion

$\quad = A_2 + A_3$

Writing two symbols together as though multiplied represents "and". Thus AB represents "A and B". Of course $AB = BA$. If we want to indicate the influence of time in this notation we shall have to introduce it explicitly.

All problems require a context. We shall use the solidus, "|", to mean "given the truth of" which is the same as saying "in this context".   Thus when we ask Robby about **AB|C** we are asking him (or it) to discuss **A** and **B** given that **C** is true where **A** and **B** and their denials have been unambiguously defined and stored in the computer.

UNIVERSALITY

Universality requires that we be able to compare apples and oranges.   This means we shall wish to associate real numbers with Boolean symbols.   Only real numbers allow universal comparibility, that is, plot on the real number line and permit us to compare positions.

Thus far we have decided that Robby will have

> Two registers, counters, wheels or other devices to represent each of propositions **A, B, C, AB**, etc. and their contradictories.

> The registers will contain real numbers representing the truth value of the propositions.

The user's language will require that associated with every statement and its context there be a real number, which we shall represent, tentatively, by $\phi(\textbf{AB|C})$. ϕ is the real number representing what "Robby knows about **A** and **B** given the truth of **C**".   It is an arbitrary choice to say that increasing ϕ means increasing "truth", "credibility" or "correctness" of **AB** given **C**.

CONSISTENCY

Now let us turn to the third desideratum, <u>consistency</u>. We expect that in reasoning about **A, B** and **AB** robby will have separate registers for **A|C, B|C, A|BC, B|AC** and **AB|C**. These 5 registers must be connected somehow. The designer's problem is to decide how to connect them. If, for example, there is some functional relation between **AB|C** and the other four we shall represent it in this way:

$$\phi(\textbf{AB|C}) = F[\phi(\textbf{A|BC}), \ \phi(\textbf{B|AC}), \ \phi(\textbf{A|C}), \phi(\textbf{B|C})]$$

Of course this is not the only way to design the machine. We might have chosen a different functional relation, such as

$$\phi(\textbf{AB|C}) = F[\phi(\textbf{A|BC}), \ \phi(\textbf{B|AC})]$$

There only 15 possible functions which display a dependence of $\phi(\mathbf{AB}|\mathbf{C})$ upon the other 4 registers. There are, therefore, only 15 different designs for wiring up Robby.   Not all of them will work satisfactorily.   For example, if we had chosen the functional relation in the above equation and then told Robby $\mathbf{C}=$"$\mathbf{A}$ and $\mathbf{B}$ are the same statement", Robby would have reasoned this way:

$\phi(\mathbf{A}|\mathbf{BC})$ =t since if $\mathbf{BC}$ is true and $\mathbf{C}$ says that $\mathbf{A}$ and $\mathbf{B}$ are the same statement, $\mathbf{A}$ must be true.

$\phi(\mathbf{B}|\mathbf{AC})$ =t since if $\mathbf{AC}$ is true and $\mathbf{C}$ says that $\mathbf{A}$ and $\mathbf{B}$ are the same statement, $\mathbf{B}$ must be true.

$\phi(\mathbf{AB}|\mathbf{C})$ = F(t,t),   therefore $\mathbf{AB}$ is true.

However, saying that $\mathbf{A}$ and $\mathbf{B}$ are the same statement does not make $\mathbf{AB}$ true.   Therefore, if we choose to wire up Robby according to this functional relationship, Robby will soon start to make some pretty silly inferences. We conclude, therefore that the above function would violate one of our desiderata.

I have shown elsewhere that of the 15 possible solutions, only one of them cannot be shown to lead to undesireable results.   This does not prove that the Robby reasons correctly.   It only demonstrates that incorrectness has not been proven.   The one acceptable function out of the 15 possible ones is, as Jaynes and Cox have demonstrated:

$\phi(\mathbf{AB}|\mathbf{C}) = F[\phi(\mathbf{A}|\mathbf{BC}), \phi(\mathbf{B}|\mathbf{C})]$ (1)

Because interchanging the meanings of A and B should not alter the machine or its logic, we also have:

$\phi(\mathbf{AB}|\mathbf{C}) = F[\phi(\mathbf{B}|\mathbf{AC}), \phi(\mathbf{A}|\mathbf{C})]$ (2)

This functional equation has a long history.   The first analysis of it seems to have been made in 1881 in connection with the problem of designing the most general form of slide rule (Abel, <u>Oeuvres Kompletes de Niels Henrik Abel)</u>   The equation has been studied extensively by Aczel of Waterloo University in Canada and is discussed in his book. (<u>Lectures or Functional Equations and Their Applications</u>).

I shall not dwell on how one goes about solving such an equation. Note only that we are searching for an operator that takes the two real numbers $\phi(A|BC)$ and $\phi(B|C)$ and converts them into $\phi(AB|C)$. Either of the two operations, addition or multiplication will satisfy the given functional equation. That is, we may use the function ϕ with the addition operation:

$$\phi(AB|C) = \phi(A|BC) + \phi(B|C) \tag{3}$$

or with the multiplication operation:

$$\phi(AB|C) = \phi(A|BC)\phi(B|C) \tag{4}$$

Since we may convert one solution to the other by use of a logarithmic transformatiion, these two solutions are really the same. Furthermore, if we now impose a monotonic transformation on the values of ϕ we shall not have changed the relationships. For example, we could take the square roots of both sides of the last equation or raise the terms to the same power without destroying the relationship. Inside the machine we can amplify or attenuate signals at will, transform them from linear to logarithmic scaling as we wish. All that is required is that we keep in mind the fundamental equation which links the registers together and use only monotonic transformations.

There is another functional requirement, this time that the operations which link a statement to its contradictory be unique and satisfy some simple rules. For example, whatever makes the truth value of a statement increase should decrease the truth value of its contradictory. Furthermore, the operation should cancel itself. That is, if Z is the operator that transforms $\phi(A|C)$ into $\phi(a|C)$ then we should find not only that

$$\phi(a|C) = Z[\phi(A|C)] \tag{5}$$

but also that

$$\phi(A|C) = Z^2[\phi(A|C)] \tag{6}$$

that is, applying the denial operator twice should bring the same number.

Without going into detail, here is the result of the second functional requirement (equation 6):

$$\phi^k(A|C) + \phi^k(a|C) = 1 \qquad\qquad k \neq 0 \qquad\qquad (7)$$

Robby the Robot, therefore, reasons according to the two equations, (4) and (7). Raise both sides of (4) to the power k and make the transformation of

$$p = \phi^k$$

which leads to the familiar equations:

The Multiplication Rule for combining probabilities

$$p(AB|C) = p(A|BC)p(B|C) \qquad\qquad (8)$$

The Denial Rule for relating contradictories:

$$p(A|C) + p(a|C) = 1 \qquad\qquad (9)$$

We have laid our equations on what appears to be a secure foundation, as indicated in this diagram.

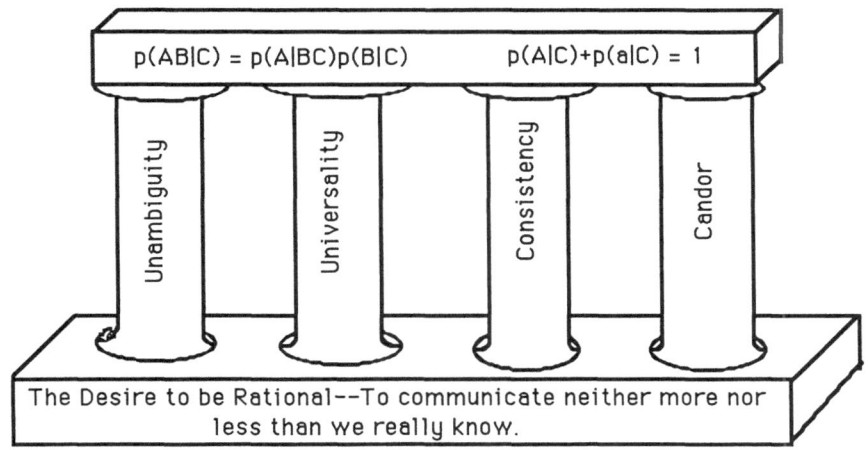

We can build upon this result very easily. It provides a strong basis.

From the symmetry of equation (4) we have:

$$p(AB|C) = p(A|BC)\, p(B|C) = p(B|AC)\, p(A|C)$$

and therefore

$$p(A|BC) = \frac{p(B|AC)p(A|C)}{p(B|C)} \qquad (10)$$

This last equation is the famous Bayes equation. If we ask Robby about **A** given that **B** and **C** are true, he goes into his memory banks and computes the value of $p(A|C)$, i.e., what was known about **A** in the absence of knowledge of **B**. Robby then multiplies this by the ratio $p(B|AC)/p(B|C)$, which is the likelihood of seeing **B**. The equation represents how Robby "learns".

Bayes equation makes it clear that Robby is not "discovering" values for probabilities. They are being assigned according to the the information furnished. Now the structure appears as follows:

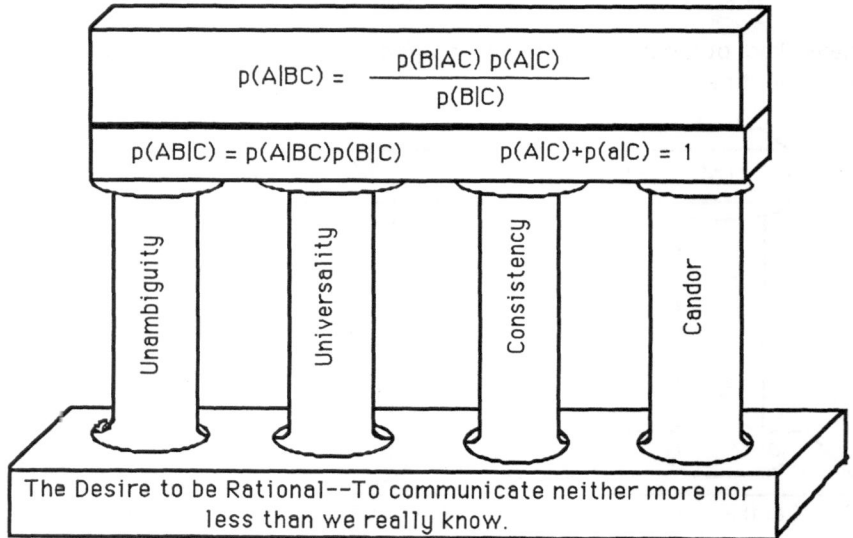

ON THE MEANING OF "PROBABILITY"

This derivation from Cox, Jaynes and Aczel, provides a new way to think about the meanings of the word "probability". In the derivation, the real number, p = p(**A**|**BC**) emerges as a unique "encoding" of *incomplete information* about **A,** *conditional* on the truth of **BC**. It is part of a code for information *transmission* . If this information is transmitted to a logical processor, the processor will operate on the given information to produce inferences according to the two equations which represent the *multiplication rule* and the **denial rule**.

In the derivation there has been no reference to frequencies or fractions. Hypothetical populations, lot sizes or superlots were not invoked in the process. Such constructs are not barred, of course. They may occur as part of the *given* data. The statements **BC** could pertain to such classes of information. The derivation in no way depends upon such constructs.

As an observer I have often been puzzled by the fact that some people insist on drawing a sharp distinction between "probability" and "statistics". I suppose one can be a probabilist without paying heed to data, but I cannot ever imagine it the other way around. It seems to me that the central problem of statistics is to see how to apply probability theory to the analysis of data, i.e. , to statistics. According to this derivation, the task is to learn how to *encode* various kinds of information (i.e., different **BC**) into numerical assignments, p_i, associated with definite propositions, $\mathbf{A}_i$, that is, to assign values to a distribution p_i:

$$p_i = p(\mathbf{A}_i|\mathbf{BC})$$

We feed data (**BC**) into the machine, along with a well formed set of propositions {$\mathbf{A}_i$} and let Robby carry out the indicated operations. Lacking Robby, we carry them out ourselves using pencil and paper.

HOW TO GET STARTED--THE ROLE OF PRIOR PROBABILITIES

The terms in Bayes equation have specific interpretations for Robby:

$$p(A|BC) = p(A|C) \, \frac{p(B|AC)}{p(B|C)} \qquad\qquad (12)$$

Knowledge about **A** if **B** is not known. ↑   ↑ Likelihood ratio: for **B** if **A** is and is not known to be true.

The term $p(A|C)$ is often called the "prior probability" which would signify that time has something to do with it. This is an unfortunate connotation. It introduces something akin to an arrow for time and this should be avoided.

The problem posed by Bayes equation is how to get started. That is, what to use for a prior probability.

There is no difficulty if the input information is taken to be a set of probabilities on the set of propositions, $\{A_i\}$. In that case we merely feed Robby with the set of prior probabilities and Robby carries out the operations.

For many people this has been the end of the design process. For many problems they say they start by introducing a "personalistic" probability. The starting point for them is how the analyst "feels about the subject". I for one do not like to start this way because it makes the topic too close to psychiatry I have no objection to psychiatry, but I do not like to have a logical method become a psychological method.

There is a way out of this dilemma. We note that the design has used only three out of the four pillars. These pillars assure us (insofar as any assurance can be given) that we are dealing with problems stated unambiguously, that we are using a universal method and that we have internal consistency. However, we have not used the last pillar, CANDOR.

CANDOR--ENTER THE MAXIMUM ENTROPY PRINCIPLE

It is at this point that the genius in Ed Jaynes' contribution comes to the fore. We began this discussion by considering the entropy measure introduced by Claude Shannon in connection with communication theory. We saw that the equation had meaning only insofar as we could figure out what the symbol p_i meant. We have now given a meaning to the symbol p_i. That meaning now gives a meaning to Shannon's entropy.

Entropy measures what we do not know when we have encoded our knowledge in a probability distribution. It measures what is left to learn when you are uncertain.

The subscript "i" on the probability distinguishes the propositions one from another. S then measures what is left to be learned about the truth of a set of well defined propositions.

To get started, therefore, we make use of Jaynes' principle of minimum prejudice:

THE PRINCIPLE OF MINIMUM PREJUDICE
(THE MAXIMUM ENTROPY PRINCIPLE)

Assign to the probabilities those values which maximize the entropy and which are consistent with whatever is known.

The maximum entropy principle allows Robby to attack a class of problems not usually undertaken by statisticians. The usual problems involve information which falls into one of three distinct categories.

Category I: The probabilities are given. In this case the input information is already encoded in probabilities and the only task is to apply the two relations, one for joint probabilities and the other for denials. Sometimes the combinaqtorial aspects of the problem create a difficulty, but this difficulty is purely mathematical, not conceptual.

Category II: The information given is symmetric. By symmetric we mean that an interchange of labeling carries no special information. In such cases the probabilities are treated as equal to one another. This is the case for drawing balls from an urn, cards from a deck or parts from a lot. The technique of randomization is an effort to make the information symmetric. Incidenta ly, Robby doesn't understand the idea of "randomization" as a physical phenomenon. To Robby it merely means making the information symmetr cal.

Category III: The information is given in terms of frequencies of past occurrences. It is an easily demonstrated consequence of the maximum entropy principle that if you have seen a large amount of data in the past for which the frequency of occurrence of A_i if f_i, then in the limit of a large amount of data, we make the assignment, $p_i = f_i$. [Note however, that we may use the probability assignments, $\{p_i\}$ to compute the *probability of a frequency* but we do compute the frequency of a probability. The concepts of **probability** and **frequency** are kept distinct.]

Category IV: Information is available only on <u>averages</u>. Here is an example of such a problem.

There are three products which are sold in great numbers. They are warranted according to the following schedule:

Part	Cost. per part
1	$50.00
2	7.50
3	1.00

If, on the average, there are 275 warranty payments per month and the total cost averages $800 per month, what is the probable failure rate of part #2?

This is a class of problems for which the maximum entropy formalism was developed. That development has lead to understanding of a host of other problems.

The general solution of this class of problems is as follows. We maximize the entropy subject to what is known. That is, we maximize:

$$S = -k \sum_i p_i \ln p_i \qquad (13a)$$

Subject to:

$$\sum_i p_i = 1 \tag{14a}$$

$$\sum_i p_i \, g_r(x_i) = <g_r> \qquad r=1,2,,... \tag{14b}$$

The constraints are all given in the form of various averages. $p_i = p(x_i|Z)$, where

x_i = "The value of x is the number x_i"

Z = "The data are in the form of averages as given in equation 14"

Using Lagrange's method of undetermined multipliers, this extremum problem is readily solved with the result:

$$p_i = \exp(-\lambda_o - \sum_r \lambda_r \, g_r(x_i)) \tag{15a}$$

$$\lambda_o = \ln \sum_i \exp(-\sum_r \lambda_r \, g_r(x_i)) \tag{15b}$$

It is seen that there are r+1 Lagrange multipliers, corresponding to the r equations of constraint (14b) and the normalization requirement (14a).

The maximum entropy solution has a number of interesting properties of which only a few are given below:

$$\frac{\partial \lambda_o}{\partial \lambda_r} = -<g_r> \tag{16a}$$

$$\frac{\partial^2 \lambda_o}{\partial \lambda_r^2} = \text{variance}(g_r) \tag{16b}$$

$$\frac{\partial^2 \lambda_o}{\partial \lambda_j \, \partial \lambda_k} = \text{covar}(g_j, g_k) \tag{16c}$$

THE STRUCTURE OF INDUCTIVE LOGIC

With the addition of the maximum entropy principle, the structure looks like this:

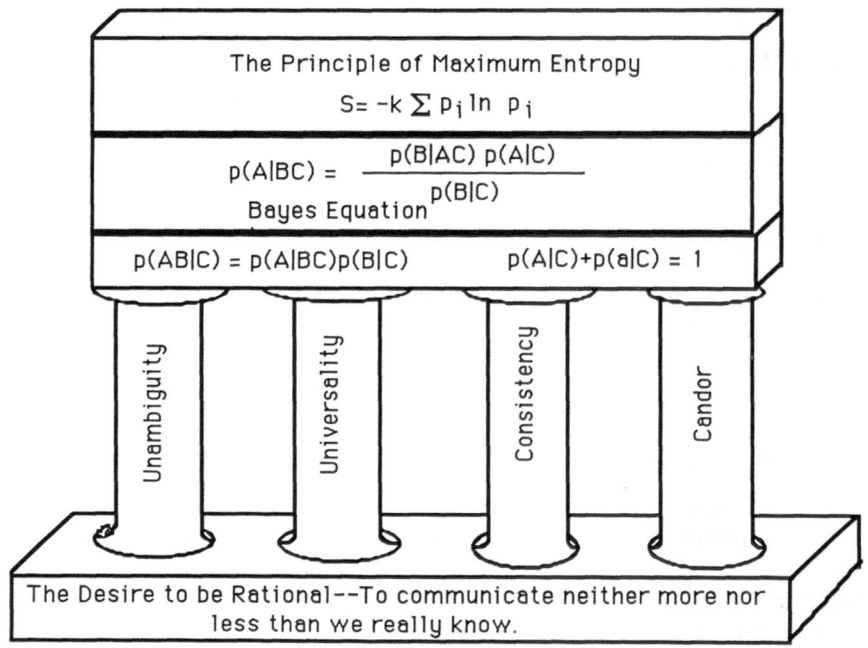

According to this principle of maximum entropy, the least prejudiced assignment of probabilities is that which maximizes Shannon's measure and agrees with the given information. With this principle our machine is completed.

SIMULATING THE MACHINE'S BEHAVIOUR

Of course we do not have to wait until someone builds the machine for us. We can simulate its behaviour by simply using the same equations in some other computer. We can even mimic its behaviour by writing and solving the equations without a computer. Returning to the problem of the warranties, we organize the input information like this:

Let $p(n_i|Z)$ = "The probability that there are n_i warranties paid for part #i, given Z, the facts of the problem as stated above"

The given information, using probability encoding is :

$$\sum_i \sum_{n_i} p(n_i|Z) \, n_i = <N> \tag{17}$$

$$\sum_i \sum_{n_i} p(n_i|Z) \, n_i \, c_i = <C> \tag{18}$$

Maximizing the entropy,which in this case is given by:

$$S = -k \sum_i \sum_{n_i} p(n_i|Z) \ln p(n_i|Z) \tag{19}$$

we find:

$$p(n_i|Z) = \exp(-\lambda_{0,i} - \lambda_1 n_i - \lambda_2 n_i c_i) \tag{20}$$

where $\lambda_{0,i}$ is adjusted to provide for normalization and the multipliers λ_1 and λ_2 are adjusted to agreee with the constraints ($<N>$ = 275 and $<C>$ = $800).

Using the above results we compute that the normalization lagrangian multiplier is given by:

$$\lambda_{o,i} = \ln \sum_{n_i=0}^{\infty} \exp(-\lambda_1 - \lambda_2 c_i) = -\ln(1 - \exp(-\lambda_1 - \lambda_2 c_i)) \tag{21}$$

and from equation 16a we find:

$$-\frac{\partial \lambda_{o,i}}{\partial \lambda_1} = <n_i> = \frac{1}{\exp(\lambda_1 + \lambda_2 c_i) - 1} \tag{22}$$

$$-\frac{\partial \lambda_{o,i}}{\partial \lambda_2} = <c_i> = \frac{c_i}{\exp(\lambda_1 + \lambda_2 c_i) - 1} \tag{23}$$

Therefore, we have:

$$\sum_{i=1}^{3} \frac{1}{\exp(\lambda_1 + \lambda_2 c_i) - 1} = <N> \qquad (24)$$

$$\sum_{i=1}^{3} \frac{c_i}{\exp(\lambda_1 + \lambda_2 c_i) - 1} = <C> \qquad (25)$$

Using $\{c_1, c_2, c_3\} = \{50, 7.5, 1\}$ as given in the problem statement and taking n=275 and C = 800, we find, using computer search techniques, λ_1=0.002 and λ_2=0.003. From these values we compute:

$\lambda_{0,1}$ = - ln(1-exp(-0.002-0.003*50)) = 1.95891

$\lambda_{0,2}$ = - ln (1-exp(-0.002-0.003*7.5))=3.72131

$\lambda_{0,3}$ = - ln (1-exp(-0.002-0.003*1))=5.11899

With these values we find:

$<n_1> = [\exp(\lambda_1 + 50^*\lambda_2) - 1]^{-1} = 6$

$<n_2> = [\exp(\lambda_1 + 7.5^*\lambda_2) - 1]^{-1} = 40$

$<n_3> = [\exp(\lambda_1 + 1.0^*\lambda_2) - 1]^{-1} = 200$

The failure rate of part #2 is 40/month. Its distribution is exponential.

I have dwelt on this method because it opens up a new line of inquiry in the field of statistics. Other speakers on this morning's program will discuss the more familiar topic of how to use Bayes' equation to assign values to parameters of statistical distributions.

BUILDING ON THE FOUNDATIONS--NEW AREAS OF APPLICATION

A small but aggressive number of new people is entering the field of statistics. They are not concentrating on the old familiar problems. They are using these new methods in such fields as photo interpretation, statistical physics, engineering design, reliability and decision analysis. They are erecting new structures. In general these people are not classically trained statisticians.

The new structure looks like this:

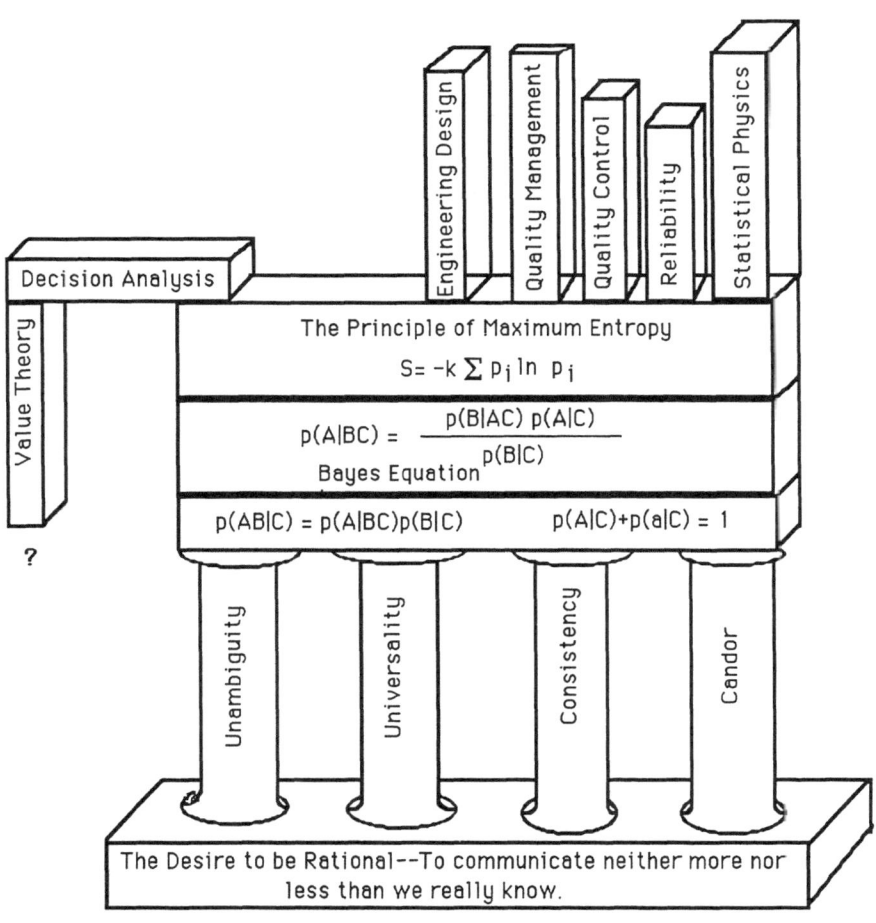

In the fields of statistical physics, reliability, quality control and similar applications the structures are on as solid a footing as we know how to devise.

In the field of decision analysis, on the other hand, there is a dependency not only on the system of inductive logic but also on the field of value analysis. Value analysis, I believe, does not have much of a foundation, so I have depicted it as hanging in mid-air.

This is fortunate, for it gives those generations which follow us something interesting to do.

In this paper I have only discussed four categories of input information for Robby. There are other equally interesting and important categories to be developed. As the number of applications increases I predict that the field of statistical inference will change to build more and more on these new techniques. Already there are young people entering the field. They will change it if you do not. Time and logic are on their side.

REFERENCES

I list here various books in Bayesian inference in no particular order of importance.

1. Jeffreys, Harold "Theory of Probability", Oxford Press, 1961 Third Edition.
(This is the first comprehensive treatment, using a unique and powerful notation with many examples. An excellent reference for those who are mathematically able)

2. Cox, Richard "The Algebra of Probable Inference", Johns Hopkins University Press, Baltimore, MD 1961. (This short book updates and extends Professor Cox's original 1946 paper. It provides the basis for a system of inference devoid of the usual reliance on populations, etc. but never loosing sight of the original goal, which is to establish a system of inference.)

3. Levine, R and Tribus, M (editors) "The Maximum Entropy Formalism", MIT Press, Cambridge, MA 1979. (This book contains a collection of 16 papers by different authors active in Bayesian inference and maximum entropy methods.)

4. Jaynes, Edwin T. "Where do we Stand on Maximum Entropy" in reference 3 above, pp 15-117. (This paper is a review of the foundations, applications and attacks upon the maximum entropy formalism. It contains numerous references worth reading in their own right.)

5. Justice, James H. (editor) "Maximum Entropy and Bayesian Methods in Applied Statistics" Cambridge University Press, England. 1986 (This book contains 17 papers in applied probability including an updating by Jaynes of attacks upon and defenses of the methods, applications to photo analysis, estimates of the Earth's density distribution and applications in seismology. Each paper contains many references in the field)

6. Tribus, Myron "Rational Descriptions, Decisions and Designs", Pergamon Press, 1969 (This book contains many of the derivations referred to in the body of this paper. It is out of print and hard to get, probably deservedly so.)

BAYESIAN INDUCTIVE INFERENCE AND MAXIMUM ENTROPY

Stephen F. Gull
Mullard Radio Astronomy Observatory
Cavendish Laboratory
Madingley Road
Cambridge CB3 0HE, United Kingdom

ABSTRACT

The principles of Bayesian reasoning are reviewed and applied to problems of inference from data sampled from Poisson, Gaussian and Cauchy distributions. Probability distributions (priors and likelihoods) are assigned in appropriate hypothesis spaces using the Maximum Entropy Principle, and then manipulated via Bayes' Theorem. Bayesian hypothesis testing requires careful consideration of the prior ranges of any parameters involved, and this leads to a quantitive statement of Occam's Razor. As an example of this general principle we offer a solution to an important problem in regression analysis; determining the optimal number of parameters to use when fitting graphical data with a set of basis functions.

INTRODUCTION

At the Calgary meeting two years ago Ed Jaynes gave a tutorial introduction (Jaynes 1986) that provides the historical and philosophical background to the principles of Bayesian inference. Like that paper, which the reader is strongly encouraged to study, the aim here is to provide a tutorial guide to Bayesian methods. A short resumé of the basic principles is presented, but the emphasis of this paper is more technical, showing the application of the method to a selection of problems that are solved in detail. We then take a glimpse of the Frontiers of the subject, where (following Jeffreys) a quantitive statement of Occam's Razor is offered. Finally, we turn to the problem of curve-fitting, a state-of-the-art example of Bayesian methods.

THE GROUND RULES

I want to distinguish clearly three stages that together make up my Bayesian view of probability theory and statistics. I believe that all three stages are essential to the process of inductive reasoning.

G. J. Erickson and C. R. Smith (eds.),
Maximum-Entropy and Bayesian Methods in Science and Engineering (Vol. 1), 53–74.
© 1988 by Kluwer Academic Publishers.

1) <u>Bayes'</u> <u>Theorem</u>.
In its simplest form this elementary theorem relates the
probabilities of two events or hypotheses A and B. It states
that the <u>joint</u> probability distribution function (p.d.f.) of
A <u>and</u> B can be expressed in terms of the <u>marginal</u> and
<u>conditional</u> distributions:

$$pr(A,B) = pr(A) \, pr(B|A) = pr(B) \, pr(A|B).$$

Bayes' theorem is merely a re-arrangement of this
decomposition, which itself follows from the requirement of
consistency for the manipulation of probabilities (Cox
1946). Of course, anyone can prove this theorem, but people
who believe it and use it are called <u>Bayesians</u>. However,
before anyone, even Bayesians, can use it, the joint p.d.f.
has to be assigned. Because Bayes' theorem is simply a rule
for manipulating probabilities, it cannot by itself help us
to assign them in the first place, and for that we have to
look elsewhere.

2) <u>Maximum</u> <u>Entropy</u>.
The Maximum Entropy principle (MaxEnt) is a variational
principle for the <u>assignment</u> of probabilities under certain
types of constraint called <u>Testable</u> <u>Informatation</u>. These
constaints are ones that refer to the probability
distribution directly: e.g. for a discrete p.d.f. $\{p_i\}$, the
ensemble average of a quantity r $<r> = \sum_i r_i p_i$ constitutes

testable information. MaxEnt states that the probabilities
are given by maximising the <u>Entropy</u>

$$S = -\sum_i p_i \log p_i/m_i \text{ under the constraints } \sum_i p_i = 1 \text{ and}$$

$<r>$ given, where $\{m_i\}$ is a suitable measure over the space
of possibilities (hypothesis space). The MaxEnt rule can be
justified as the only consistent variational principle for
the assignment of probability distributions (Shore & Johnson
1980, Gull & Skilling 1984, Skilling 1988). It can also be
derived in a multitude of other ways (Jaynes 1986). In the
simplest case there is no additional information other than
normalisation: MaxEnt then gives equal probabilities to all
events, in accordance with Laplace's "principle of
indifference". In fact, I believe that MaxEnt is the only
logical method we have for the assignment of probabilities
- but it is so powerful that it may be all we need. Of
course, MaxEnt is rule for assigning probabilities once the
hypothesis space has been defined; to choose the hypothesis
space we have again to look elsewhere.

3) <u>Choosing the hypothesis space</u>
The real art is to choose an appropriate "space of
possibilities", and to date we have no systematic way of

generating it. Transformation group arguments can often help
us (Jaynes 1968) in problems involving physical quantities;
the appropriate measure space is often uniform (location
parameters) or uniform in the logarithm (scale parameters).
MaxEnt will then assign a uniform "prior" probability
distribution over this space. However, in many problems one
has no guarantee that our choice is right in any final
sense, and this feeling of ambiguity has led to much soul-
searching. I feel (along with Jaynes, 1986) that our aims
should be different. We should not seek a "final truth" in
our hypothesis space, but use our common sense to capture
enough structure of the real problem being solved so that we
can make useful predictions. If the predictions are useful,
then that is an indication the the hypothesis space is good
enough for now, without prejudice to the possibility of
revising it later. If the predictions are not good, this is
not a disaster, for we then have learnt that the hypotheses
have to be reformulated and the ways in which our
predictions are wrong may help us to do this. In any case we
simply have nothing to lose by choosing an interim
hypothesis space and proceeding with the calculation.

Of course, not everyone sees it that way, but once you are
used to the process there is nothing more painful than the
sight of grown men being psychologically unable to make a
simple Bayesian calculation just because they might be
wrong. They could agonise forever about the hypothesis space
or prior, but unless they make that calculation they will
never know!

WHY I AM A BAYESIAN

I am ashamed to have to admit that, when I was a
physics student, I thought that the lectures on probability
theory and statistics were an unnecessary distraction from
"real physics". Whatever my motives at the time, the result
was that I had an open mind when confronted some years later
by Bayesian statistics. Whilst observing the radio sky (see
example 3) I met Geoff Daniell in a pub to discuss the
analysis of the data. In the course of that evening he
proved Bayes' theorem to my satisfaction by drawing on a
beer-mat a circle that had two lines across it, and gave me
a few examples. The following is the first example I did for
myself when I returned to the telescope.

Poisson distribution - the radioactive solid
Suppose there is a sample of a radioactive solid that
produces, on average, α decays per second. You have observed
N decays in T seconds: what is α? The Likelihood or sampling
distribution for the Poisson process is well-known as a
limiting form of Binomial distribution:

$$pr(N|\alpha,T) = (\alpha T)^N \exp{-\alpha T} / N!.$$

This distribution can also be derived by MaxEnt (see for example Skilling & Gull 1984) with a constraint on $<N> = \alpha T$, using as a measure the form $Q^N/N!$ for the number ways of distributing N objects in a large number Q of cells.

A Bayesian analysis should start with the joint p.d.f.:

$$pr(\alpha,N) = pr(\alpha) pr(N|\alpha) \quad (T \text{ is always known}).$$

To complete the assignment of this joint distribution we have to specify the hypothesis space sufficiently to determine the prior distribution $pr(\alpha)$. We can do this by noting that α is a scale parameter: if we were totally ignorant of the amount of radioactivity, we would be just as ignorant if there was twice (or half) as much. This leads to a uniform prior in $\log\alpha$, but to be quite complete we should specify some limits $[\alpha_{min}, \alpha_{max}]$ so that the prior can be normalised. Let us define a "sensible" range of: α_{min} = Hubble's constant / Avagadro's number (1 decay per gram molecule in the age of the Universe) and $\alpha_{max} = 10^{-4}$ of a lethal radiation level (or you can find another experimenter).

We now use Bayes' theorem by writing the joint p.d.f. in its alternative form:

$$pr(\alpha,N) = pr(N) pr(\alpha|N).$$

Renormalising, we get the posterior distrubuton for α:

$$pr(\log\alpha|N) = (\alpha T)^N \exp{-\alpha T} / (N-1)!.$$

The re-normalisation is possible over an infinite range of α if T>0 and N>0. This is entirely reasonable: if N=0 then we don't yet know it is radioactive, and if T=0 we haven't started looking.

This is a very simple, but highly instructive example. Figure 1 shows the Likelihood and the posterior distribution for the case $T = N = 5$, $\alpha = 1$. It is the same function of the two variables (α,N), but plotted on different axes, so that there is a remarkable switch of meaning. The Likelihood gives the probability of different numbers of decays for a constant value of α, whereas the posterior gives the relative probability of different parameter values for the single value of N that was actually observed. These are completely different concepts and it is only through Bayes' theorem that there is any relationship between them.

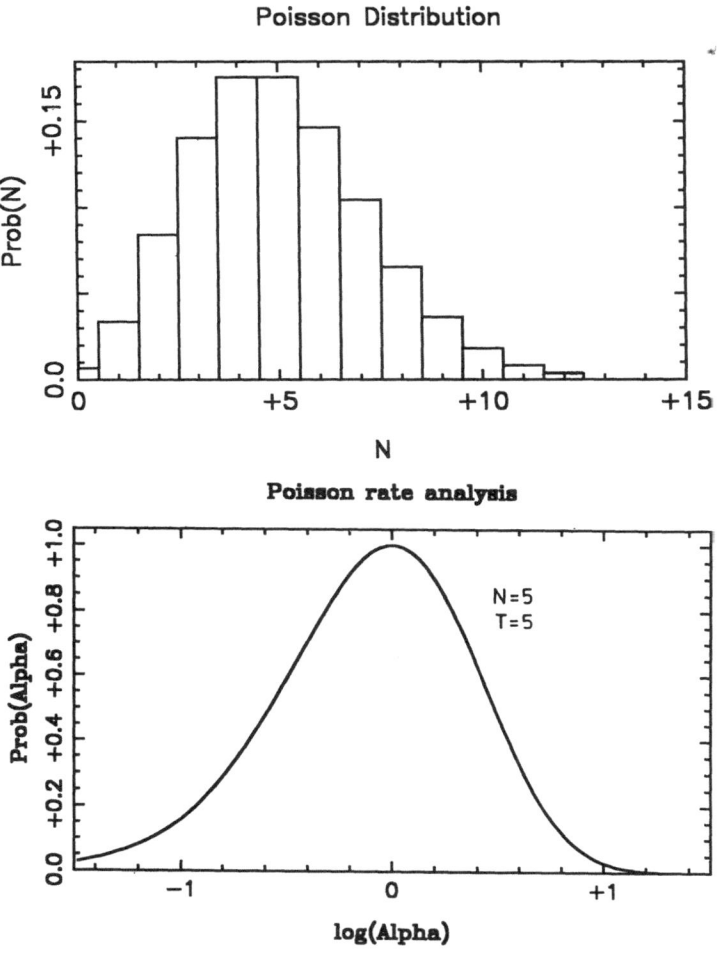

Figure 1. Likelihood and posterior probability
distribution for a Poisson process with α=1, N=T=5.

I think we should not lose sight of this; the Bayesian
rationale given above is, of course, entirely consistent
with the "Maximum Likelihood" method - in fact it provides a
justification for that method. But when we use the ML method
a natural misunderstanding arises by the word-play inherent
in the very name "Maximum Likelihood" - it makes you think
that the answer you get is the "most likely" one. Not so:
you get the parameter for which the observed datum had the
greatest Likelihood. It is only by confronting Bayes'
theorem that one can see that this is indeed (under many
circumstances) the "most likely".

Poisson rate analysis

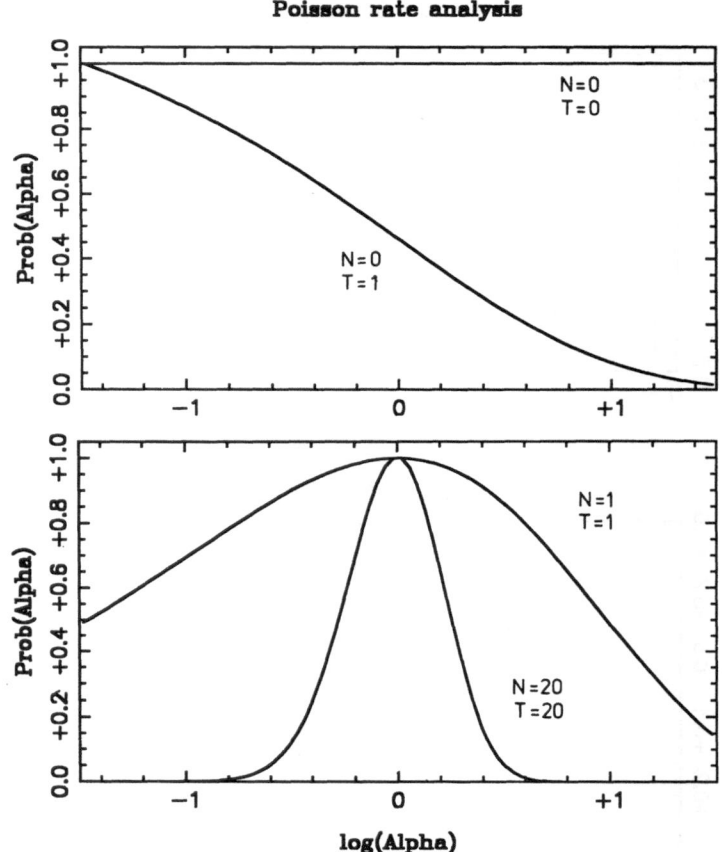

<u>Figure</u> <u>2</u>. Evolution of the posterior p.d.f. of the rate
parameter α of a Poisson process as more data become
available.

Additional features

1) Figure 2 shows some results generated by a Poisson
process on my PC. At T=0 the pr(logα) is uniform, and as
T>0, but N remains 0, high values of α becomes less likely,
and the p.d.f. can be allowed to extend over an infinite
upper interval. When the first decay occurs the lower limit
can also be extended to zero.

2) The moments of the posterior distribution are easily
calculated in terms of Gamma functions:

$$\langle\alpha\rangle = N/T,$$
$$\langle\alpha^2\rangle = N(N+1)/T.$$

As T and N increase, this leads to a width $\delta\alpha = N^{1/2}/T$ that shrinks like $T^{-1/2}$ as expected.

3) Another useful technique I should mention is to expand the logarithm of the p.d.f. near its maximum. A Taylor series about this point will yield an estimator and a width:

$$\log(pr(\log\alpha)) = \text{const.} - \alpha T + N\log\alpha T.$$

If we differentiate with respect to $\log\alpha$ we get an unbiased estimate:

$$\partial(\log pr)/\partial\log\alpha = -\alpha T + N \qquad \text{(zero at maximum)}$$
$$\partial^2(\log pr)/\partial\log\alpha^2 = -\alpha T.$$

This yields a maximum probability at $\alpha = N/T$ and an approximate width of $\delta\log\alpha = N^{-1/2}$.

These features were sufficient to convince me of the usefulness of Bayesian methods. I was, and still am, impressed by the way the beautiful result $\langle\alpha\rangle = N/T$ depends on the careful consideration of the prior for α. The next example is even more staightforward, but is still the cause of heated debate with non-Bayesians in my department.

Cauchy distribution - the lighthouse problem
(Taken from a Cambridge Part 1A examples sheet). A lighthouse is somewhere off a piece of straight coastline at position x_0 along the coast and a distance y out to sea. It emits a series of short, highly collimated flashes at random intervals and hence at random azimuths. These pulses are intercepted on the coast by photo-detectors that record only the fact that a flash has occurred, but not the azimuth from which it came. N Flashes have so far been recorded at positions $\{x_i, i=1,N\}$. Where is the lighthouse?

For any one sample the likelihood can be written in terms of the azimuthal angle θ , where $y \tan \theta = x - x_0$:

$$pr(x|x_0,y) \, dx = pr(\theta) \, d\theta = d\theta / \pi.$$

This gives the Cauchy distribution:

$$pr(x|x_0,y) = y / (\pi (y^2 + (x - x_0)^2)).$$

Different pulses are independent so that the total likelihood can be written as a product:

$$pr(\{x_i\}|x_0,y) = (y/\pi)^N \prod_i (y^2 + (x_i - x_0)^2)^{-1}.$$

Use the joint p.d.f. again, taking a uniform prior probability for the position (x_0,y) as they are location

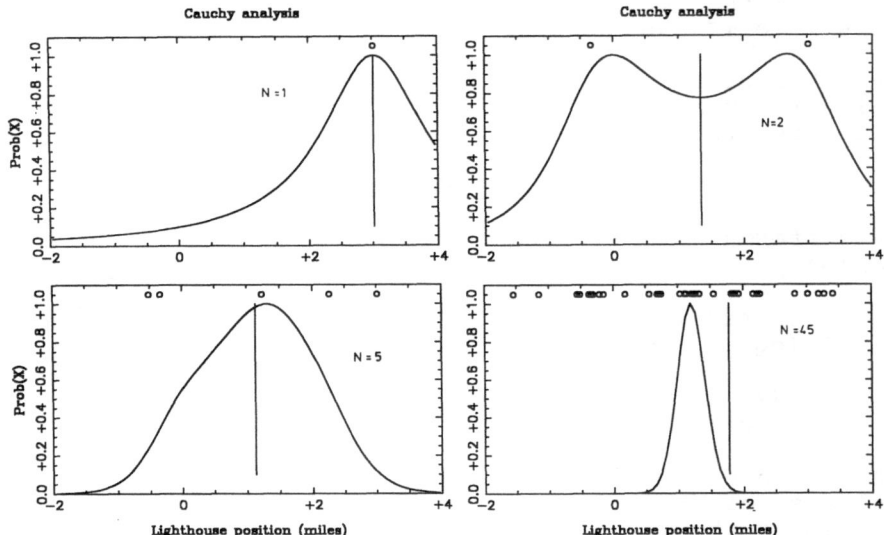

<u>Figure 3</u>. One-dimensional posterior p.d.f. of the
lighthouse position for various data samples. Note that
the distribution can be multi-modal. The vertical bar
shows the postion of the sample mean. The correct
position was at $x_0 = 1$.

parameters:

$$pr(\{x\},x_0,y) = pr(\{x\}|x_0,y)\ pr(x_0,y)$$
$$= pr(x_0,y|\{x\})\ pr(\{x\})$$

and obtain: $pr(x_0,y|\{x\}) \propto pr(\{x\}|x0,y)$.

This formula is illustrated by computer example for two
cases:

Case 1: The lighthouse is known to be 1 mile off the coast,
so that we have a one-dimensional probability distribution.

Case 2: No such restriction, so that there is two-
dimensional plot.

The figures are very revealing.

1) The Cauchy distribution has very wide wings, i.e. there
are many more "bad" data points than, for example in a
Gaussian distribution. For the 1-dimensional case (Figure 3)
this can lead to the posterior distribution being bi-modal
if the first few points are sufficiently discordant.

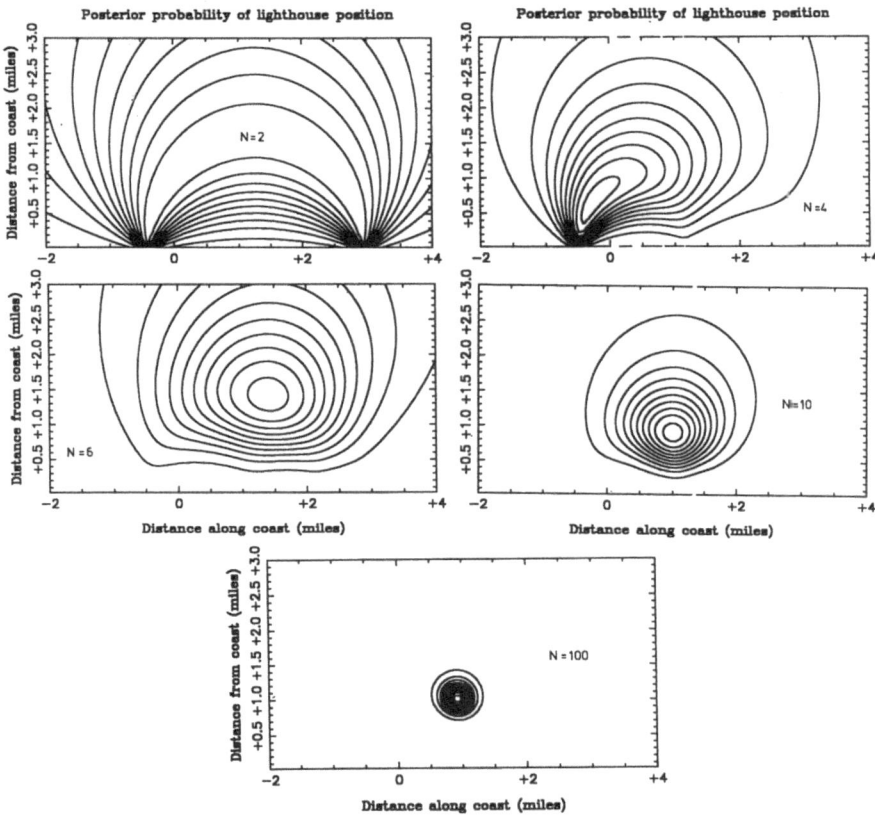

<u>Figure 4</u>. Two-dimensional plot of lighthouse position   as
function of x_0 and y. The correct position was at $x_0=y=1$.

2) Nevertheless the bulk of "good" data eventually overwhelm
the bad and the allowed range of (x_0,y) shrinks (Figure 4).

3) The sample mean $\Sigma x/N$ is not a good statistic for this
problem, and does not approach the value of x_0 any more
closely as N increases. For an excellent discussion of this
see Jaynes (1976).

<u>Gaussian distribution</u>
Meanwhile, back at the telescope, I was observing a patch of
sky repeatedly in an attempt to detect a putative "hole" in
the temperature of the Cosmic Microwave Background
Radiation. The depth of this hole is about 0.5mK, and
individual 1 minute measurements had a variance of about
10mK (and cost about $3 each). (Patience has now been
rewarded with 3 results of ≈10σ after 10 years (Birkinshaw
et al. 1985)). Suppose that we model the data collection as

a Gaussian process with mean μ and standard deviation σ. You have N samples $\{x_i\}$: what are μ and σ? This simple problem is worth solving here because it illustrates quite a few of the mathematical subtleties that will appear later in the section on Bayesian curve-fitting.

The single-sample likelihood for the Gaussian distribution can be derived by MaxEnt, using constraints on the first two moments of $\mathrm{pr}(x)$: $<x> = \mu$ and $<(x - \mu)^2> = \sigma^2$, and a uniform measure $m(x)$. The MaxEnt likelihood for multiple samples $\{x_i\}$ is then independent:

$$\mathrm{pr}(\{x_i\}|\mu\sigma) = (2\pi\sigma^2)^{-N/2} \exp{-\sum_i(x_i-\mu)^2/2\sigma^2}.$$

To manipulate this expression it is best to re-write the exponential as: $-(1/2\sigma^2) [N\mu^2 - 2\mu\Sigma x_i + \Sigma x_i^2].$

Now complete the square, defining the sample mean and variance $\bar{x} = \Sigma x/N$ and $V = \Sigma x^2 - N\bar{x}^2$:

$$-(1/2\sigma^2) [N(\mu - \bar{x})^2 + V].$$

We are now ready for Bayes' theorem using the joint p.d.f. again:

$$\mathrm{pr}(\{x_i\},\mu\sigma) = \mathrm{pr}(\{x_i\}|\mu\sigma) \mathrm{pr}(\mu\sigma) = \mathrm{pr}(\{x_i\}) \mathrm{pr}(\mu\sigma|\{x_i\}).$$

The prior for μ and σ has been much discussed: σ is a scale parameter and should have a uniform prior in $\log\sigma$; μ is a location parameter and should have a uniform prior. At the time that this talk was presented I followed conventional wisdom that this implied a uniform prior $\mathrm{pr}(\mu,\log\sigma)$. But if we start by allocating the prior in $\log\sigma$ over some range $[\sigma_{min}, \sigma_{max}]$ then we clearly have to assign the range in μ <u>given</u> the <u>knowledge of</u> σ. Perhaps the range in μ should be proportional to σ, which would lead rather surprisingly to $\mathrm{pr}(\mu,\log\sigma) \propto 1/\sigma^2$. Yoel Tikochinsky has another argument based on a transformation group that yields the same result. Although I will now assume $\mathrm{pr}(\mu,\log\sigma) = \mathrm{constant}$, I think that Yoel has a good point and that there still seems to be some life in this old argument!

Write the posterior distribution:

$$\mathrm{pr}(\mu,\log\sigma|\{x_i\}) \propto \sigma^{-N} \exp{-[N(\mu-\bar{x})^2+V]/2\sigma^2}.$$

The marginal distributions are interesting: the distribution for μ is a "Student-t" with N-1 degrees of freedom.

$$\mathrm{pr}(\mu|\{x_i\}) = \int d\log\sigma\, \mathrm{pr}(\mu,\log\sigma|x\{i\})$$

$$\propto [N(\mu-\bar{x})^2 + V]^{-N/2}.$$

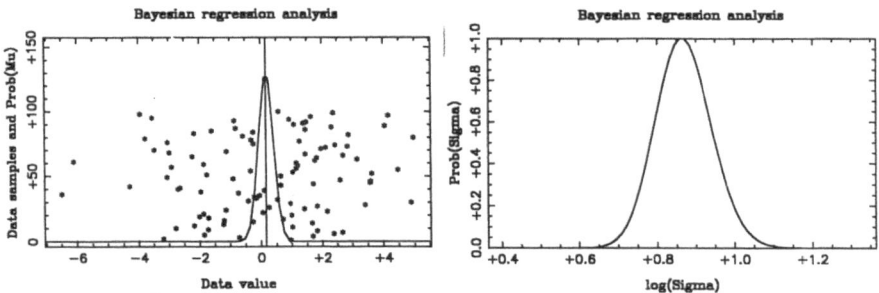

Figure 5. Marginal posterior distributions of mean and standard deviation for a Gaussian distribution. There were 100 samples with $\mu=0$ and $\sigma=2$.

Marginalising the other way we find:

$$pr(\log\sigma \mid \{x_i\}) \propto \sigma^{-N+1} \exp\text{-}V/2\sigma^2,$$

and defining $X=V/\sigma^2$:

$$pr(X \mid \{x_i\}) \propto X^{(N-3)/2} \exp\text{-}X/2.$$

In more conventional language this says that V/σ^2 is distributed like χ^2 with N-1 degrees of freedom. A good estimator is therefore $\sigma^2 \approx V/(N-1)$. This can also be seen by differentiating $\log(pr(\log\sigma))$ with respect to $\log\sigma$. Some results are plotted as Figure 5.

Note that in these examples I have treated scale parameters systematically by taking logarithms. This is good practice, because the prior is uniform in the logarithm, corresponding to the suggestion that we use log graph paper to plot the distribution. If we insist on plotting the parameter itself, then then prior is $1/\sigma$, for example. This looks a bit mysterious, even to a practising Bayesian like myself. But we don't need to confuse - take logarithms.

BAYESIAN HYPOTHESIS TESTING

The story of Mr. A and Mr. B
Why do we prefer theories with only a few parameters? The principle proposed by William of Occam - that there is more intrinsic merit in simpler theories - is universally accepted by scientists. But why? The following argument, due to Harold Jeffreys (1939, Chapter 5) explains that a simple theory can become more probable than a complicated one when confronted with data.

Suppose we have two competing theories to explain the data

D. The theory proposed by Mr. B has a parameter λ, which has to be known before the data can be predicted, but Mr. A's theory has none, and predicts the data directly. An example that occurred in physics some time ago was the Brans-Dicke scalar field theory that included a ratio ω of the strength of scalar and tensor fields (Brans & Dicke 1961). If there was no scalar component ($\omega = 0$) then the theory reduced to Einstein's General Relativity. The data in question were the classical tests of G.R., along with some new measurements of solar oblateness.

Write the likelihoods $pr(D|A)$ and $pr(D|B,\lambda)$. There is presumably a value of λ that fits the data best - call it λ_0. Let us suppose for the sake of illustration that for our particular case the likelihood is a Gaussian with width $\Delta\lambda$. Also, we must suppose that Mr. B's extra parameter allows him to fit the data better than Mr. A's inflexible one, which may or may not be a special case of Mr. B's. (The Brans-Dicke camp might say G.R. was just a special case of their theory, but the other side might retort that no such parameter existed!)

The question to be asked is then: how much bigger should $pr(D|B,\lambda_0)$ be than $pr(D|A)$ for Mr. B's theory to be preferred? We need a _quantitative_ statement of Occam's Razor. Let us try to calculate the relative probabilities of Mr. A and Mr. B's theories in the light of the data.

$$\frac{pr(A|D)}{pr(B|D)} = \frac{pr(A|D)}{\int d\lambda\, pr(B,\lambda|D)} = \frac{pr(A)\, pr(D|A)}{\int d\lambda\, pr(B,\lambda)\, pr(D|B,\lambda)}$$

$$\frac{pr(A)}{pr(B)} \times \frac{pr(D|A)}{\int d\lambda\, pr(\lambda|B)\, pr(D|B,\lambda)}$$

The difficult term is the prior for the parameter λ in Mr. B's theory. Let us take it as uniform in some range $[\lambda_{min}, \lambda_{max}]$ specified by Mr. B. Then, using the assumption of a Gaussian likelihood we find:

$$\frac{pr(A|D)}{pr(B|D)} = \frac{pr(A)}{pr(B)} \times \frac{pr(D|A)}{pr(D|B,\lambda_0)} \times \frac{(\lambda_{max} - \lambda_{min})}{(2\pi)^{\frac{1}{2}}\, \Delta\lambda}.$$

The first term in this product is a prior prejudice in favour of Mr. A or Mr. B that has nothing to do with the theory being tested. It might be taken as unity, or might even reflect their past performances. The second term is the best-case likelihood ratio, that is expected to favour Mr.

B. The third term is the "Occam factor" we are looking for and is due to the posterior collapse of Mr. B's hypothesis space. If A and B were equally probable to start with, then Mr. B has to spread his share of probability over a bigger space from λ_{min} to λ_{max}. When the data are given, many of these possible parameter values perish, and only the range $\Delta\lambda$ survive.

This analysis is the same as that given by Jeffreys, he then says that there are difficulties, which indeed there are, because λ_{min} and λ_{max} are left in an unsatisfactorily ambiguous state: what stops us taking an infinite range? That gives an infinite penalty for the parameter, which is just as bad as having no penalty at all. We have to be fair to both Mr. A and Mr. B. However, when stated in the abstract as here, I think that this ambiguity is inevitable - there can be no panacea to solve all such problems. On the other hand we can certainly make progress for many specific problems, when our prior information, whilst still vague, is not actually zero.

A further note of interest is that the decomposition of the posterior probability is precisely the same (if you take the logarithm) as that given by Peter Cheeseman and others in their "minimum message-length" approach.

BAYESIAN CURVE-FITTING

Suppose that you are given a graph consisting of N pairs of {x,y} values, and that the values of the ordinate {y_i} are subject to a constant, but unknown, amount of noise σ. The task is to fit a set of M parameters {a_j} so that the {y_i} can be adequately represented in terms of a set of basis functions {$f_j(x)$}:

$$y(x_i) \approx \hat{y}(x_i) = \sum_{j=1}^{M} a_j f_j(x_i),$$

or: $\hat{y} = f.a$, where f is an (N x M) matrix. The functions {f} might, for example, be a set of polynomials, or a Fourier series.

I must emphasise that this model problem is one where we suppose that measurement noise σ is added to an exact underlying relation $\hat{y} = f.a$ and that the {a_j} are unknown, but with no intrinsic variation from sample to sample of {y_i}. Another scenario for this sort of problem is the case where there is very little measurement noise, but the data {y_i} relate to individual objects that have a spread of {a_j} values. An example of this latter type of problem is the colour-luminosity relation for main-sequence stars (the Hertzprung-Russell diagram). Although this other case is interesting, it is _not_ the problem addressed here.

The ultimate goal of our analysis is to answer the question of how many parameters M we should use. However, we start with the relatively straightfoward task of determining the parameters $\{a_j\}$ when M and σ are known in advance. Write the joint p.d.f. as:

$$pr(\{y\},\{a\}|\sigma,M) = pr(\{a\}|\sigma,M)\ pr(\{y\}|\{a\},\sigma,M).$$

(In all of what follows the values of $\{x_i\}$ and N are known as well, but will be omitted to avoid cluttering the conditioning statements.) For the moment take the prior $pr(\{a\})$ as uniform over some large hypervolume $\delta^M a$. The likelihood (which can be derived by MaxEnt) can be taken as an independent Gaussian:

$$pr(\{y\}|\{a\},\sigma,M) = (2\pi\sigma^2)^{-N/2}\ exp\ -\sum_i(y_i - \sum_j f_{ij}a_j)^2/2\sigma^2 .$$

We can make life easier by a little rearrangement of the exponential; write it as $- V/2\sigma^2$, where:

$$V = \Sigma(y - fa)^2 = V_0 - 2\ a^t.B + a^t.A.a,$$

$$V_0 = \sum_i y_i^2, \quad B_j = \sum_i y_i f_j(x_i), \quad A_{jl} = \sum_i f_j(x_i)f_l(x_i) = f^t f.$$

The (MxM) A matrix tells us about the structure of the space spanned by the $\{f\}$; it is strictly positive definite if the $\{f\}$ are linearly independent. If the basis functions are dependent or M>N then we don't really deserve to solve the problem. A further definition we will need is the Maximum Likelihood estimator $\hat{a}$, which is the solution of the equation: $A.\hat{a} = B$. We then have:

$$V = V(M) + (a - \hat{a})^t.A.(a - \hat{a}),$$

$$V(M) = V_0 - \hat{a}^t.B .$$

The minimum of V, namely V(M), occurs at $a = \hat{a}$.

We now use Bayes' Theorem to obtain the posterior distribution:

$$pr(\{a\}|\{y\},\sigma,M) \propto exp\text{-}V(M)/2\sigma^2\ exp\text{-}(a-\hat{a})^t A(a-\hat{a})/2\sigma^2 .$$

This is a multivariate Gaussian distribution for the variables $\{a\}$, with best estimator $<a> = \hat{a}$ and covariance:

$$<\delta a_j \delta a_l> = \int d^M a\ \delta a_j \delta a_l\ pr(\{a\}|\{y\}) = \sigma^2\ [A^{-1}]_{jl},$$

where $\delta a = a - \hat{a}$.

It is worth dwelling for a while on these formulae, particularly that for the covariance. This result is "obvious" if you "diagonalise" the matrix A (in imagination only!), because the individual dimensions in the integral then separate. Another thing we can do with the formula is to use it for <u>prediction</u> of $\hat{y}(x)$:

$$\langle\hat{y}\rangle = f(x) \cdot \hat{a},$$

$$\langle\delta\hat{y}^2\rangle = \sigma^2 \sum_{jl} [A^{-1}]_{jl} \, f_j(x) \, f_l(x).$$

A final note is that the practical solution of the equation for $\hat{a}$ is best done by least-squares solution of $|B-f.a|^2$, not by solving the normal equations directly. The numerical conditioning of the least-squares solution is determined by the singular values of f, which are the square roots of the eigenvalues of A itself.

The <u>determination</u> of σ

We now turn to the next easiest problem: M is given but the noise σ is unknown. To do this we expand the hypothesis space to include σ as a parameter, taking a uniform prior for $\log\sigma$ because σ is a scale parameter. To be definite we take this over a range $[\sigma_{min}, \sigma_{max}]$, but the range will not matter if N>M. This procedure is equivalent to "forgetting" σ - the posterior distribution for σ will then single out the most likely value of σ consistent with the data. We now should be careful about all the factors of σ that appear in the normalisation of the joint p.d.f, and find:

$$pr(\log\sigma,\{a\}|\{y\},M) \propto \sigma^{-N} \, \exp\text{-}V(M)/2\sigma^2 \, \exp\text{-}\delta a^t A \delta a /2\sigma^2.$$

We now integrate this over $\{a\}$ to find the marginal distribution:

$$pr(\log\sigma|\{y\}) = \int d^M a \, pr(\log\sigma,a|\{y\})$$

$$\propto \sigma^{M-N} \, \exp\text{-}V(M)/2\sigma^2.$$

This means that the posterior distribution of $V(M)/\sigma^2$ is like χ^2 with N-M degrees of freedom. If we have to choose a single value of σ, then $\sigma^2 = V(M)/(N-M)$ is a pretty good guess. As might be expected, we lose a degree of freedom for each parameter estimated.

The <u>real</u> <u>problem:</u> <u>which</u> <u>M</u> <u>to</u> <u>use?</u>

Figure 6 shows some data kindly provided to me by colleagues as a blind test. It is (I was assured) a polynomial, with added noise, though not much. Also plotted is the fit for

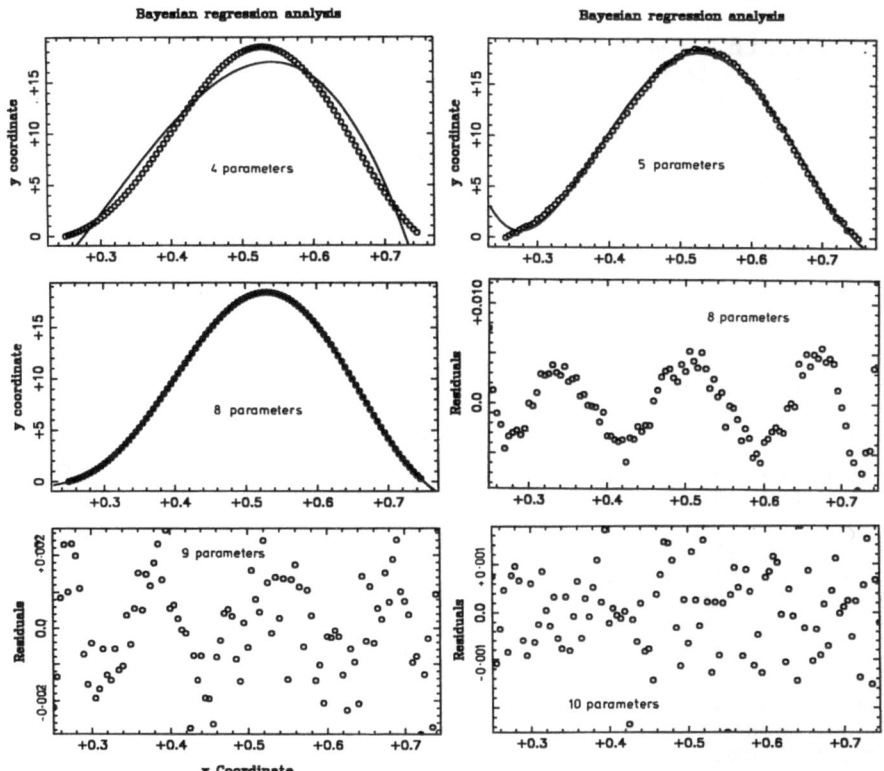

<u>Figure 6</u>. Best-fitting polynomial curves of order 4,5 and
8 compared with a sample of 100 {x,y} values. The
residuals are shown for polynomial orders 8,9 and 10.

M=4,5,8 and the residuals for M=8,9,10. The graph of the
minimum Variance V(M) against polynomial order M is shown in
Figure 7. We see now the real problem: the V(M) curve
decreases monotonically, quickly at first, but then more
slowly. But how much decrease of V(M) must we have before it
is worth adding a new parameter? We need to be fair: if we
accept <u>any</u> decrease, then we approach the dreaded "Sure
Thing" Theory (Copyright (c) E.T. Jaynes), if we are over-
cautious we will miss true structure. In Bayesian terms,
this problem is related to the hypervolume $\delta^M a$ associated
with any M. We must complete the assignment of priors:

$$pr(\{a\},M) = pr(M) \, pr(\{a\}|M).$$

The final prior pr(M) scarcely matters and we take it as
constant in $1 < M < M_{max}$.

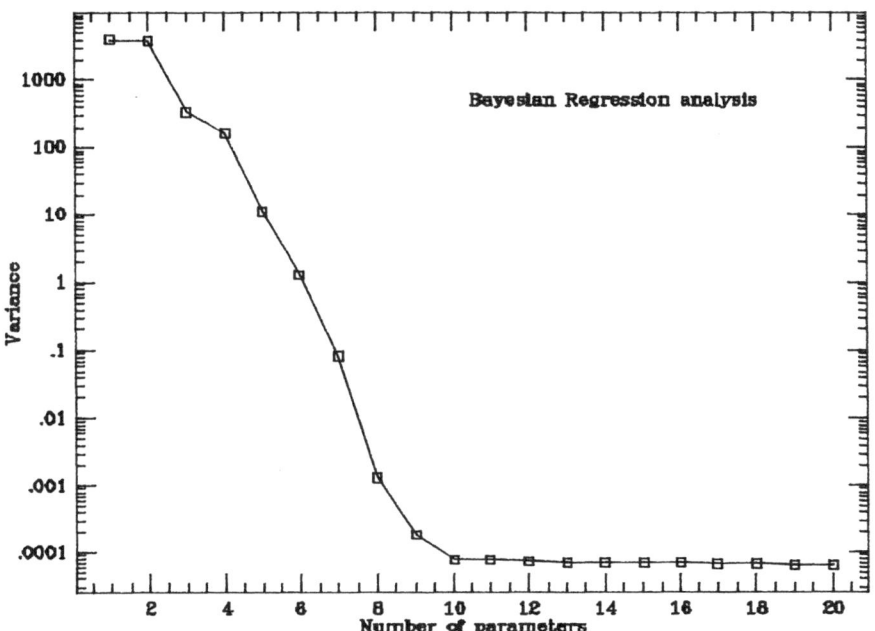

Figure 7. Minimum variance V(M) as a function of polynomial order for the example of Figure 6.

The important factor is, of course, $pr(\{a\}|M)$; we need a prior that ties together the different values of M. A possible way in which we can do this is by referring everything to the N-dimensional space of the ordinates $\{\hat{y}(x_i)\}$. Suppose that the points $\{x,\hat{y}\}$ are drawn on a piece of graph-paper, and that the $\hat{y}$-axis extends from -R to +R. We know that the ordinates will be somewhere on this graph-paper (and so are the samples $\{y_i\}$ to within the noise σ), but we want to encode this information gently, in a way that does not prejudice the <u>shape</u> of the curve. <u>If we knew</u> R, we could accomplish this rather neatly by an ensemble average constraint of the variance:

$$<\Sigma\hat{y}^2> = N\ R^2.$$

Note that this is a statement only about the average variance, we are not constraining the variance itself. But

$$<\Sigma\hat{y}^2> = \int d^Ma\ \sum_i\ (\sum_j f_j(x_i)a_j)^2\ pr(\{a\}|R)$$

is <u>testable</u> information; it relates directly to $pr(\{a\})$. We can therefore derive $pr(\{a\})$ using MaxEnt, maximising

$$S = -\int da \; pr(a) \; \log(p(a)/m(a))$$

over some large measure space m(a) (taken as constant because the {a} are location parameters in a M-dimensional vector space). We use the Partition function:

$$Z(\beta) = \int da \; \exp-(\beta/2)a^t A a \; = (2\pi/\beta)^{M/2} \; (\det A)^{-\frac{1}{2}},$$

where the constraint is satified by: $NR^2 = M/\beta$. This leads to the prior:

$$pr(\{a\}|M,R) = (\beta/2\pi)^{M/2} \; (\det A)^{\frac{1}{2}} \; \exp -(\beta/2)a^t A a.$$

This prior neatly incorporates all the properties of orthogonality and normalisation of the basis functions, and relates everthing to the same N-dimensional hypervolume NR^2. In that sense we are being fair. But what should be be the size of this hypervolume? We face the same problem as before: if the hypervolume is to big, we pay too great a price for a new parameter and will miss real structure; if the hypervolume is too small we take too many parameters and have the additional disadvantage that the prior biases the answer too much. The hypervolume has to be "just right", which means $NR^2 \approx \Sigma y^2$ (of the data set). We can show this by expanding our hypothesis space yet further to include different values of R (or β). R and β are scale parameters, so we take uniform prior in logR or logβ. By doing this we essentially "forget" the size of the graph-paper (which we didn't know anyway!), yet retain the "fairness" property between the different values of M. The posterior distribution of pr(logR) will then automatically select the best hypervolume for our purposes, just as previously happened for the case of pr(logσ). We could, of course, simply integrate R out of the problem here and now, and obtain a nice-looking prior:

$$pr(a|M) \propto (a^t A a)^{-M/2}.$$

The proportionality warns us that this is an improper prior, still depending on the limits of logR, with weak (logarithmic) singularities at both large and small values of a. However, this integration would be counter-productive as far as practical manipulation is concerned; we will keep the Gaussian distributions around as long as possible, because we can always integrate them exactly. The difficult functional forms are those involving R and σ, and we will delay their determination until last. For the moment we note that the consequence of the prior (at fixed β) is to change our estimate of the parameters:

$$<a> = k \; \hat{a},$$

with $k = 1/(1 + \sigma^2\beta)$, the fractional weight of the data versus the prior. There is thus a (small) bias of the parameters towards zero (very small in the example given).

Our final formula for the posterior distribution is:

$$pr(M, \log\sigma, \log\beta \mid \{y_i\}) \propto$$

$$\beta^{M/2} \sigma^{-N} (\beta + 1/\sigma^2)^{-M/2} \exp[-V(M)/2\sigma^2 - (\beta k/2)B^t.\hat{a}].$$

This formula is the basis of the computer program that produced the figures. For each M, the maximum posterior probability was found and a numerical "steepest descents" integration performed to get the marginal distribution $pr(M \mid \{y\})$. However, with the caveat explained in the next section, we can proceed further analytically for the limiting case $V(M) \ll V_0$ (i.e. good data!). For this case can set $k = 1$ and $B^t.\hat{a} \approx V_0$, and find that the integral over β and σ separates into two Gamma function integrals:

$$pr(\log\sigma, \log\beta, M) \propto \beta^{M/2} \exp{-\beta V_0/2} \quad \sigma^{M-N} \exp{-V(M)/2\sigma^2}.$$

This implies chi-squared distributions for β and $1/\sigma^2$ and estimators:

$$<\sigma^2> \approx V(M)/(N-M)$$

as before, and

$$<1/\beta> \approx V_0/M,$$

leading to $V_0 \approx NR^2$ as predicted.

Further, integrating over $\log\sigma$ and $\log\beta$ we find:

$$\log pr(M) = const. + \log(\Gamma(M/2)) + \log(\Gamma((N-M)/2)) + \ldots$$

$$\ldots + (N-M)/2 \log (V_0/V(M)).$$

The error in this formula is $O(MV(M)/V_0)$, which is small for $V(M) \ll V_0$.

The performance of this formula can be judged by Figure 8, which shows the posterior distribution as a function of M. It is instructive to look again at the residuals shown in Figure 6: most people agree that there is clear evidence for a new parameter at M=9, but it would be a brave man that suggested one at M=10.

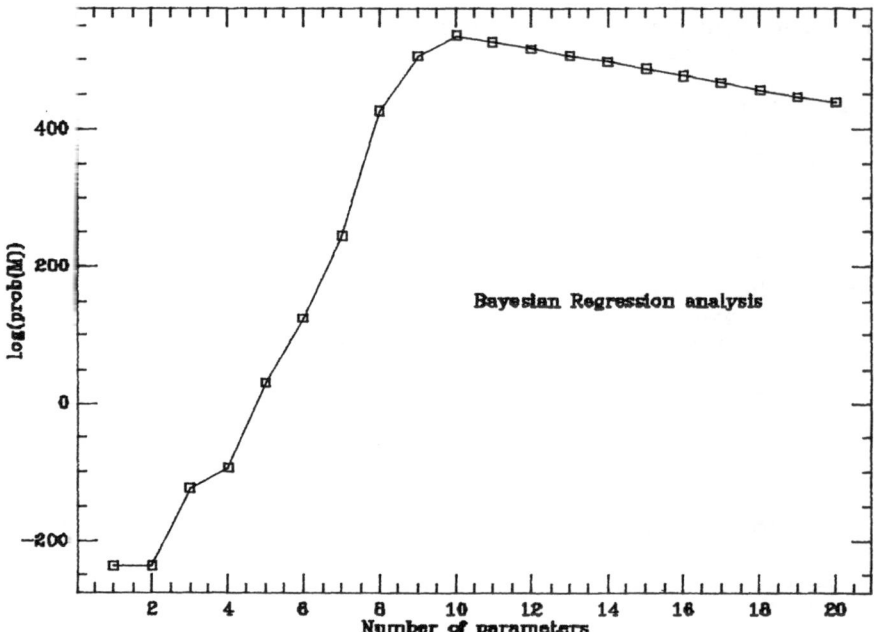

Figure 8. Posterior probability distribution of the polynomial expansion order for the example of Figure 6. There is a maximum at M = 10.

Sermon on the spike

When we perform the integral over β more carefully, by changing to kβ as a new variable, we encounter a singularity:

$$\int_0^{1/\sigma^2 - \epsilon} d(k\beta) \, (k\beta)^{M/2-1} \exp(-k\beta B^t.\hat{a}/2) \, (1 - k\beta\sigma^2)^{-1}$$

where σ is related to the maximum allowed prior range of β: $\epsilon = 1/\sigma^2\beta_{max}$. There is, therefore, a tiny "spike" which gives a logarithmically divergent contribution to the integral. This behaviour is related to the fact that we took the range parameter β (or R) to be a scale parameter, and is again a warning that some aspect of that prior assumption remains relevant in the posterior distribution. Rather than being frightened of these spikes that occur in problems of this type, let us instead make a simple calculation to see just how relevant our prior information is, after the data have arrived. We can do this by calculating what the cutoff β_{max} would have to be in order to make a 50 per cent contribution to the integral. The main part of the integral is approximately $\Gamma(M/2) \, (2/V_0)^{M/2-1}$, and the spike involves the value of the cutoff. Making suitable approximations we find that the fraction in the spike

$$\sim (V_0/2\sigma^2)^{M/2-1} \exp(-V_0/2\sigma^2) \log(\sigma^2\beta_{max}) \; / \; \Gamma(M/2)$$

For our example, with $V_0 = 4 \times 10^3$, $\sigma = 10^{-4}$, M=10, we find that the spike is important if:

$$\sigma^2\beta_{max} \sim \exp(\exp(2 \times 10^{11})),$$

where we are justified in ignoring a factor of 10^{45} as "small"! Such values of β_{max} are, of course, quite incomprehensible even to an astronomer, and indicates that our integral does indeed converge for all practical purposes.

However, let us think for a moment about the origin of this divergence. The data provide us with likelihood factors of about $\exp(-V_0/2\sigma^2)$, which are certainly large (see above), but nevertheless <u>finite</u>. The prior contains the range parameter β as a scale parameter, and in particular allows us to think of the limit $\beta_{max} \Longrightarrow \infty$, which corresponds to reducing the allowed range to zero. Eventually we arrive at the case where the prior is so sure that a = 0 that it is incapable of learning from the data. If this situation is permitted without limit, then the finite likelihood factors will not be able to overwhelm the prior. The purpose of the above calculation was to show just how pig-headed one would have to be in order to ignore the data completely. In that respect it is telling us something useful. We note, finally, that there is no corresponding problem with the limit $\beta_{min} \Longrightarrow 0$.

CONCLUSIONS

We have given a selection of examples that illustrate the simplicity and power of Bayesian methods. Bayes' rule is used to manipulate probabilities in the light of experimental data; MaxEnt is used to assign probability distributions given testable information. However, it is up to us to choose a hypothesis space that is suitable for our problem, and this not only requires us to assign an appropriate measure in the space of possibilities, but to define a range of allowed values for any parameters involved.

The collapse of hypothesis space hyper-volume leads to a penalty for introducing a new parameter. This was first described by Jeffreys, but it is a very general phenomenon that deserves to be better known.

A tentative solution has been offered to the problem of

determining the optimal number of parameters in regression analysis. The essential feature of this solution is the attempt to treat all expansion orders equally, by relating their available parameter-space hyper-volumes to a common range parameter.

ACKNOWLEDGMENTS

Thanks are due to Yoel Tikochinsky and Larry Bretthorst for helpful comments, and to Gary Erickson for patience beyond the call of duty. The graph-fitting problem was motivated by a debate at the Laramie 1985 meeting between Peter Cheesman and Jorma Rissanen.

REFERENCES

Birkinshaw, M., Gull, S.F. & Hardebeck, H. (1984). Nature, 309, 34-35.

Brans, C. & Dicke, R.H. (1961). Phys. Rev., 124, 925.

Cox, R.P. (1946). Probability, Frequency and Reasonable Expectation. Am. Jour. Phys. 17, 1-13.

Gull, S.F. & Skilling, J. (1984). Maximum entropy method in image processing. IEE Proc.,131(F), 646-659.

Jaynes, E.T. (1968). Prior probabilities. Reprinted in E.T. Jaynes: Papers on Probability, Statistics and Statistical Physics, ed. R. Rosenkrantz, 1983 Dordrecht: Reidel.

Jaynes, E.T. (1976). Confidence intervals versus Bayesian Intervals. Reprinted in E.T. Jaynes: Papers on Probability, Statistics and Statistical Physics, ed. R. Rosenkrantz, 1983. Dordrecht: Reidel.

Jaynes, E.T. (1986). Bayesian Methods - an Introductory Tutorial. In Maximum Entropy and Bayesian Methods in Applied Statistics. ed. J.H. Justice. Cambridge University Press.

Jeffreys, H. (1939). Theory of Probability, Oxford University Press. Later editions 1948, 1961,1983.

Shore, J.E. & Johnson, R.W. (1980). Axiomatic derivation of the principle of maximum entropy and the principle of minimum cross-entropy. IEEE Trans.Info.Theory, IT-26, 26-39 and IT-29, 942-943.

Skilling, J. (1986). The Axioms of Maximum Entropy. Presented at 1986 Maximum Entropy conference, Seattle, Washington (this volume).

Skilling, J. & Gull, S.F. (1984). The entropy of an image. SIAM Amer. Math. Soc. proc. Appl. Math.,14, 167-189

Excerpts from
Bayesian Spectrum Analysis and Parameter Estimation[1]

G. Larry Bretthorst

Department of Physics
Washington University
St. Louis, MO. 63130

ABSTRACT

Bayesian spectrum analysis is still in its infancy. It was born when E. T. Jaynes derived the periodogram[2] as a sufficient statistic for determining the spectrum of a time sampled data set containing a single stationary frequency. Here we extend that analysis and explicitly calculate the joint posterior probability that multiple frequencies are present, independent of their amplitude and phase, and the noise level. This is then generalized to include other parameters such as decay and chirp. Results are given for computer simulated data and for real data ranging from magnetic resonance to astronomy to economic cycles. We find substantial improvements in resolution over Fourier transform methods.

TABLE OF CONTENTS

G. J. Erickson and C. R. Smith (eds.),
Maximum-Entropy and Bayesian Methods in Science and Engineering (Vol. 1), 75–145.
© *1988 by Kluwer Academic Publishers.*

I. INTRODUCTION.

Experiments are performed in three general steps: first, the experiment must be designed; second, the data must be gathered; and third, the data must be analyzed. These three steps are highly idealized and no clear boundary exists between them. The problem of analyzing the data is one that should be faced early in the design phase. Gathering the data in such a way as to learn the most about a model is what doing an experiment is all about. It will do an experimenter little good to obtain a set of data that does not bear directly on the model to be tested.

In many experiments it is essential that one does the best possible job in analyzing the data. This could be true because no more data can be obtained, or one is trying to discover a very small effect. Furthermore, thanks to modern computers, sophisticated data analysis is far less costly than data acquisition, so there is no excuse for not doing the best job of analysis that we can. Unfortunately, the theory of optimum data analysis, which takes into account not only the raw data but also the prior knowledge that one has to supplement the data, is almost nonexistent. We hope to show the advantage of such a theory by developing a little of it and applying the results to some real data.

In Section I we outline the calculation procedure used in this paper. The spectrum estimation problem is approached using probability theory and Bayes' theorem to remove the nuisance parameters.

In Section II, we analyze a time series which contains a single stationary harmonic signal plus noise, because it contains most of the points of principle that must be faced in the more general problem. In particular, we derive the probability that a signal of frequency ω is present, regardless of its amplitude, phase, and the variance of the noise. An example is given of numerical analysis of real data illustrating these principles.

In Section III, we discuss the types of model equations used, introduce the concept of an orthonormal model, and derive a transformation which will take any nonorthonormal model into an orthonormal model. Using these orthonormal models, we then generalize the simple harmonic analysis to arbitrary model equations and discuss a number of surprising features to illustrate the power and generality of the method.

In Section IV, we collect technical discussions of several side issues that are necessary for completeness from the standpoint of the expert. Here we calculate a number of expectation values including the estimated amplitude of the signal, the variance of the data, and the power spectral density.

In Section V, we specialize the discussion to spectral estimates. In particular we discuss the estimation of multiple harmonic frequencies and their power spectra. We will then generalize the frequency and spectrum estimation problem to frequencies and spectra which are not stationary.

In Section VI, we apply the theory to a number of real time series including Wolf's relative sunspot numbers, some NMR data containing multiple close frequencies with decay, and to an economic time series which has a large trend. These analyses will give the reader a better feel for the types of applications and complex phenomena which can be investigated easily using Bayesian techniques.

The basic reasoning used in this work will be a straightforward application of Bayes' theorem: denoting by $P(A|B)$ the conditional probability that proposition A is true, given that proposition B is true, Bayes' theorem is

$$P(H|DI) = P(H|I)\frac{P(D|HI)}{P(D|I)}. \tag{1}$$

It is nothing but the probabilistic statement of an almost trivial fact: Aristotelian logic is commutative. That is, the propositions:

$$HD = \text{"Both } H \text{ and } D \text{ are true"}$$

$$DH = \text{"Both } D \text{ and } H \text{ are true"}$$

say the same thing, so they must have the same truth value in logic and the same probability, whatever our information about them. In the product rule of probability theory, we may then interchange H and D:

$$P(HD|I) = P(H|DI)P(D|I) = P(H|I)P(D|HI)$$

which is Bayes' theorem. In our problems, H is any hypothesis to be tested, D is the data, and I is the prior information. In the terminology of current statistical literature, $P(H|DI)$ is called the posterior probability of the hypothesis, given the data and the prior information. This is what we would like to compute for several different hypotheses concerning what systematic "signal" is present in our data. Bayes' theorem tells us that to compute it we must have three terms: $P(H|I)$ is the prior probability of the hypothesis (given only our prior information), $P(D|I)$ is the prior probability of the data (this term will always be absorbed into a normalization constant and will not change the distribution), and $P(D|HI)$ is called the direct probability of the of the data, given the hypothesis and the prior information. The direct probability is called the "sampling distribution" when the hypothesis is held constant and one considers different sets of data, and it is called the "likelihood function" when the data are held constant and one varies the hypothesis. Often, a prior probability distribution is called simply a "prior".

In a specific Bayesian probability calculation, we need to "define our model"; i.e. to enumerate the set $(H_1, H_2, \cdots)$ of hypotheses concerning the systematic signal that is to be tested by the calculation. A serious weakness of all Fourier transform methods is that they do not consider this aspect of the problem. In the widely used Blackman-Tukey[3] method of spectrum analysis, for example, there is no mention of any model or any systematic signal at all. From the standpoint of probability theory, the class of problems for which the Blackman-Tukey method is appropriate has never been defined. In the problems we are considering, specification of a definite model (i.e. stating just what prior information we have about the phenomenon being observed) is essential; the information we can extract from the data depends crucially on which model we analyze.

In the following section we consider the simplest nontrivial model and analyze it in some depth to show some elementary but important points of principle in the technique of using probability theory with nuisance parameters and "uninformative" priors.

II. SINGLE STATIONARY SINUSOID PLUS NOISE.

We begin the analysis by constructing the direct probability. We think of this as the likelihood of the parameters, because it is the dependence of the likelihood function on the model parameters which concerns us here. The time series $y(t)$ we are considering is postulated to contain a single stationary harmonic signal $f(t)$ plus noise $e(t)$. The basic model is always: we have recorded a discrete data set $D = \{d_1, \cdots, d_N\}$; sampled from $y(t)$ at discrete times $\{t_1, \cdots, t_N\}$; with a model equation

$$d_i = y(t_i) = f(t_i) + e_i, \quad (1 \le i \le N).$$

Different models correspond to different choices of the signal $f(t)$. We repeat the analysis originally done by Jaynes[2] using a different, but equivalent, set of model functions. We repeat this analysis for two reasons: first, by using a different formulation of the problem we can see how to generalize to multiple frequencies and more complex models; and second, to introduce a different prior probability for the amplitudes. This different prior simplifies the calculation but has almost no effect on the final result. The model we are considering in this section is

$$f(t) = A_1\cos(\omega t) + A_2\sin(\omega t)$$

which has three parameters (A_1, A_2, ω) that may be estimated from the data. The model used by Jaynes[2] was the same, but expressed in polar coordinates:

$$f(t) = A\cos(\omega t + \theta)$$
$$A = \sqrt{A_1^2 + A_2^2}$$
$$\tan\theta = -\frac{A_2}{A_1}$$

$$dA_1\, dA_2\, d\omega = A\, dA\, d\theta\, d\omega.$$

It is the factor A in the volume elements which is treated differently in the two calculations. Jaynes used a prior probability that initially considered equal intervals of A and θ to be equally likely, while we shall use a prior that initially considers equal intervals of A_1 and A_2 to be equally likely.

Of course, neither choice fully expresses all the prior knowledge we are likely to have in a real problem. This means that the results we find are conservative, and in a case where we have quite specific prior information about the parameters, we would be able to do somewhat better than in the following calculation. However, the differences arising from different prior probabilities are small provided we have a reasonable amount of data. A good rule of thumb is that one more power of A^{-1} in the prior has about the same effect on our conclusions as having one more data point.

A. The likelihood function.

To construct the likelihood we take the difference between the model function, or "signal", and the data. If we knew the true signal, then this difference would be just the noise. We wish to assign a noise prior probability density which is consistent with the available prior information. The prior should be as uninformative as possible to prevent us from "seeing" things in the data which are not there. To derive this prior probability for the noise is a simple application of the principle of maximum entropy, or if the noise is known to be the result of many small independent effects, the central limit theorem of probability theory leads to the Gaussian form independently of the fine details. Regardless; the prior probability assignment will be the same:

$$P(e_i) = \frac{1}{\sqrt{2\pi\sigma^2}} \exp\left(-\frac{e_i^2}{2\sigma^2}\right).$$

Next we apply the product rule from probability theory to obtain the probability of a set of noise values $\{e_1, \cdots e_N\}$ given by

$$P(e_1, \cdots, e_N) = \prod_{i=1}^{N} \left[\frac{1}{\sqrt{2\pi\sigma^2}} \exp\left(-\frac{e_i^2}{2\sigma^2}\right)\right]. \tag{2}$$

For a detailed discussion of why and when a Gaussian distribution should be used for the noise probability, see the original paper by Jaynes.[2] Additionally, the book of Jaynes' collected papers contains a discussion of the principle of maximum entropy and much more.[4]

The probability that we should obtain the data $D=\{d_1 \cdots d_N\}$ given the parameters is

$$P(D|H,I) \propto L(A_1,A_2,\omega,\sigma) = \prod_{i=1}^{N} \sigma^{-1} \exp\left\{-\frac{1}{2\sigma^2}[d_i - f(t_i)]^2\right\}$$

$$L(A_1,A_2,\omega,\sigma) = \sigma^{-N} \times \exp\left\{-\frac{1}{2\sigma^2}\sum_{i=1}^{N}[d_i - f(t_i)]^2\right\}. \tag{3}$$

The usual way to proceed is to fit the sum in the exponent. Finding the parameter values which minimize this sum is called least squares. The (in the Gaussian case) equivalent procedure of finding parameter values that maximize $L(A_1,A_2,\omega,\sigma)$ is called "maximum likelihood". The maximum likelihood procedure is more general than least squares: it has theoretical justification when the likelihood is not Gaussian. The departure of Jaynes was to use (3) instead in Bayes' theorem (1), and then to remove the phase and amplitude from further consideration by integration over these parameters. To do this we first expand (3)

$$L(A_1,A_2,\omega,\sigma) \propto \sigma^{-N} \exp\left\{-\frac{N}{2\sigma^2}\left[\overline{d^2} - \frac{2}{N}[A_1P(\omega)+A_2Q(\omega)] + \tfrac{1}{2}(A_1^2+A_2^2)\right]\right\} \tag{4}$$

where

$$P(\omega) = \sum_{i=1}^{N} d_i\cos(\omega t_i)$$

$$Q(\omega) = \sum_{i=1}^{N} d_i\sin(\omega t_i)$$

are the sine and cosine transforms of the data and

$$\overline{d^2} = \frac{1}{N}\sum_{i=1}^{N} d_i^2$$

is the observed mean-square data value. For a simplified preliminary discussion we have assumed the data have zero mean value (any nonzero average value has been subtracted from the data), and we simplified the quadratic term as follows:

$$\sum_{i=1}^{N}\cos^2(\omega t_i) = \frac{N}{2} + \tfrac{1}{2}\sum_{i=1}^{N}\cos(2\omega t_i) \approx \frac{N}{2}.$$

The neglected term is of order one, and is assumed small compared to N except for the isolated special case of $\omega \approx 0$. We have specifically eliminated this special case from consideration by subtracting off the constant term. A similar simplification occurs with the sine squared term. In addition, the cross term, $2A_1A_2\sum_{i=1}^{N}\cos(\omega t_i)\sin(\omega t_i)$, is at most of the same order as the terms we just ignored; therefore, this term is also ignored.

The assumption that this cross term is zero is equivalent to assuming the sine and cosine functions are orthogonal on the discrete time sampled region. Indeed, this is the actual case for uniformly spaced time intervals; however, even without uniform spacing this is a good assumption provided N is large. The assumption that the cross terms are zero by orthogonality will prove to be the key to generalizing this problem to more complex models, and eventually the assumptions that we are making now will become exact by a change of variables.

B. Elimination of nuisance parameters.

In a harmonic analysis one is usually interested only in the frequency ω. Then if the amplitude, phase, and the variance of the noise are unknown, they are referred to as nuisance parameters. The principles of probability theory uniquely determine how nuisance parameters should be eliminated. Suppose ω is a parameter of interest, and θ is a nuisance parameter. What we want is $P(\omega|D,I)$, the posterior probability (density) of ω. This may be calculated as follows: first calculate the joint posterior probability (or probability density) of ω and θ by Bayes' theorem:

$$P(\omega,\theta|D,I) = P(\omega,\theta|I)\frac{P(D|\omega,\theta,I)}{P(D|I)}$$

and then integrate out θ, obtaining the marginal posterior probability density for ω:

$$P(\omega|D,I) = \int d\theta P(\omega,\theta|D,I)$$

which expresses what the data and prior information have to tell us about ω, regardless of the value of θ.

Usually, the prior probabilities are independent:

$$P(\omega,\theta|I) = P(\omega|I)P(\theta|I) .$$

But even if they are not, the prior can be factored as

$$P(\omega,\theta|I) = P(\omega|I) P(\theta|\omega,I)$$

so the calculation can always be organized as follows: calculate the "quasi-likelihood" of ω;

$$L(\omega) = \int d\theta P(D|\omega,\theta,I) P(\theta|\omega,I) \tag{5}$$

then, to within a normalization constant, the desired distribution for ω is

$$P(\omega|D,I) \propto P(\omega|I)L(\omega).$$

If we had prior information about the nuisance parameters (such as: they had to be positive, they could not exceed an upper limit, or we had independently measured values for them) then equation (5) would be the place to incorporate that information into the calculation. We assume no prior information about the amplitudes A_1 and A_2 and assign them a prior probability which indicates "complete ignorance of a location parameter". This prior is a uniform, flat, prior density; it is called an improper prior probability because it is not normalizable. In principle, we should approach an improper prior as the limit of a sequence of proper priors. However, in this problem there are no difficulties with the use of the uniform prior because the Gaussian cutoff in the likelihood function ensures convergence in (5), and the result is the same.

Upon multiplying and integrating the likelihood (4) with respect to A_1 and A_2 one obtains the joint quasi-likelihood of ω and σ:

$$L(\omega,\sigma) \propto \sigma^{-N+2} \times \exp\left\{-\frac{N}{2\sigma^2}\left[\overline{d^2} - \frac{2C(\omega)}{N}\right]\right\} \tag{6}$$

where

$$C(\omega) \equiv \frac{1}{N}\left[P^2(\omega) + Q^2(\omega)\right]$$

the Schuster periodogram $C(\omega)$,[5] has appeared in a very natural way. If one knows the variance σ from some independent source and has no additional prior information about ω, then the problem is completed. The posterior probability density for ω is proportional to

$$P(\omega|D,\sigma,I) \propto \exp\left(\frac{C(\omega)}{\sigma^2}\right). \tag{7}$$

Because we have assumed no prior information about A_1, A_2, and ω this probability density will yield the most conservative estimate one can make from probability theory of ω and its probable accuracy.

C. Resolving power.

To obtain the (mean) ± (standard deviation) approximation for the frequency ω we expand $C(\omega)$ about the peak

$$C(\omega) = C(\omega_{max}) - \frac{b^2}{2}(\omega - \omega_{max})^2 + \cdots$$

where

$$b^2 \equiv - C''(\omega_{max}) > 0$$

we have a Gaussian approximation

$$<\hat{p}(\omega)> \approx 2C(\omega_{max}) \exp\left\{- \frac{b^2(\omega - \omega_{max})^2}{2\sigma^2}\right\}$$

from which we would estimate of the frequency

$$\omega_{est} = \omega_{max} \pm \frac{\sigma}{b}.$$

The accuracy depends on the curvature of $C(\omega)$ at its peak. For example, if the data are composed of a single sine wave plus noise $e(t)$ of standard deviation σ

$$d_t = A_1\cos(\hat{\omega}t) + e_t$$

and $\sigma \ll A_1$, then as found by Jaynes:[2]

$$\omega_{max} \approx \hat{\omega}$$

$$C(\omega_{max}) \approx \frac{NA_1^2}{4}$$

$$\omega_{est} \approx \hat{\omega} \pm \frac{\sigma}{A_1}\sqrt{48/N^3} \tag{8}$$

which indicates, as common sense would lead us to expect, that the accuracy depends on the signal-to-noise ratio, and quite strongly on how much data we have.

However, before comparing these results with experience we need to note that we are here using dimensionless units, since we took the data sampling interval to be 1. Converting to ordinary physical units, let the sampling interval be Δt seconds, and denote by f the frequency in Hz. Then the total number of cycles in our data record is

$$\frac{\hat{\omega}(N-1)}{2\pi} = (N-1)\hat{f}\Delta t = \hat{f}T$$

where $T = (N-1)\Delta t$ seconds is the duration of our data run. So the conversion of dimensionless ω to f in physical units is

$$f = \frac{\omega}{2\pi\Delta t} \text{ Hz}.$$

The frequency estimate (8) becomes

$$f_{est} = \hat{f} \pm \delta f \text{ Hz}$$

where now, not distinguishing between N and $(N-1)$,

$$\delta f = \frac{\sigma}{2\pi A_1 T} \sqrt{48/N} = 1.1 \frac{\sigma}{A_1 T \sqrt{N}} \text{ Hz}. \tag{9}$$

For example, if we have an RMS signal-to-noise ratio $= A_1/\sqrt{2}\sigma = 1$, and we take data every $\Delta t = 10^{-3}$ sec. for $T = 1$ second, thus getting $N = 1000$ data points, the theoretical accuracy for determining the frequency of a single steady sinusoid is

$$\delta f = \frac{1.1}{\sqrt{2000}} = 0.025 \text{ Hz} \tag{10}$$

while the Nyquist frequency for the onset of aliasing is $f_N = (2\Delta t)^{-1} = 500$ Hz, greater by a factor of 20,000.

To some, this result will be quite startling. Indeed, had we considered the periodogram itself to be a spectrum estimator, we would have calculated instead the width of its central peak. A noiseless sinusoid of frequency $\hat{\omega}$ would have a periodogram proportional to

$$C(\omega) \propto \frac{\sin^2\{N(\omega-\hat{\omega})/2\}}{\sin^2\{(\omega-\hat{\omega})/2\}}$$

thus the half-width at half amplitude is given by $|N(\hat{\omega}-\omega)/2| = \pi/4$ or $\delta\omega = \pi/2N$. Converting to physical units, the periodogram will have a width of about

$$\delta f = \frac{1}{4N\Delta t} = \frac{1}{4T} = 0.25 \text{ Hz} \tag{11}$$

just ten times greater than the value (10) indicated by probability theory. This factor of ten is the amount of narrowing produced by the exponential peaking of the periodogram in (7), even for unity signal-to-noise ratio.

But some would consider even the result (11) to be a little overoptimistic. The famous Rayleigh criterion[6] for resolving power of an optical instrument supposes that the minimum resolvable frequency difference corresponds to the peak of the periodogram of one sinusoid coming at the first zero of the periodogram of the second. This is twice (11):

$$\delta f_{\text{Rayleigh}} = \frac{1}{2T} = 0.5 \text{ Hz}. \tag{12}$$

There is widely believed "folk-theorem" among theoreticians without laboratory experience, which seems to confuse the Rayleigh limit with the Heisenberg uncertainty principle, and holds that (12) is a fundamental irreducible limit of resolution. Of course there is no such theorem, and workers in high resolution NMR have been routinely determining line positions to an accuracy that surpasses the Rayleigh limit by an order of magnitude, for thirty years.

The misconception is perhaps strengthened by the curious coincidence that (12) is also the minimum half-width that can be achieved by a Blackman-Tukey spectrum analysis[3] (even at infinite signal-to-noise ratio) because the "Hanning window" tapering function that is applied to the data to suppress side-lobes (the secondary maxima of $[\sin(x)/x]^2$) just doubles the width of the periodogram. Since the Blackman-Tukey method has been used widely by economists, oceanographers, geophysicists, and engineers for many years, it has taken on the appearance of an optimum procedure.

According to E. T. Jaynes, Tukey himself acknowledged[7] that his method fails to give optimum resolution, but held this to be of no importance because "real time series do not have sharp lines." Nevertheless, this misconception is so strongly held that there have been attacks on the claims of Bayesian/Maximum Entropy spectrum analysts to be able to achieve results like (10) when the assumed conditions are met. Some have tried to put such results in the same category with circle squaring and perpetual motion machines. Therefore we want to digress to explain the premise in very elementary physical terms why it is the Bayesian result (9) that does correspond to what a skilled experimentalist can achieve.

Suppose first that our only data analysis tool is our own eyes looking at a plot of the raw data of duration $T = 1$ sec., and that the unknown frequency f in (10) is 100Hz. Now anyone who has looked at a record of a sinusoid and equal amplitude wide-band noise, knows that the cycles are quite visible to the eye. One can count the total number of cycles in the record confidently (using interpolation to help us over the doubtful regions) and will feel quite sure that the count is not in error by even one cycle. Therefore by raw eyeballing of the data and counting the cycles, one can achieve an accuracy of

$$\delta f \approx \frac{1}{T} = 1 \text{ Hz.}$$

But in fact, if one draws the sine wave that seems to fit the data best, he can make a quite reliable estimate of how many quarter-cycles were in the data, and thus achieve

$$\delta f \approx \frac{1}{4T} = 0.25 \text{ Hz}$$

corresponding just to the periodogram width (11). Then the use of probability theory needs to surpass the naked eye by another factor of ten to achieve the Bayesian width (10).

What probability theory does is essentially to average out the noise in a way that the naked eye cannot do. If we repeat some measurement N times, any randomly varying component of the data will be suppressed relative to the systematic component by a factor of $N^{-1/2}$, the standard rule.

In the case considered, we assumed $N = 1000$ data points. If they were all independent measurements of the same quantity with the same accuracy, this would suppress the noise by about a factor of 30. But in our case not all measurements are equally cogent for estimating the frequency. Data points in the middle of the record contribute very little to the result; only data points near the ends are highly relevant for determining the frequency, so the effective number of observations is less than 1000. The probability analysis leading to (25) indicates that the "effective number of observations" is only about $N/10 = 100$; thus the Bayesian width (25) that results from the exponential peaking of the periodogram now appears to be, if anything, somewhat conservative. Indeed, that is what Bayesian analysis always does when we use smooth, uninformative priors for the parameters, because then probability theory makes allowance for all possible values that they might have. As noted before, if we had any cogent prior information about ω and expressed it in a narrower prior, we would be led to still better results; but they would not be much better unless the prior range became comparable to the width of the likelihood $L(\omega)$.

D. Elimination of the noise level σ.

The above analysis is valid whenever the noise variance (or power) is known. Frequently one has no independent prior knowledge of the noise. The noise variance σ^2 then becomes a nuisance parameter. We eliminate it in much the same way as the amplitudes were eliminated. Now σ is restricted to positive values and additionally it is a scale parameter. The prior which indicates "complete ignorance" of a scale parameter α is $d\alpha/\alpha = d\log\alpha$. This prior was first suggested by Sir Harold Jeffreys [8] some 50 years ago. It has since been derived by several different methods [9,10] as being the only consistent prior which indicates "complete ignorance" of a scale parameter, by several different criteria of "consistent". Multiplying equation (6) by the Jeffreys prior and integrating over all positive values gives

$$P(\omega|D,I) \propto \left[1 - \frac{2C(\omega)}{N\overline{d^2}}\right]^{\frac{2-N}{2}} \tag{13}$$

This is called a "student t-distribution" for historical reasons, although it is expressed here in very nonstandard notation. In our case it is the posterior probability density that a stationary harmonic frequency ω is present in the data when we have no prior information about σ.

This simple result shows explicitly why the discrete Fourier transform tends to peak at the location of a frequency when the data are noisy. Namely, the discrete Fourier transform is directly related to the probability that a simple harmonic frequency is present in the data, even when the noise level is unknown. Additionally, zero padding a time series (i.e. adding zeros at its end to make a longer series) and then taking the discrete Fourier transform of the padded series, is equivalent to calculating the Schuster periodogram at smaller frequency intervals. If the signal one is analyzing is a simple harmonic frequency plus noise, then the maximum of the periodogram will be the best estimate of the frequency in the absence of prior information about it.

If the signal is other than a single sinusoid, then the above analysis does not apply and the discrete Fourier transform may peak at the "incorrect" frequencies: i.e. frequencies different from those we wish to estimate. This occurs, not because the discrete Fourier transform is "wrong", but because it is answering what we should then regard as the "wrong" question. Put differently, the discrete Fourier transform is by definition the spectrum of the noisy data; but we are trying to use it to estimate a frequency in a particular model. If that model is other than a simple harmonic model (i.e. if there are several signals present, or the variation is periodic but not sinusoidal, or there is decay or chirp), there is no reason to expect the discrete Fourier transform to be a reasonable data analysis method for our different model. For each model, we must re-examine what probability theory has to say.

To apply these procedures to more complex signals we must generalize the formalism, this is done in Section III; for now we apply the simple result (13) to Wolf's relative sunspot numbers.

E. An example: Wolf's relative sunspot numbers.

Wolf's relative sunspot numbers are, perhaps, the most analyzed set of data in all of spectrum analysis. These numbers (defined as: $R = k[10g + f]$, where g is the number of sunspot groups, f is the number of individual sunspots, and k is used to scale different telescopes onto a common scale) have been collected on a yearly basis since 1700, and on a monthly basis since 1748.[11] The exact physical mechanism which generates the sunspots is unknown and no complete theory exists. Different analyses of these numbers have been published more or less regularly since their tabulation began. Here we will analyze the sunspot numbers with a number of different models including the simple harmonic analysis just completed; even though we know this analysis is too simple to be realistic for these numbers. We have plotted the time series from 1700 to 1985, Fig. 1(A). A cursory examination of this time series does indeed show a cyclic variation with a period of about 11 years. Next we computed, Fig. 1(B) the Schuster periodogram (continuous curve) and the discrete Fourier transform (open circles); these clearly show a maximum with a period near 11 years. It is a theorem that the discrete Fourier transform contains all the information that is in the periodogram; but one sees that the information is much more apparent to the eye in the continuous periodogram. We then computed the "student t-distribution" (13), Fig. 1(C), to determine the accuracy of the frequency estimate without making any assumption about σ. Now because of the processing in equation (13) all details in the periodogram have been

Figure 1.

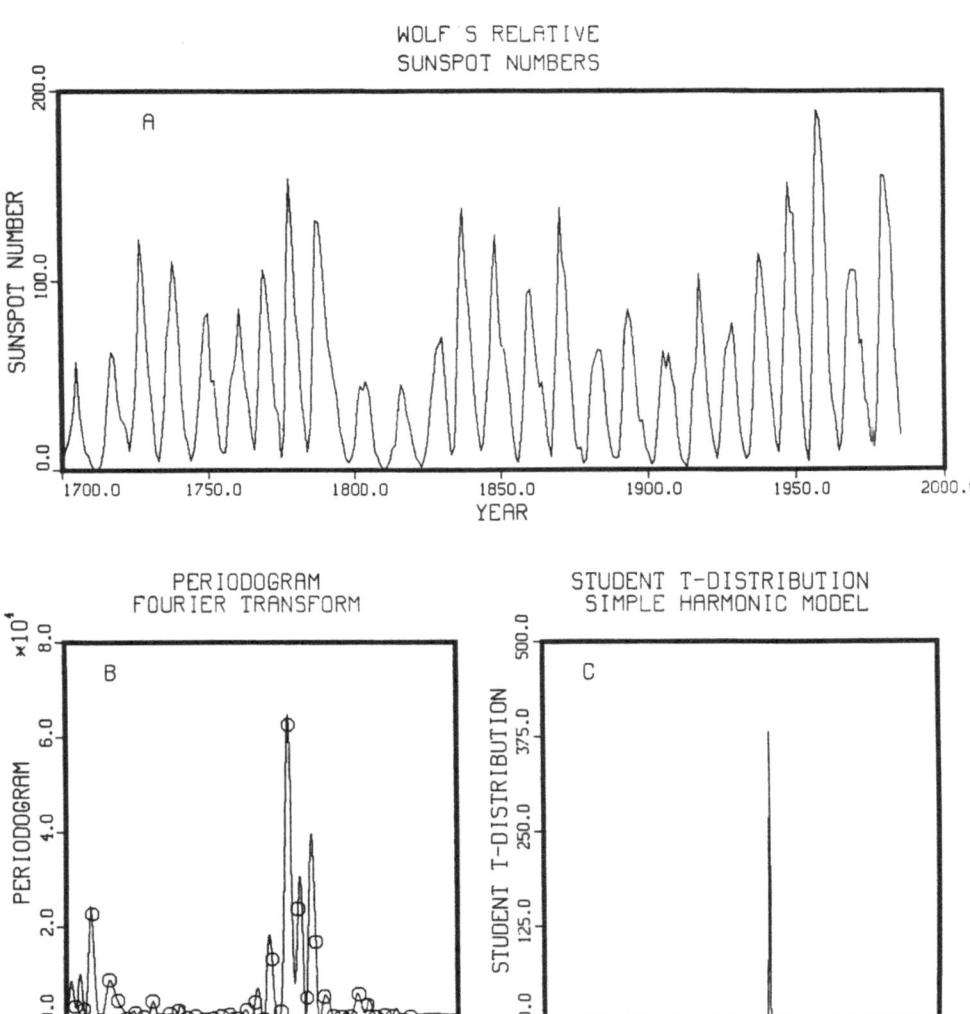

Wolf's relative sunspots Fig. 1(A) have been collected on a yearly basis since 1700. The periodogram Fig. 1(B) contains evidence of several complex phenomena. However, using the single frequency model the "student t-distribution", Fig. 1(C), picks out the 11.04 year cycle to an estimated accuracy of ±10 days.

suppressed and only the peak at 11 years remains.

We determined the accuracy of the frequency estimate as follows: We located the maximum of the "student t-distribution", integrated about a symmetric interval, and recorded the enclosed probability at a number of points. This gives:

period in years		accuracy in years	probability enclosed
11.04	±	0.012	.50
	±	0.015	.62
	±	0.020	.75
	±	0.026	.90

as the error estimates. According to this, there is not one chance in ten that the true period differs from 11.04 years by more than ten days. At first glance, this appears too good to be true.

But what does raw eye-balling of the data give? In 285 years, there are about $285/11 = 26$ cycles. If we can count these to an accuracy of $\pm 1/4$ cycle, our period estimate would be about

$$(f)_{est} = 11 \text{ years} \pm 39 \text{ days}.$$

Probability averaging of the noise, as discussed above, would reduce this uncertainty by about a factor of $\sqrt{285/10} = 5.3$, giving

$$(f)_{est} = 11 \text{ years} \pm 7.3 \text{ days}, \quad \text{or} \quad (f)_{est} = 11 \pm 0.02 \text{ years}$$

which corresponds nicely with the result of the probability analysis.

These results came from analyzing the data by a model which said there is nothing present but a single sinusoid plus noise. Probability theory, given this model, is obliged to consider everything in the data that cannot be fit to a single sinusoid to be noise. But a glance at the data shows clearly that there is more present than our model assumed: therefore, probability theory must estimate the noise to be quite large.

This suggests that we might do better by using a more realistic model which allows the "signal" to have more structure. Such a model can be fit to the data more accurately, therefore it will estimate the noise to be smaller. This should permit a still better period estimate!

III. THE GENERAL MODEL EQUATION PLUS NOISE.

These simple results already represent progress toward the more general spectral analysis problem because we were able to remove consideration of the amplitude, phase and noise level, and find what probability theory has to say about the frequency alone. In addition, it has given us an indication about how to proceed to more general problems. If we had used a model where the quadratic term in the likelihood function did not simplify, we would have a more complicated analytical solution. Although any multivariate Gaussian integral can be done, the key to being able to remove the nuisance parameters easily, and above all, selectively was that the likelihood factored into independent parts. In the full spectrum analysis problem worked on by Jaynes,[2] the nuisance parameters were not independent, and the explicit solution required the diagonalization of a matrix that could be quite large. To understand an easier approach to complex models, suppose we have a model of the form

$$d_i = f(t_i) + e_i$$

$$f(t) = \sum_{j=1}^{m} B_j G_j(t). \tag{14}$$

The model functions, $G_i(t)$, are themselves functions of other parameters which we collectively label $\{\omega\}$ (these parameters might be frequencies, chirp rates, decay rates, or any other quantities one could encounter). Now if we substitute this model into the likelihood (3) the simplification that occurred in (4) does not take place:

$$L(\{B\},\{\omega\},\sigma) \propto \sigma^{-N} \times \exp\left\{-\frac{N}{2\sigma^2}\left[\overline{d^2} - \frac{2}{N}\sum_{j=1}^{m}\sum_{i=1}^{N}B_jd_iG_j(t_i) + \frac{1}{N}\sum_{j=1}^{m}\sum_{k=1}^{m}g_{jk}B_jB_k\right]\right\} \quad (15)$$

$$g_{jk} = \sum_{i=1}^{N}G_j(t_i)G_k(t_i). \quad (16)$$

If the desired simplification is to take place the matrix g_{jk} must be diagonal.

A. The orthonormal model equations.

For the matrix g_{jk} to be diagonal the model functions G_j must be made orthogonal. This can be done by taking appropriate linear combinations of them. But care must be taken; we do not desire a set of orthogonal functions of a continuous variable t, but a set of vectors which are orthogonal when summed over the discrete sampling times t_i. It is the sum over t_i appearing in the quadratic term of the likelihood which must simplify.

To accomplish this, consider the real symmetric matrix g_{jk} defined above (16). Since for all $\sum x_j^2 > 0$,

$$\sum_{j,k=1}^{m}g_{jk}x_jx_k = \sum_{i=1}^{N}\left(\sum_{j=1}^{m}x_jG_j(t_i)\right)^2 \geq 0$$

g_{jk} is positive definite if it is of rank m. If it is of rank $r<m$, then the model functions $G_j(t)$ and/or the sampling times t_i were poorly chosen. That is, if a linear combination of the $G_j(t)$ is zero at every sampling point:

$$\sum_{j=1}^{m}x_jG_j(t_i) = 0 , \quad (1 \leq i \leq N)$$

then at least one of the model functions $G_j(t)$ is redundant and can be removed from the model without changing the problem.

We suppose that redundant model functions have been removed, so that g_{jk} is positive definite and of rank m in what follows. Let e_{kj} represent the j'th component of the k'th normalized eigenvector of g_{jk}; i.e. $\sum_{k=1}^{m}g_{jk}e_{lk} = \lambda_le_{lj}$, where λ_l is the l'th eigenvalue of g_{jk}. Then the functions $H_j(t)$, defined as

$$H_j(t) = \frac{1}{\sqrt{\lambda_j}}\sum_{k=1}^{m}e_{jk}G_k(t), \quad (17)$$

have the desired orthonormality condition,

$$\sum_{i=1}^{N}H_j(t_i)H_k(t_i) = \delta_{jk}. \quad (18)$$

The model equation can now be rewritten in terms of these orthonormal functions as

$$f(t) = \sum_{k=1}^{m}A_kH_k(t).$$

The amplitudes B_k are linearly related to the A_k by

$$B_k = \sum_{j=1}^{m}\frac{A_je_{jk}}{\sqrt{\lambda_j}} \quad \text{and} \quad A_k = \sqrt{\lambda_k}\sum_{j=1}^{m}B_je_{kj}. \quad (19)$$

The volume elements are given by

$$dB_1 \cdots dB_m = \left| \frac{e_{lj}}{\sqrt{\lambda_j}} \right| dA_1 \cdots dA_m. \qquad (20)$$

The Jacobian is a function of the $\{\omega\}$ parameters and is a constant so long as we are not integrating over these $\{\omega\}$ parameters. At the end of the calculation the linear relations between the A's and B's can be used to calculate the expected values of the B's from the expected value of the A's and the same is true of the second posterior moments

$$<B_k> = \sum_{j=1}^{m} \frac{<A_j> e_{jk}}{\sqrt{\lambda_j}} \qquad (21)$$

$$<B_k B_l> = \sum_{i=1}^{m} \sum_{j=1}^{m} \frac{e_{ik} e_{jl} <A_i A_j>}{\sqrt{\lambda_i \lambda_j}}. \qquad (22)$$

The two operations of making a transformation on the model functions and a change of variables will transform any nonorthonormal model of the form (14) into an orthonormal model (18). We still have a matrix to diagonalize, but this is done once at the beginning of the calculation. It is not necessary to carry out the inverse transformation if we are interested only in estimating the $\{\omega\}$ parameters, since these parameters are transferred into the $H_j(t)$ functions.

B. Elimination of the nuisance parameters.

We are now in a position to proceed as before. Because the calculation is essentially identical to the single harmonic calculation we will proceed very rapidly. The likelihood can now be factored into a set of independent likelihoods for each of the A_j. It is now possible to remove the nuisance parameters easily. Using the joint likelihood (15) we make the change of function (17) and the change of variables (19) to obtain the joint likelihood of the new parameters

$$L(\{A\},\{\omega\},\sigma) \propto \sigma^{-N} \times \exp\left\{ -\frac{N}{2\sigma^2} \left[\overline{d^2} - \frac{2}{N} \sum_{j=1}^{m} A_j h_j + \frac{1}{N} \sum_{j=1}^{m} A_j^2 \right] \right\} \qquad (23)$$

$$h_j \equiv \sum_{i=1}^{N} d_i H_j(t_i), \qquad (1 \le j \le m). \qquad (24)$$

Here h_j is just the projection of the data onto the orthonormal model function H_j. In the simple harmonic analysis performed in Section II, the $P(\omega)$ and $Q(\omega)$ functions are the analogues of these h_j functions. However, the h_j functions are more general: we did not make any approximations in deriving them. The orthonormality of the H_j functions was used to simplify the quadratic term. This simplification makes it possible to complete the square in the likelihood and to integrate over the A_j's, or any selected subset of them.

As before, if one has prior information about these amplitudes, then here is where it should be incorporated. We will assume that no prior information is available, and thus obtain the most conservative estimates by assigning the amplitudes a uniform prior. Then performing the m integrations one obtains

$$L(\{\omega\},\sigma) \propto \sigma^{-N+m} \times \exp\left\{ -\frac{N\overline{d^2} - m\overline{h^2}}{2\sigma^2} \right\} \qquad (25)$$

where

$$\overline{h^2} \equiv \frac{1}{m} \sum_{j=1}^{m} h_j^2 \qquad (26)$$

is the mean-square of the observed projections. This equation is the analogue of equation (6) in the simple harmonic calculation. Although it is exact and far more general, it is actually simpler in structure and gives us a better intuitive understanding of the problem, as we will see in the Bessel inequality below. In a sense $\overline{h^2}$ is a generalization of the periodogram to arbitrary model functions. In its dependence on the parameters $\{\omega\}$ it is a sufficient statistic for all of them.

Now if σ is known, then the problem is again completed provided we have no additional prior information. The joint posterior probability of the $\{\omega\}$ parameters, conditional on the data and our knowledge of σ, is

$$P(\{\omega\}|D,\sigma,I) \propto \exp\left\{\frac{m\,\overline{h^2}}{2\sigma^2}\right\}. \tag{27}$$

But if there is no prior information available about the noise, then σ is a nuisance parameter and can be eliminated as before. Using the Jeffreys prior $1/\sigma$ and integrating (25) over σ gives

$$P(\{\omega\}|D,I) \propto \left[1 - \frac{m\,\overline{h^2}}{N\,\overline{d^2}}\right]^{\frac{m-N}{2}}. \tag{28}$$

This is again of the general form of the "student t-distribution" that we found before (13). But one may be troubled by the negative sign [in the big brackets (28)], which suggests that (28) might become singular. We pause to investigate this possibility by Bessel's famous argument.

C. The Bessel inequality.

Suppose we wish to approximate the data vector $\{d_1, \cdots, d_N\}$ by the orthogonal functions $H_j(t_i)$:

$$d_i = \sum_{j=1}^{m} a_j H_j(t_i) + \text{"error"}, \qquad (1 \le i \le N).$$

What choice of $\{a_1, \cdots, a_m\}$ is "best"? If our criterion of "best" is the mean-square error, we have

$$0 \le \sum_{i=1}^{N} \left[d_i - \sum_{j=1}^{m} a_j H_j(t_i)\right]^2$$

$$= N\overline{d^2} + \sum_{j=1}^{m} (a_j^2 - 2a_j h_j)$$

$$= N\overline{d^2} - m\,\overline{h^2} + \sum_{j=1}^{m} (a_j - h_j)^2$$

where we have used (22) and the orthonormality (18). Evidently, the "best" choice of the coefficients is

$$a_j = h_j, \qquad (1 \le j \le m)$$

and with this best choice the minimum possible mean-square error is given by the Bessel inequality

$$\overline{d^2} - \frac{m}{N}\overline{h^2} \ge 0 \tag{29}$$

with equality if and only if the approximation is perfect. In other words, (28) becomes singular somewhere in the parameter space if and only if the model

$$f(t) = \sum_{j=1}^{m} A_j H_j(t)$$

can be fitted to the data exactly. But in that case we know the parameters by deductive reasoning, and probability theory becomes superfluous. Even so, probability theory is still working correctly, indicating an infinitely greater probability of the true parameter values than for
any others

D. An intuitive picture.

This gives us the following intuitive picture of the meaning of equations (25-28). The
data $\{d_j, \cdots, d_N\}$ comprise a vector in an N-dimensional linear vector space S_N. The
model equation

$$d_i = \sum_{j=1}^{m} A_j H_j(t_i) + e_i, \qquad (1 \le i \le N)$$

supposes that these data can be separated into a "systematic part" $f(t_i)$ and a white Gaussian
"random part" e_i. Estimating the parameters of interest $\{\omega\}$ that are hidden in the model
functions $H_j(t)$ amounts essentially to finding the values of the $\{\omega\}$ that permit $f(t)$ to make
the closest possible (by the mean-square criterion) fit to the data. Put differently, probability
theory tells us that the most likely values of the $\{\omega\}$ are those that allow a maximum amount
of the mean-square data $\overline{d^2}$ to be accounted for by the systematic term; from (29), those are
the values that maximize $\overline{h^2}$.

However, we have N data points and only m model functions to fit to them. Therefore,
to assign a particular model is equivalent to supposing that the systematic component of the
data lies only in an m-dimensional subspace S_m of S_N. What kind of data should we then
expect?

Let us look at the problem backwards for a moment. Suppose someone knows (never
mind how he could know this) that the model is correct, and he also knows the true values of
all the model parameters $(\{A\}, \{\omega\}, \sigma)$; call this the Utopian state of knowledge U; but he
does not know what data will be found. Then the probability density that he would assign to
any particular data set $D=\{d_1, \cdots, d_N\}$ is just our original sampling distribution (15):

$$P(D|U) = (2\pi\sigma^2)^{-\frac{N}{2}} \exp\left\{-\frac{1}{2\sigma^2}\sum_{i=1}^{N}[d_i - f(t_i)]^2\right\}.$$

From this he would find the expectations and covariances of the data:

$$<d_i> = f(t_i) \qquad (1\le i \le N)$$

$$<d_i d_j> - <d_i><d_j> = (2\pi\sigma^2)^{-\frac{N}{2}} \int d^N x x_i x_j \exp\left[-\frac{1}{2\sigma^2}\sum_{i=1}^{N}x_i^2\right] = \sigma^2 \delta_{ij}$$

therefore he would "expect" to see a value of $\overline{d^2}$ of about

$$<\overline{d^2}> = \frac{1}{N}\sum_{i=1}^{N}<d_i^2> \tag{30}$$

$$= \frac{1}{N}\sum_{i=1}^{N}(<d_i>^2 + \sigma^2)$$

$$= \frac{1}{N}\sum_{i=1}^{N}f^2(t_i) + \sigma^2,$$

but from the orthonormality (18) of the $H_j(t_i)$ we have

$$\sum_{i=1}^{N}f^2(t_i) = \sum_{l=1}^{N}\sum_{j,k=1}^{m}A_j A_k H_j(t_i)H_k(t_i) = \sum_{j=1}^{m}A_j^2.$$

So that (30) becomes

$$<\overline{d^2}> = \frac{m}{N}\overline{A^2} + \sigma^2.$$

Now, what value of $\overline{h^2}$ would he expect the data to generate? This is

$$<\overline{h^2}> = \frac{1}{m}\sum_{j=1}^{m} <h_j^2> \tag{31}$$

$$= \frac{1}{m}\sum_{j=1}^{m}\left[\sum_{i,k=1}^{N} <d_i d_k> H_j(t_i) H_j(t_k)\right]$$

$$= \frac{1}{m}\sum_{j=1}^{m}\left[\sum_{i,k=1}^{N} (<d_i><d_k> + \sigma^2\delta_{ik}) H_j(t_i) H_j(t_k)\right].$$

But

$$\sum_{i=1}^{N} <d_i>H_j(t_i) = \sum_{i=1}^{N}\sum_{l=1}^{m} A_l H_l(t_i) H_j(t_i) = \sum_{l=1}^{m} A_l \, \delta_{lj} = A_j$$

and (31) reduces to

$$<\overline{h^2}> = \overline{A^2} + \sigma^2.$$

So he expects the left-hand side of the Bessel inequality (29) to be approximately

$$<\overline{d^2}> - \frac{m\overline{h^2}}{N} \approx \frac{N-m}{N}\sigma^2. \tag{32}$$

This agrees very nicely with our intuitive judgment that as the number of model functions increases, we should be able to fit the data better and better. Indeed, when $m=N$, the $H_j(t_i)$ become a complete orthonormal set on S_N, and the data can always be fit exactly, as (32) suggests.

E. A simple diagnostic test.

If σ is known, these results give a simple diagnostic test for judging the adequacy of our model. Having taken the data, calculate $(N\overline{d^2} - m\overline{h^2})$. If the result is reasonably close to $(N - m)\sigma^2$, then the validity of the model is "confirmed" (in the sense that the data give no evidence against the model). On the other hand, if $(N\overline{d^2} - m\overline{h^2})$ turns out to be much larger than $(N - m)\sigma^2$, the model is not fitting the data as well as it should: it is "underfitting" the data. That is evidence either that the model is inadequate to represent the data (we need more model functions), or our supposed value of σ^2 is too low. The next order of business would be to investigate these possibilities.

It is also possible, although unusual, that $(N\overline{d^2} - m\overline{h^2})$ is far less than $(N - m)\sigma^2$; the model is "overfitting" the data. That is evidence either that our supposed value of σ is too large (the data are actually better than we expected), or that the model is more complex than it needs to be. By adding more model functions we can always improve the apparent fit, but if our model functions represent more detail than is really in the systematic effects at work, part of this fit is misleading: we are "fitting the noise".

A test to confirm this would be to repeat the whole experiment under conditions where we know the parameters should have the same values as before, and compare the parameter estimates from the two experiments. Those parameters that are estimated to be about the same in the two experiments are probably real systematic effects. If some parameters are estimated to be quite different in the two experiments, they are almost surely spurious: i.e. not real effects but only artifacts of fitting the noise. The model should then be simplified, by removing the spurious parameters.

Unfortunately, a repetition is seldom possible with geophysical or economic time series, although one may split the data into two parts and see if they make about the same estimates. But repetition is usually easy and standard practice in the controlled environment of a physics experiment. Indeed, the physicist's common-sense criterion of a real effect is its reproducibility. Probability theory does not conflict with good common-sense judgment; it only sharpens it and makes it quantitative. A striking example of this is given in the scenario below.

Consider now the case that σ is completely unknown, where probability theory led us to (28). As we discussed in Section II, integrating σ out of the problem as a nuisance parameter is much like estimating σ from the data, and using that estimate in our equations; if σ is actually well determined by the data, the two procedures are essentially equivalent. We can see what estimate of σ is being made in (28) by comparing it to (27). Using the fact that if $x \ll 1$ and $N \gg 1$, $(1 - x)^{-N} \approx \exp(Nx)$, (28) is crudely approximated by

$$P(\{\omega\}|D,I) \approx \exp\left\{\frac{N - m}{2} \frac{m \overline{h^2}}{N \overline{d^2}}\right\}$$

which corresponds to (27) with the variance σ^2 replaced with the estimate given by

$$(\sigma^2)_{est} = \frac{N}{N-m} \overline{d^2} = \frac{1}{N-m} \sum_{i=1}^{N} d_i^2 . \tag{33}$$

In effect, probability theory tells us that we should suppose the first m degrees of freedom to be fit by the m model functions, and apportion the observed $\sum d_i^2$ to the remaining $(N - m)$ noise degrees of freedom. But this approximation is good only when $(N - m) \gg 1$ and $m \overline{h^2} \ll N \overline{d^2}$; i.e. there are many noise degrees of freedom and the fit to the data is poor. We shall presently find the exact mean value estimate of σ^2, which turns out to be [equations (40), (41)]

$$<\sigma^2> = \frac{N}{N-m-2}\left[\overline{d^2} - \frac{m \overline{h^2}}{N}\right] \tag{34}$$

and agrees with (33) in this limit.

More interesting is the opposite extreme when (28) approaches a singular value. Consider the following scenario. You have obtained some data which are recorded automatically to six figures and look like this: $D = \{d_1=1.42316, d_2=1.50977, d_3=1.59638, \cdots \}$. But you have no prior knowledge of the accuracy of those data; for all you know, σ may be as large as 0.1 or even larger, making the last four digits garbage. But you plot the data, to determine a model function that best fits them. Suppose, for simplicity, that the model function is linear: $d_i = a + si + e_i$. On plotting d_i against i, you are astonished and delighted to see the data falling exactly on a straight line (i.e. to within the six figures given). What conclusions do you draw from this?

Intuitively, one would think that the data must be far "better" than had been thought; you feel sure that $\sigma < 10^{-5}$, and that you are therefore able to estimate the slope s to an accuracy considerably better than $\pm 10^{-5}$, if the amount of data N is large. It may, however, be hard to see at first glance how probability theory can justify this intuitive conclusion that we draw so easily.

But that is just what (28) and (34) tell us; Bayesian analysis leads us to it automatically and for any model functions. Even though you had no reason to expect it, if it turns out that the data can be fit almost exactly to a model function, then from the Bessel inequality (29) it follows that σ^2 must be extremely small and, if the other parameters are independent, they can all be estimated almost exactly.

IV. ESTIMATING THE NUISANCE PARAMETERS.

When the models had been rewritten in terms of these orthonormal model functions we were able to remove the nuisance parameters $\{A\}$ and σ. The integrals performed in removing the nuisance parameters were all Gaussian; therefore, one can always compute the moments of these parameters.

There are a number of reasons why these moments are of interest: the first moments of the amplitudes are needed if one intends to reconstruct the original model function $f(t)$; the second moments are related to the energy carried by the signal; the estimated noise variance σ^2 and the energy carried by the signal can be used to estimate the signal-to-noise ratio of the data. Thus the parameters $\{A\}$ and σ are not entirely "nuisance" parameters; it is of some interest to estimate them.

A. The expected amplitudes $<A_j>$.

To begin we will compute the expected amplitudes $<A_j>$ in the case where the variance is assumed known. Now the likelihood (23) is a function of the $\{\omega\}$ parameters and to estimate the $<A_j>$ independently of the $\{\omega\}$'s, we should integrate over these parameters. Because we have not specified the model functions we cannot do this once and for all. But we can obtain the estimated $<A_j>$ as functions of the $\{\omega\}$ parameters. This gives us what would be the "best" estimate of the amplitudes if we knew the $\{\omega\}$ parameters. The expected amplitudes are given by

$$<A_j(\{\omega\})> = \frac{\int_{-\infty}^{+\infty} dA_1 \cdots dA_m \, A_j L(\{\omega\},\{A\},\sigma)}{\int_{-\infty}^{+\infty} dA_1 \cdots dA_m \, L(\{\omega\},\{A\},\sigma)}.$$

We will carry out the first integration in detail to illustrate the procedure, and later just give results. Using the likelihood (23) and having no prior information about A_j we assign a uniform prior and integrate over the $\{A_j\}$. Because the joint likelihood is a product of their independent likelihoods, all of the integrals except the one over A_j cancel:

$$<A_j(\{\omega\})> = \frac{\int_{-\infty}^{+\infty} dA_j \, A_j \exp\left\{-\frac{1}{2\sigma^2}\left[A_j^2 - 2A_j h_j\right]\right\}}{\int_{-\infty}^{+\infty} dA_j \, \exp\left\{-\frac{1}{2\sigma^2}\left[A_j^2 - 2A_j h_j\right]\right\}}.$$

A simple change of variables $u_j=(A_j-h_j)/\sqrt{2\sigma^2}$ reduces the integrals to

$$<A_j(\{\omega\})> = \frac{\int_{-\infty}^{+\infty} du_j \, \left\{\sqrt{2\sigma^2}u_j + h_j\right\} \exp[-u^2]}{\int_{-\infty}^{+\infty} du_j \, \exp[-u^2]}.$$

The first integral in the numerator is zero by symmetry and the second gives

$$<A_j(\{\omega\})> = h_j(\{\omega\}). \tag{35}$$

This is the result one would expect. After all, we are expanding the data on an orthonormal set of vectors. The expansion coefficient is just the projection of the data onto the expansion vectors and that is what we find.

We can use these expected amplitudes $<A_j>$ to calculate the expectation values of the amplitudes $<B_k>$ in the nonorthonormal model. Using (21), these are given by

$$<B_k(\{\omega\})> = \sum_{j=1}^{m} \frac{h_j e_{jk}}{\sqrt{\lambda_j}}.$$

Care must be taken in using this formula, because the dependence of the $<B_k>$ on the $\{\omega\}$ is hidden. The functions h_j, the eigenvectors e_{kj} and the eigenvalues λ_j are all functions of the $\{\omega\}$ parameters. If one wishes to integrate over the $\{\omega\}$ parameters to obtain the best estimate of the B_k, then the integrals must be done over $<B_k(\{\omega\})>$ times the probability density of the $\{\omega\}$ parameters, including the Jacobian (20).

We would like to compute $<A_j>$ when the noise variance σ^2 is unknown to see if obtaining independent information about σ will affect these results. To do this we need the likelihood $L(\{A\},\{\omega\})$; as a function of $\{A\}$ and $\{\omega\}$ this is given by

$$L(\{\omega\},\{A\}) \propto \left[\overline{d^2} - \frac{m\overline{h^2}}{N} + \frac{1}{N}\sum_{i=1}^{m}(A_j - h_j)^2\right]^{-\frac{N}{2}}. \tag{36}$$

Using equation (36) and repeating the calculation for $<A_j>$ one obtains the same result. Apparently it does not matter if we know the variance or not. We will make the same estimate of the amplitudes regardless. As with some of the other results discovered in this calculation, this is what one's intuition might have said; knowing σ affects the accuracy of the estimates but not their actual values. Indeed, the first moments were independent of the value of σ when the variance was known; it is hard to see how the first moments could suddenly become different when it is unknown.

B. The second posterior moments $<A_jA_k>$.

The second posterior moments $<A_jA_k>$ cannot be independent of the noise variance σ^2, for that is what limits the accuracy of our estimates of the A_j. The second posterior moments when the variance is assumed known are given by

$$<A_jA_k> = \frac{\int_{-\infty}^{+\infty} dA_1 \cdots dA_m \, A_jA_kL(\{\omega\},\{A\},\sigma)}{\int_{-\infty}^{+\infty} dA_1 \cdots dA_m \, L(\{\omega\},\{A\},\sigma)}.$$

Performing the integrals gives

$$<A_jA_k> = h_jh_k + \sigma^2\delta_{jk} \tag{37}$$

or, in view of (35), the posterior covariances are

$$<A_jA_k> - <A_j><A_k> = \sigma^2\delta_{jk}.$$

The A_j parameters are uncorrelated [we defined the model functions $H_j(t)$ to ensure this], and each one is estimated to an accuracy $\pm\sigma$. Intuitively, we might anticipate this but we would not feel very sure of it.

The expectation value $<A_jA_k>$ may be related back to the expectation value for the original model amplitudes by using equation (22):

$$<B_kB_l> - <B_k><B_l> = \sigma^2\sum_{i=1}^{m} \frac{e_{ik}e_{il}}{\lambda_i}. \tag{38}$$

These are the explicit Bayesian estimates for the posterior covariances for the original model. These are the most conservative estimates (in the sense discussed before) one can make.

We can repeat these calculations for the second posterior moments in the case when σ is assumed unknown to see if obtaining explicit information about σ is of use. Of course, we expect the results to differ from the previous result since (38) depends explicitly on σ. Performing the required calculation gives

$$<A_jA_k> = h_jh_k + \left(\frac{N}{N-2}\right)\left(\frac{2N-5}{2N-5-2m}\right)\left(\frac{2N-7}{2N-7-2m}\right)\left(\overline{d^2} - \frac{m\overline{h^2}}{N}\right)\delta_{jk}.$$

Comparing this with (37) shows that obtaining independent information about σ will affect the estimates of the second moments.

C. The power spectral density $\hat{p}(\{\omega\})$.

Although not explicitly stated, we have calculated an estimate of the total energy of the signal. The estimated total energy of the signal is just $\sum <f^2(t_i)>$, which in our orthonormal model is given by $<\sum A_j^2>$. Now we have computed this expectation value as a function of the $\{\omega\}$ parameters. We would like to express the total energy carried as a density. This is easily done, the power spectral density $\hat{p}(\{\omega\})$ is given by

$$\hat{p}(\{\omega\}) = \left[m\sigma^2 + m\,\overline{h^2}\right] \frac{P(\{\omega\}|DI\sigma)}{\int d\{\omega\}\,P(\{\omega\}|DI\sigma)}. \tag{39}$$

This function is the estimated energy carried by the signal (not the noise) per unit $\{\omega\}$.

That term of $m\sigma^2$ in (39) might be a little disconcerting to some; if (39) estimates the energy carried by the "signal" why does it include the noise power σ^2? If $\overline{h^2}\gg\sigma^2$ then the term is of no importance. But in the unlikely event $\overline{h^2}\ll\sigma^2$, then what is this term telling us? When these equations were formulated we essentially put in the fact that there is present noise of variance σ^2 and a signal in a subspace of m model functions. But then if $\overline{h^2}\ll\sigma^2$, there is only one explanation: the noise is such that its components on those m model functions just happened to cancel the signal. But if the noise just cancels the signal, the power carried by the signal must be equal to the power $m\sigma^2$ carried by the noise in those m functions; and that is exactly the answer one obtains. This is an excellent example of the sophisticated subtlety of Bayesian analysis.

D. The estimated variance σ.

One of the things that is of interest in an experiment is to estimate the noise power σ^2. We can obtain the expected value of σ as a function of the $\{\omega\}$ parameters; however, we can just as easily obtain $<\sigma^s>$ for any power s. Using equation (25), and the Jeffreys prior $1/\sigma$ we integrate:

$$<\sigma^s> = \frac{\int_0^{+\infty} \frac{d\sigma}{\sigma}\,\sigma^s\,L(\{\omega\},\sigma)}{\int_0^{+\infty} \frac{d\sigma}{\sigma}L(\{\omega\},\sigma)}$$

to obtain

$$<\sigma^s> = \Gamma\left(\frac{N-m-s}{2}\right)\Gamma\left(\frac{N-m}{2}\right)^{-1}\left\{\frac{N}{2}\left[\overline{d^2} - \frac{m\,\overline{h^2}}{N}\right]\right\}^{\frac{s}{2}}. \tag{40}$$

For $s = 2$ this gives the estimated variance as

$$<\sigma^2> = \frac{N}{N-m-2}\left[\overline{d^2} - \frac{m\,\overline{h^2}}{N}\right]. \tag{41}$$

The estimate depends on the number m of expansion functions used in the model. The more model functions we use the smaller the last factor in (41), because by the Bessel inequality (29) the larger models fit the data better and $(\overline{d^2} - mN^{-1}\overline{h^2})$ decreases. But this should not decrease our estimate of σ^2 unless that factor decreases by more than we would expect from fitting the noise. The factor $N/(N-m-2)$ takes this into account; another example of sophisticated subtlety.

E. The estimated signal-to-noise ratio.

These results may be used to estimate the signal-to-noise ratio of the data. We define this as the square root of the (power carried by the signal) divided by the (mean power carried by the noise):

$$\frac{\text{Signal}}{\text{Noise}} = \left[<\sum_{j=1}^{m} A_j^2> / N\sigma^2 \right]^{\frac{1}{2}}.$$

This may be obtained from equations (37)

$$\frac{\text{Signal}}{\text{Noise}} = \left\{ \frac{m}{N} \left[1 + \frac{\overline{h^2}}{\sigma^2} \right] \right\}^{\frac{1}{2}}. \tag{42}$$

A similar signal-to-noise ratio may be obtained when the noise variance σ is unknown by replacing σ in (44) by the estimated noise variance (42).

V. SPECTRAL ESTIMATION.

The previous sections surveyed the theory in generality. In this section we will specialize the analysis to frequency and spectrum estimates. Our ultimate aim is to derive explicit Bayesian estimates of the power spectrum and other parameters when multiple nonstationary frequencies are present. We will do this by proceeding through several stages beginning with the simplest spectrum estimation problem. We do this because as was shown by Jaynes[2] when multiple, well-separated frequencies are present $[|\omega_j - \omega_k| \gg 2\pi/N]$, the spectrum estimation problem essentially separates into independent single-frequency problems. It is only when multiple frequencies are close together that we will need to use more general models.

A. The Simple Harmonic Spectrum.

The simplest frequency estimation problem one can discuss is the single frequency problem presented in Section II. For this problem, when the data are uniformly sampled in time the model can be written

$$f(t) = B_1 \cos\omega l + B_2 \sin\omega l$$

where l is an index running over a symmetric time interval $(-T \leq l \leq T)$ and $(2T+1=N)$. The matrix g_{ij} becomes

$$g_{ij} = \begin{bmatrix} \displaystyle\sum_{l=-T}^{l=T} \cos^2\omega l & \displaystyle\sum_{l=-T}^{l=T} \cos\omega l \sin\omega l \\ \displaystyle\sum_{l=-T}^{l=T} \cos\omega l \sin\omega l & \displaystyle\sum_{l=-T}^{l=T} \sin^2\omega l \end{bmatrix}.$$

For uniform time sampling the off diagonal terms are zero and the diagonal term may be summed explicitly to obtain

$$g_{ij} = \begin{bmatrix} c & 0 \\ 0 & s \end{bmatrix}$$

where c and s are given by

$$c = \frac{N}{2} + \frac{\sin(N\omega)}{2\sin(\omega)}$$

$$s = \frac{N}{2} - \frac{\sin(N\omega)}{2\sin(\omega)}.$$

Then the orthonormal model functions may be written as

$$H_1(t) = \frac{\cos(\omega t)}{\sqrt{c}}$$

$$H_2(t) = \frac{\sin(\omega t)}{\sqrt{s}} \ .$$

The posterior probability of a frequency ω in a uniformly sampled data set, independent of the signal amplitude, and phase, and the noise level, is given by equation (28). Substituting these model functions gives this as

$$P(\omega|D,I) \propto \left[1 - \frac{P(\omega)^2/c + Q(\omega)^2/s}{N \overline{d^2}} \right]^{\frac{2-N}{2}} \tag{43}$$

where $P(\omega)$ and $Q(\omega)$ are the squares of the real and imaginary parts of the discrete Fourier transform (7, 8). Notice, when $N \gg 1$ the normalization constants c and s reduce to $N/2$ and (43) reduces to equation (13) found eralier.

We would like to use the posterior probability to derive an estimate of the power spectral density $\hat{p}(\omega)$. We caution the reader again that the terms "power spectrum" or "spectral density" are used in the literature with several different meanings. Our meaning was defined previously as the expected power, over the joint posterior probability distribution of all the parameters, carried by the signal (not the noise), during the observation time. We made such an estimate in Section IV, but those estimates assumed the noise variance σ^2 was known. When the variance is unknown, the desired quantity is easily obtained from equation (39)

$$\hat{p}(\omega) \approx \left[\frac{P^2(\omega)}{c} + \frac{Q^2(\omega)}{s} \right] \frac{P(\omega|D,I)}{\int d\omega \, P(\omega|D,I)}$$

where we have dropped a term which is essentially the estimated variance of the noise. The estimated variance term can be neglected provided it is small compared to maximum of $\overline{h^2}$. This will occur whenever $\sum \langle A_j^2 \rangle \gg \sigma^2$. In practice this approximation is good when one has a few hundred data points and a signal-to-noise ratio larger than about one. But if the number of data points is large, then this equation can be further simplified to obtain

$$\hat{p}(\omega) = C(\omega) \frac{d\omega \, P(\omega|D,I)}{\int d\omega \, P(\omega|D,I)} \tag{44}$$

$$P(\omega|D,I) \approx \left[1 - \frac{2C(\omega)}{N \overline{d^2}} \right]^{\frac{2-N}{2}} \ .$$

In N is large $P(\omega|D,I)$ is effectively a delta function; the peak value of $c(\omega)$ is approximately the total energy carried by the signal.

To obtain a better understanding of the use of this power spectral estimate, we have prepared an example: the data consist of a single harmonic frequency plus Gaussian white noise, Fig. 2. We generated these data from the following equation

$$d_j = 0.001 + \cos(0.3j + 1) + e_j$$

where j is a simple index running over the symmetric interval $-T$ to T in half integer steps $(2T+1=512)$, and e_i was a Gaussian distributed random number with unit variance. After generating the time series we computed its average value and subtracted it from each data point: this insures the data have zero mean value. Figure 2(A) is a plot of this computer simulated time series, Fig. 2(B) is a plot of the Schuster periodogram (continuous curve) with the discrete Fourier transform marked with open circles. The periodogram and the discrete Fourier transform have spurious side lobes, but these do not appear in the plot of the power

Figure 2.

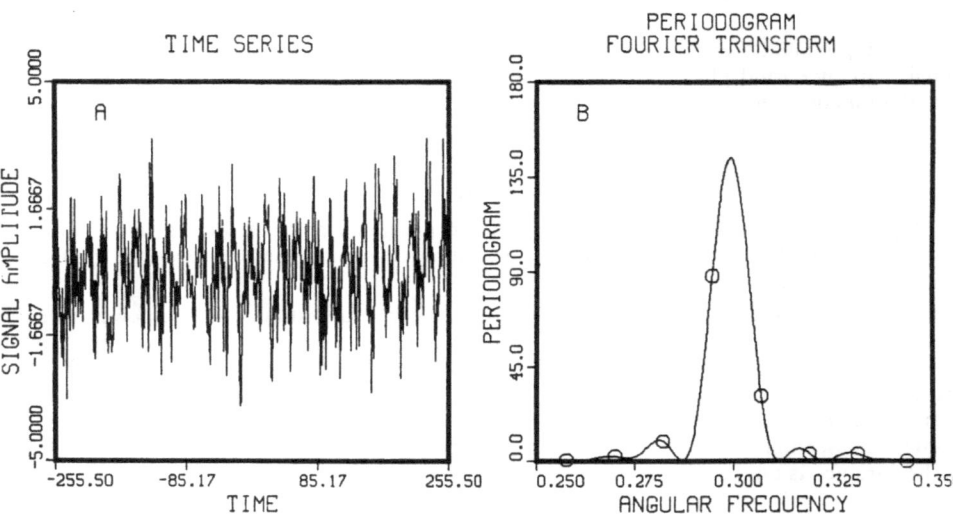

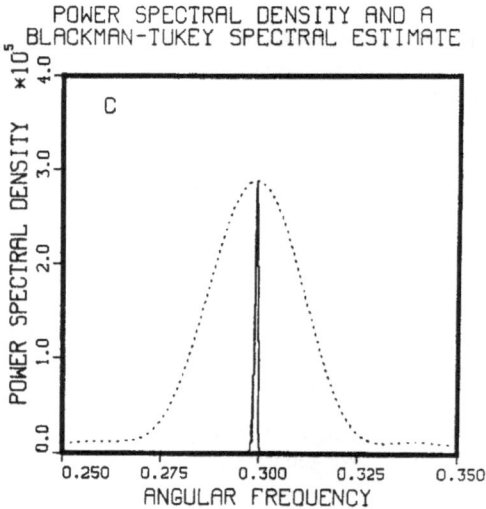

The data in Fig. 2(A) contain a single harmonic frequency plus noise. There are 512 data points in the signal with $S/N \approx 1$. The Schuster periodogram, Fig. 2(B) solid curve, and the discrete Fourier transform, open circles, clearly show a single sharp peak plus side lobes. These side lobes do not show up in the Bayesian power spectral density, Fig. 2(C), because $\hat{p}(\omega) \approx 2C(\omega) P(\omega|DI)$; the normalized posterior probability is very sharply peaked around the maximum of the periodogram. The dotted line in Fig. 2(C) is a Blackman-Tukey spectrum with a Hanning window and 256 lag coefficients. If we had used a 1/10 lag as Tukey suggests the BT spectrum would have been nearly a flat line on this scale.

spectral density Fig. 2(C) because, the processing in (39) will effectively suppress all but the very highest peak in the periodogram. This just illustrates numerically what we already knew analytically; it is only the very highest part of the periodogram that is important for estimation of a single frequency.

We have included a Blackman-Tukey spectrum estimate (dotted line) in Fig. 2(C) for comparison. The dotted line is a Blackman-Tukey spectrum using a Hanning window. The Blackman-Tukey spectrum has removed the side lobes at the cost of half the resolution in the discrete Fourier transform. The maximum lag was set at 256, i.e. over half the data. Had we used a lag of one-tenth as Tukey[3] advocates, the Blackman-Tukey spectrum would look nearly like a horizontal straight line on the scale of this plot.

Of course, the peak of the periodogram and the peak of the power spectral density occur at the same frequency. Indeed, for a simple harmonic signal the peak of the periodogram is the optimum frequency estimator. But in our problem (i.e. our model), the periodogram is not even approximately a valid estimator of the power spectrum, as Schuster supposed it to be. Consequently, even though these techniques give nearly the same frequency estimates, they give very different power spectral estimates.

B. The Simple Harmonic Signal with Lorentzian Decay.

The simple harmonic frequency problem just discussed may be generalized easily to include Lorentzian or Gaussian decay. We assume, for this discussion, that the decay is Lorentzian the generalization to other types of decay will become more obvious as we proceed. For a uniformly sampled interval the model we are considering is

$$f(l) = [\, B_1 \cos(\omega l) + B_2 \sin(\omega l) \,]\, e^{-\alpha l} \qquad (45)$$

where l is restricted to values ($1 \leq l \leq N$). We now have four parameters to estimate: the amplitudes B_1, B_2; the frequency ω; and the decay rate α. The solution to this problem is a straight forward application of the general procedures. The matrix g_{ij} (16) is given by

$$g_{ij} = \begin{bmatrix} \sum\limits_{l=1}^{N} \cos^2(\omega l) e^{-2\alpha l} & \sum\limits_{l=1}^{N} \cos\omega l \sin\omega l\, e^{-2\alpha l} \\ \sum\limits_{l=1}^{N} \cos\omega l \sin\omega l^{-2\alpha l} & \sum\limits_{l=1}^{N} \sin^2(\omega l) e^{-2\alpha l} \end{bmatrix} .$$

These sums may be done explicitly or approximated in any number of ways. We will approximate them as follows:

$$c \equiv \sum_{l=1}^{N} \cos^2(\omega l) e^{-2l\alpha} \approx \sum_{l=1}^{N} \sin^2(\omega l) e^{-2l\alpha} \approx \frac{1}{2} \sum_{l=1}^{N} e^{-2l\alpha} = \frac{1}{2} \left[\frac{1 - e^{-2N\alpha}}{e^{2\alpha} - 1} \right]. \qquad (46)$$

The off diagonal terms are at most the same order·as the ignored terms; these terms are therefore ignored. The matrix g_{ij} can be written as

$$g_{ij} \approx \begin{bmatrix} c & 0 \\ 0 & c \end{bmatrix} .$$

The orthonormal model functions may then be written as

$$H_1(l) = c^{-\frac{1}{2}} \cos(\omega l) e^{-\alpha l} \qquad (47)$$

$$H_2(l) = c^{-\frac{1}{2}} \sin(\omega l) e^{-\alpha l} \qquad (48)$$

The projections of the data onto the orthonormal model functions (24) are given by

$$h_1 \equiv c^{-\frac{1}{2}} P(\omega, \alpha) = c^{-\frac{1}{2}} \sum_{l=1}^{N} d_l \cos(\omega l) e^{-\alpha l}$$

$$h_2 \equiv c^{-\frac{1}{2}}Q(\omega,\alpha) = c^{-\frac{1}{2}}\sum_{l=1}^{N} d_l \sin(\omega l)e^{-\alpha l}$$

and the posterior probability of a frequency ω and a decay rate α is given by

$$P(\omega,\alpha|DI) \propto \left[1 - \frac{P(\omega,\alpha)^2 + Q(\omega,\alpha)^2}{Nc\,\overline{d^2}}\right]^{\frac{2-N}{2}}. \tag{49}$$

This approximation is valid provided there is plenty of data $N \gg 1$, and there is no evidence of a low frequency, there is no restriction on the range of α: if $\alpha > 0$ the signal is decaying with increasing time, if $\alpha < 0$ the signal is growing with increasing time, and if $\alpha = 0$ the signal is stationary. This equations is exactly analogous to (13) and reduces to (43) in the limit $\alpha \to 0$.

We would like to derive an estimate of the accuracy of the frequency and decay parameter estimates. To do this we can approximate the probability distribution $P(\omega,\alpha|D,I,\sigma)$ by a Gaussian. This may be done readily by assuming a form of the data, and then expanding $\overline{h^2}$ around the maximum of the probability distribution (49) as was done in Section II. From the second derivative we may obtain the desired (mean) $\pm$ (standard deviation) estimates. We take as the data

$$d(t) = A_1 \cos(\hat{\omega}t)e^{-\hat{\alpha}t} \tag{50}$$

where $\hat{\omega}$ is the true frequency of oscillation and $\hat{\alpha}$ is the true decay rate. We have assumed only a cosine component to effect some simplifications in the discussion. It will be obvious at the end of the calculation that the result for an arbitrary signal phase can be obtained by replacing the amplitude A_1^2 by the squared magnitude $A^2 \equiv A_1^2 + A_2^2$.

The projection of the data (50) onto the model functions (47, 48) is:

$$h_1 = \frac{A_1}{2\sqrt{c}}\left\{\sum_{l=1}^{N}\cos(\omega-\hat{\omega})l\,e^{-(\alpha+\hat{\alpha})l} + \sum_{l=1}^{N}\cos(\omega+\hat{\omega})l\,e^{-(\alpha+\hat{\alpha})l}\right\}$$

and $h_2 \ll h_1$ and is ignored. The sums may be done explicitly using (46) to obtain

$$h_1 = \frac{A_1}{4\sqrt{c}}\left\{\frac{1-e^{-2Nv}}{e^{2v}-1} + \frac{1-e^{-2Nu}}{e^{2u}-1}\right\}$$

where

$$v = \frac{\alpha+\hat{\alpha}-i(\omega-\hat{\omega})}{2} \quad \text{and} \quad u = \frac{\alpha+\hat{\alpha}+i(\omega-\hat{\omega})}{2},$$

and $i = \sqrt{-1}$ in the above equations. Then the sufficient statistic $\overline{h^2}$ is given by:

$$\overline{h^2} = \frac{A_1^2}{16}\left[\frac{e^{2\alpha}-1}{1-e^{-2N\alpha}}\right]\left[\frac{1-e^{-2Nv}}{1-e^{2v}} + \frac{1-e^{-2Nu}}{1-e^{2u}}\right]^2$$

The region of the parameter space we are interested in is where the unitless decay rate is small compared to one, and $\exp(N\hat{\alpha})$ is large compared to one. In this region the true signal decays away in the observation time, but not before we obtain a good representative sample of it. We are not considering the case were the decay is so slow that the signal is nearly stationary, nor are we considering the case were the decay is so strong that the signal is gone within a small fraction of the observation time. Within these limits the sufficient statistic $\overline{h^2}$ is

$$\overline{h^2} \approx \frac{A_1^2\alpha}{4}\left[\frac{\alpha+\hat{\alpha}}{(\alpha+\hat{\alpha})^2 + (\omega-\hat{\omega})^2}\right]^2.$$

The first derivatives of $\overline{h^2}$ evaluated at $\omega=\hat{\omega}$ and $\alpha=\hat{\alpha}$ are zero, as they should be. The mixed second partial is also zero. This gives the second derivatives of $\overline{h^2}$ as

$$\left(\frac{\partial^2 \overline{h^2}}{\partial \omega^2}\right)_{\omega=\hat{\omega}} = -\frac{A_1^2}{4\hat{\alpha}^3} \quad \text{and} \quad \left(\frac{\partial^2 \overline{h^2}}{\partial \alpha^2}\right)_{\alpha=\hat{\alpha}} = -\frac{A_1^2}{32\hat{\alpha}^3} .$$

We can now expand $\overline{h^2}$ in a Taylor series about the maximum and normalizing the distribution gives

$$P(\omega\alpha|D,I,\sigma) \approx (4\pi\delta_\omega\delta_\alpha)^{-1}\exp\left[-\frac{(\omega-\hat{\omega})^2}{2\delta_\omega^2} - \frac{(\alpha-\hat{\alpha})^2}{2\delta_\alpha^2}\right]$$

$$(\alpha)_{est} = \hat{\alpha} \pm \delta_\alpha \quad \text{and} \quad (\omega)_{est} = \hat{\omega} \pm \delta_\omega$$

where

$$\delta_\alpha \approx \frac{5.6\sigma\hat{\alpha}^{\frac{3}{2}}}{A_1} \quad \text{and} \quad \delta_\omega \approx \frac{2\sigma\hat{\alpha}^{\frac{3}{2}}}{A_1} \tag{51}$$

The accuracy estimate δ_α for the decay parameter is almost a factor of 3 worse than the estimate δ_ω for the frequency. This result has been noted before but why it should be so was not understood. Our independent probability analysis clearly indicates that this must be the case.

How does this compare to the results obtained for the simple harmonic frequency? Converting to Hertz involves dividing these by $2\pi\Delta t$, for a signal with $N = 1000$, $\hat{\alpha} = 0.01$, $A_1/\sqrt{2}\sigma = 1$ and, including a factor of 2 to obtain the values at the full-width at half maximum we have the estimated accuracy for frequency and decay as

$$<\omega> = \hat{\omega} \pm .9\text{Hz} \quad \text{and} \quad <\alpha> = \hat{\alpha} \pm 2.5\text{Hz}.$$

This compares to 0.025Hz for a stationary signal with the same signal-to-noise ratio. This is a factor of 36 times larger and since the error varies like $N^{-\frac{3}{2}}$ we have effectively lost all but one tenth of the data. When we have reached the unitless time of $t = 100$ the signal is down by a factor of 2.7 and has all but disappeared into the noise.

We wish to plot the power spectral estimate as a function of frequency and decay. These are given by

$$\hat{p}(\omega) \approx m \frac{\int d\alpha \; \overline{h^2}P(\omega,\alpha|D,I)}{\int d\alpha d\omega \; P(\omega,\alpha|D,I)} \tag{53}$$

$$\hat{p}(\alpha) \approx m \frac{\int d\omega \; \overline{h^2}P(\omega,\alpha|D,I)}{\int d\alpha d\omega \; P(\omega,\alpha|D,I)} \tag{54}$$

where $P(\omega,\alpha|D,I)$ is taken from (28) using (45) as the model: then, $\hat{p}(\omega)$ is useful for estimating the frequency; and $\hat{p}(\alpha)$ is useful for estimating the decay rate. These integrals can be computed numerically. The computer code used to evaluate the "student t-distribution" in this paper (in fact in all of the examples in this work) is included in Appendix A. This appendix contains a general routine for evaluating the "student t-distribution" (28), the orthonormal amplitudes (36), the power spectral density (39), and the estimated variance (40) with $s = 1$. In addition there is an example of how to use this subroutine in Appendix B. In this power spectral estimate (53, 54) (and in the computer code) we have assumed the estimated noise variance σ^2 is small compared to $\overline{h^2}$ and have ignored this term.

To illustrate some of these points we have prepared another example, Fig. 3. This time series was prepared from the following equation

$$d_j = 0.001 + \cos(0.3j + 1)e^{-0.01j} + e_j \ .$$

The $N = 512$ data samples were prepared in the following manner: first, we generated the data without the noise; we then computed the average of the data, and subtracted it from each data point, thus to ensure that the average value of the data was zero; we then repeated this process on the Gaussian white noise; next, we computed the average mean-square of the signal and the noise; and scaled the data by the appropriate ratio to make the signal-to-noise ratio of the data exactly one; last, we added the noise to the data. The time series clearly shows a small signal which rapidly decays away, Fig. 3(A). Figure 3(B), the periodogram (continuous curve) and the discrete Fourier transform (open circles) clearly show the Lorentzian line shape. The noise is now significantly affecting the periodogram: the periodogram is no longer an optimum frequency estimator.

Figures 3(C) and 3(D) contain plots of the power spectral density (53, 54). In Fig. 3(C) we have treated the frequency as a nuisance parameter and have integrated it out; as was emphasized earlier this is essentially the posterior probability distribution for α normalized to a power level rather than to unity. In Fig. 3(D) we have treated the decay as the nuisance parameter and have integrated it out. This gives the power spectral estimate as a function of frequency.

The width of these curves is a measure of the uncertainty in the determination of the these parameters. We have determined full-width at half maximum (numerically) for each of these and have compared these to the theoretical "best" estimates (51) and find

$$(\omega)_{est} = 0.2998 \pm 5.3 \times 10^{-4} \text{ and } (\omega)_{best} = 0.3000 \pm 3 \times 10^{-4},$$

$$(\alpha)_{est} = 0.0109 \pm 5.5 \times 10^{-4} \text{ and } (\alpha)_{best} = 0.0100 \pm 8 \times 10^{-4} \ .$$

Converting to Hz, $5.3 \times 10^{-4}/2\pi N = 0.84$Hz. The frequency estimate compares nicely with the "best" estimate, while our decay estimate is a little better. Given that the theoretical estimates were only approximations they are in good agreement with each other.

C. The Spectrum of Two Harmonic Frequencies.

We now turn our attention to the slightly more general problem of analyzing a data set which we postulate contains two distinct harmonic frequencies. The "student t-distribution" represented by equation (28) is, of course, the general solution to this problem. Unfortunately, that equation does not lend itself readily to understanding the probability distribution. In particular we would like to know what the behavior of these equations are in three different limits: first, when the frequencies are well separated; second, when they are close but distinct; and third, when they are so close as to be, for all practical purposes, identical. To investigate these we will solve, approximately, the two frequency problem.

The model equation for the two frequency problem is a simple generalization of the single harmonic problem:

$$f(t) = B_1\cos(\omega_1 t) + B_2\cos(\omega_2 t) + B_3\sin(\omega_1 t) + B_4\sin(\omega_2 t).$$

The model functions can then be used to construct the g_{jk} matrix. On a uniform grid this is given by

$$g_{jk} = \begin{bmatrix} C_{11} & C_{12} & 0 & 0 \\ C_{12} & C_{22} & 0 & 0 \\ 0 & 0 & S_{11} & S_{12} \\ 0 & 0 & S_{12} & S_{22} \end{bmatrix}$$

where

Figure 3.

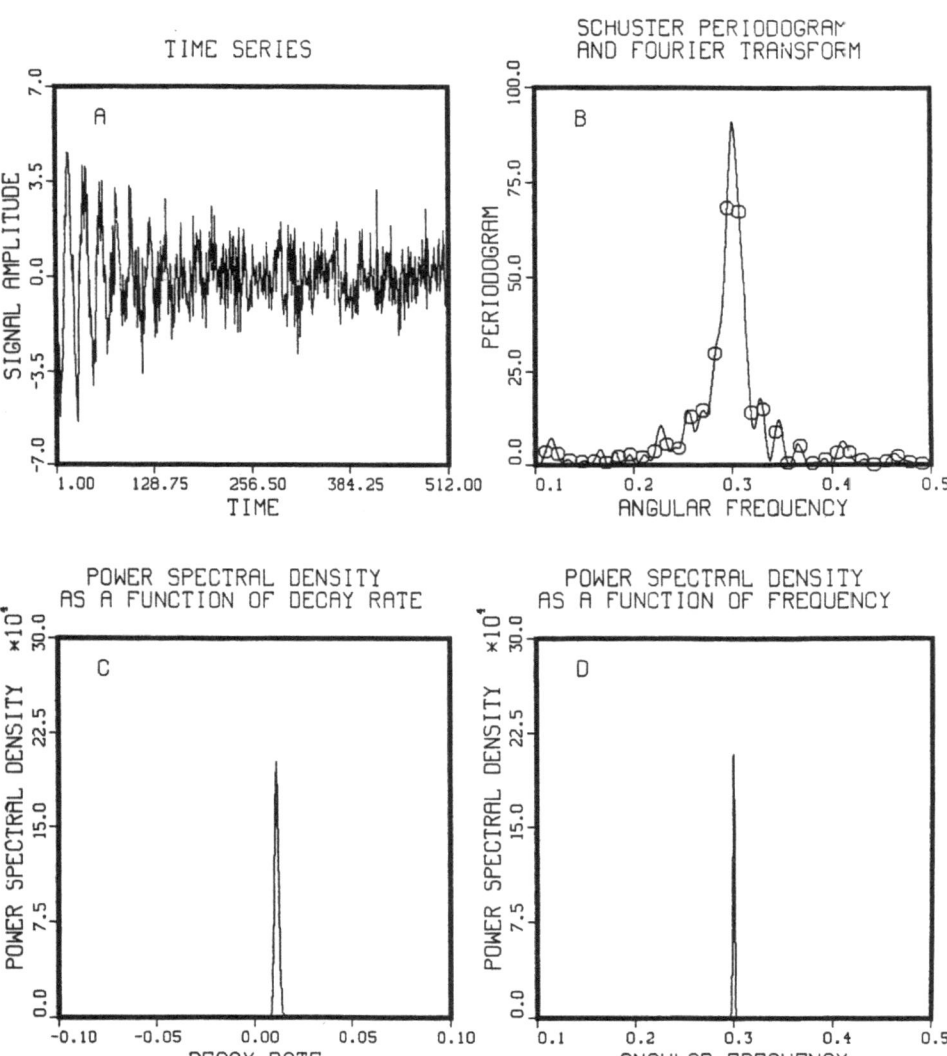

The data in Fig. 3(A) contain a simple frequency with a Lorentzian decay plus noise. In Fig. 3(B), the noise has significantly distorted the periodogram (continuous curve) and the discrete Fourier transform (open circles). The power spectral density may be computed as a function of decay rate Fig. 3(C), or as a function of frequency Fig. 3(D). Neither of these have been significantly affected by the noise.

$$C_{jk} = \sum_{l=-T}^{T} \cos(\omega_j l)\cos(\omega_k l) = \frac{\sin(\tfrac{1}{2}N\omega_+)}{2\sin(\tfrac{1}{2}\omega_+)} + \frac{\sin(\tfrac{1}{2}N\omega_-)}{2\sin(\tfrac{1}{2}\omega_-)}$$

$$S_{jk} = \sum_{l=-T}^{T} \sin(\omega_j l)\sin(\omega_k l) = \frac{\sin(\tfrac{1}{2}N\omega_-)}{2\sin(\tfrac{1}{2}\omega_-)} - \frac{\sin(\tfrac{1}{2}N\omega_+)}{2\sin(\tfrac{1}{2}\omega_+)}$$

$$w_+ = \omega_j + \omega_k, \quad (j,k=1 \text{ or } 2)$$

$$w_- = \omega_j - \omega_k.$$

The eigenvalue and eigenvectors problem for g_{jk} splits into two separate problems each involving 2×2 matrices. The eigenvalues are

$$\lambda_{1\&2} = \frac{C_{11} + C_{22}}{2} \pm \sqrt{(C_{11} - C_{22})^2 + 4C_{12}^2}$$

$$\lambda_{3\&4} = \frac{S_{11} + S_{22}}{2} \pm \sqrt{(S_{11} - S_{22})^2 + 4S_{12}^2} .$$

We can go on and obtain the exact solution to this problem but that will not be necessary. When the frequencies are well separated $|\omega_1-\omega_2| \gg 2\pi/N$, the eigenvalues reduce to $\lambda = N/2$. That is, g_{jk} goes into $N/2$ times the unit matrix. Then each of the model equations are effectively orthogonal and the sufficient statistic $\overline{h^2}$ reduces to

$$\overline{h^2} = \frac{2}{N}\left[C(\omega_1) + C(\omega_2)\right];$$

and the probability, when the variance is known, is given by

$$P(\omega_1\omega_2|D,I,\sigma) \propto \exp\left[\frac{C(\omega_1) + C(\omega_2)}{\sigma^2}\right]. \tag{55}$$

The problem has separated: one can estimate each of the frequencies separately. The maximum of the two frequency posterior probability density will be located at the two greatest peaks in the periodogram, in agreement with the common sense usage of the Fourier transform. A similar result holds for the general frequency estimation problem. Then the r frequencies, corresponding to the maximum of the joint posterior probability, are essentially the estimates obtained from the r biggest peaks in the periodogram.[2]

The labels ω_1, ω_2, etc. for the frequencies in the model are arbitrary, and accordingly their joint probability density is invariant under permutations. That means, for the two frequency problem, there is an axis of symmetry running along the line $\omega_1 = \omega_2$. We do not know from (55) what is happening along that line. This is easily investigated when $\omega_1 = \omega_2 \equiv \omega$: the eigenvalues become

$$\lambda_1 = N, \quad \lambda_2 = 0, \quad \lambda_3 = N, \quad \lambda_4 = 0.$$

The matrix g_{jk} has two redundant eigenvalues, and the probability distribution becomes

$$P(\omega|D,I,\sigma) \propto \exp\left[\frac{C(\omega)}{\sigma^2}\right]. \tag{56}$$

The probability density goes smoothly into the single frequency probability distribution along this axis of symmetry. Given that the two frequencies are equal, our estimate of them will be identical, in value and accuracy, to those of the one frequency case.

The problem of understanding the posterior probability density when there are two close but distinct frequencies must now be addressed. The matrix g_{jk} for this two frequency problem is readily diagonalized and the exact solution for the two frequency problem obtained. An approximate solution may be obtained that is valid in the same sense that the approximate solution to the single frequency problem obtained in Section II is valid. To obtain this approximate solution one needs only to examine the matrix g_{jk} and notice that the elements of this matrix consist of the diagonal elements given by:

$$C_{11} = \frac{N}{2} + \frac{\sin(N\omega_1)}{2\sin(\omega_1)} \approx \frac{N}{2},$$

$$C_{22} = \frac{N}{2} + \frac{\sin(N\omega_2)}{2\sin(\omega_2)} \approx \frac{N}{2},$$

$$S_{11} = \frac{N}{2} - \frac{\sin(N\omega_1)}{2\sin(\omega_1)} \approx \frac{N}{2},$$

$$S_{22} = \frac{N}{2} - \frac{\sin(N\omega_2)}{2\sin(\omega_2)} \approx \frac{N}{2},$$

and the off diagonal elements. The off diagonal terms are small compared to N unless the frequencies are specifically in the region of $\omega_1 \approx \omega_2$, then only the terms involving the difference $(\omega_1 - \omega_2)$ are large. We can approximate the off diagonal terms as:

$$C_{12} \approx S_{12} \approx \frac{1}{2} \sum_{l=-T}^{T} \cos\frac{1}{2}(\omega_1-\omega_2)l = \frac{1}{2} \frac{\sin\frac{1}{2}N(\omega_1-\omega_2)}{\sin\frac{1}{2}(\omega_1-\omega_2)} \equiv \frac{B}{2}. \tag{57}$$

When the two frequencies are well separated, (57) is of order one and is ignorable. When the two frequencies are nearly equal, then the off diagonal terms are large and are given accurately by (57). So the approximation is valid for all values of ω_1 and ω_2.

With this approximation for g_{jk} it is now possible to write a simplified solution for the two frequency problem. The matrix g_{jk} (16) is given approximately by

$$g_{jk} = \frac{1}{2} \begin{vmatrix} N & B & 0 & 0 \\ B & N & 0 & 0 \\ 0 & 0 & N & B \\ 0 & 0 & B & N \end{vmatrix}.$$

The orthonormal model functions (17) may now be constructed:

$$H_1(t) = \frac{1}{\sqrt{N + B}} \left\{ \cos(\omega_1 t) + \cos(\omega_2 t) \right\}, \tag{58}$$

$$H_2(t) = \frac{1}{\sqrt{N - B}} \left\{ \cos(\omega_1 t) - \cos(\omega_2 t) \right\},$$

$$H_3(t) = \frac{1}{\sqrt{N + B}} \left\{ \sin(\omega_1 t) + \sin(\omega_2 t) \right\},$$

$$H_4(t) = \frac{1}{\sqrt{N - B}} \left\{ \sin(\omega_1 t) - \sin(\omega_2 t) \right\}.$$

We can now write the sufficient statistic $\overline{h^2}$ in terms of these orthonormal model functions to obtain

$$\overline{h^2} = h_+^2 + h_-^2,$$

$$h_+^2 \equiv \frac{1}{4(N + B)} \left\{ [P(\omega_1) + P(\omega_2)]^2 + [Q(\omega_1) + Q(\omega_2)]^2 \right\},$$

$$h_-^2 \equiv \frac{1}{4(N - B)} \left\{ [P(\omega_1) - P(\omega_2)]^2 + [Q(\omega_1) - Q(\omega_2)]^2 \right\},$$

where P and Q are the sine and cosine transforms of the data as functions of the appropriate frequency. The factor of 4 comes about because for this problem there are $m = 4$ model functions. Using (26), the posterior probability that two distinct frequencies are present, given the roise variance σ^2, is

$$P(\omega_1,\omega_2|D,I,\sigma) \propto \exp\left[\frac{2\overline{h^2}}{\sigma^2}\right]. \tag{59}$$

A quick check on the asymptotic forms of this will verify that when the frequencies are well separated one has $\overline{h^2} = \frac{1}{2}[C(\omega_1) + C(\omega_2)]$, and it has reduced to (55) and, when the frequencies are the same the second term smoothly to zero, and the first term goes into $\frac{1}{2}C(\omega)$, to reduce to (56) as expected.

When the frequencies are very close or far apart we can apply the results obtained by Jaynes[2] concerning the accuracy of the frequency estimates:

$$(\omega_{est}) = (\omega_{max}) \pm \frac{\sigma}{A}\sqrt{48/N^3} . \tag{60}$$

In the region where the frequencies are close but distinct, (59) appears very different. We would like to understand what is happening in this region, in particular we would like to know just how well two close frequencies can be estimated. To understand this we will construct a Gaussian approximation similar to what was done for the case with Lorentzian decay. We Taylor expand the $\overline{h^2}$ in (59) to obtain

$$P(\omega_1,\omega_2|D,I,\sigma) \approx \exp\left(- \frac{(\omega_1-\hat{\omega}_1)^2}{2r^2\sigma^2} - \frac{(\omega_2-\hat{\omega}_2)^2}{2s^2\sigma^2} - \frac{(\omega_1-\hat{\omega}_1)(\omega_2-\hat{\omega}_2)}{2u^2\sigma^2}\right)$$

where

$$\frac{1}{r^2} = -\frac{\partial^2\overline{h^2}}{\partial\omega_1^2}\bigg|_{\substack{\omega_1=\hat{\omega}_1 \\ \omega_2=\hat{\omega}_2}}$$

$$\frac{1}{s^2} = -\frac{\partial^2\overline{h^2}}{\partial\omega_2^2}\bigg|_{\substack{\omega_1=\hat{\omega}_1 \\ \omega_2=\hat{\omega}_2}}$$

$$\frac{1}{u^2} = -\frac{\partial^2\overline{h^2}}{\partial\omega_1\partial\omega_2}\bigg|_{\substack{\omega_1=\hat{\omega}_1 \\ \omega_2=\hat{\omega}_2}}$$

where $\hat{\omega}_1$, $\hat{\omega}_2$ are the locations of the maxima of (59). If we have a uniformly sampled signal of the form

$$f_l = \hat{b}_1\cos(\hat{\omega}_1 l) + \hat{b}_2\cos(\hat{\omega}_2 l) + \hat{b}_3\sin(\hat{\omega}_1 l) + \hat{b}_4\sin(\hat{\omega}_2 l) + e_l \tag{61}$$

where $-T \leq l \leq T$, $2T+1 = N$, and $\hat{b}_1$, $\hat{b}_2$, $\hat{b}_3$, $\hat{b}_4$ are the true amplitudes, $\hat{\omega}_1$, $\hat{\omega}_2$ are the true frequencies, and $e_l \ll$ the signal, then h_1 is given by the projection of H_1 (58) onto the data (61) to obtain

$$h_1 = \frac{1}{\sqrt{N + B(\omega_1,\omega_2)}} \sum_{l=-T}^{T} \{\cos(\omega_1 l) + \cos(\omega_2 l)\} f_l$$

where

$$\frac{B(\omega_1,\omega_2)}{2} \equiv \frac{1}{2}\sum_{l=-T}^{T}\cos(\omega_1-\omega_2)l = \frac{1}{2}\frac{\sin\frac{1}{2}N(\omega_1-\omega_2)}{\sin\frac{1}{2}(\omega_1-\omega_2)} \tag{62}$$

For uniform time series these h_j may be summed explicitly using equation (62) to obtain

$$h_1 = \frac{1}{2\sqrt{N + B(\omega_1,\omega_2)}} \left\{ \hat{b}_1[B(\hat{\omega}_1,\omega_1) + B(\hat{\omega}_1,\omega_2)] + \hat{b}_2[B(\hat{\omega}_2,\omega_1) + B(\hat{\omega}_2,\omega_2)] \right\}$$

$$h_2 = \frac{1}{2\sqrt{N - B(\omega_1,\omega_2)}} \left\{ \hat{b}_1[B(\hat{\omega}_1,\omega_1) - B(\hat{\omega}_1,\omega_2)] + \hat{b}_2[B(\hat{\omega}_2,\omega_1) - B(\hat{\omega}_2,\omega_2)] \right\}$$

$$h_3 = \frac{1}{2\sqrt{N + B(\omega_1,\omega_2)}} \left\{ \hat{b}_3[B(\hat{\omega}_1,\omega_1) + B(\hat{\omega}_1,\omega_2)] + \hat{b}_4[B(\hat{\omega}_2,\omega_1) + B(\hat{\omega}_2,\omega_2)] \right\}$$

$$h_4 = \frac{1}{2\sqrt{N - B(\omega_1,\omega_2)}} \left\{ \hat{b}_3[B(\hat{\omega}_1,\omega_1) - B(\hat{\omega}_1,\omega_2)] + \hat{b}_4[B(\hat{\omega}_2,\omega_1) - B(\hat{\omega}_2,\omega_2)] \right\}.$$

We have kept terms corresponding to the differences in the frequencies. When the frequencies are close together it is only these terms which are important: the approximation is consistent with the others made.

The sufficient statistic $\overline{h^2}$ is then given by

$$\overline{h^2} = \frac{1}{4}(h_1^2 + h_2^2 + h_3^2 + h_4^2). \tag{63}$$

To obtain a Gaussian approximation for (59) one must calculate the second derivative of (63) with respect to ω_1 and ω_2. The problem is simple in principle but difficult in practice. To get these partial derivatives we Taylor expand (63) around the maximum located at $\hat{\omega}_1$ and $\hat{\omega}_2$ and then take the deratives. The intermediate steps are of little concern and were carried out using an algebra manipulation package. Terms of order one compared to N were again ignored and, we have assumed the frequencies are close but distinct: we used the small angle approximations for the sine and cosine at the end of the calculation. The local variable δ [defined as $(\hat{\omega}_2-\hat{\omega}_1)/2 \equiv \delta/N$] measures the distance between two adjacent frequencies. If δ is π then the frequencies are separated by one step apart in the discrete Fourier transform. The second partial deratives of $\overline{h^2}$ are given by:

$$\frac{\partial^2 \overline{h^2}}{\partial \omega_1^2} \Bigg|_{\substack{\omega_1=\hat{\omega}_1 \\ \omega_2=\hat{\omega}_2}} \approx -(\hat{b}_1^2+\hat{b}_3^2)N^3 \left(\frac{3\sin^2(\delta)-6\delta\cos(\delta)\sin(\delta)+\delta^2[\sin^2(\delta)+3\cos(\delta)]-\delta^4}{48\delta^3[\sin(\delta)-\delta][\sin(\delta)+\delta]} \right)$$

$$\frac{\partial^2 \overline{h^2}}{\partial \omega_2^2} \Bigg|_{\substack{\omega_1=\hat{\omega}_1 \\ \omega_2=\hat{\omega}_2}} \approx -(\hat{b}_2^2+\hat{b}_4^2)N^3 \left(\frac{3\sin^2(\delta)-6\delta\cos(\delta)\sin(\delta)+\delta^2[\sin^2(\delta)+3\cos(\delta)]-\delta^4}{48\delta^3[\sin(\delta)-\delta][\sin(\delta)+\delta]} \right)$$

$$\frac{\partial^2 \overline{h^2}}{\partial \omega_1 \partial 2_2} \Bigg|_{\substack{\omega_1=\hat{\omega}_1 \\ \omega_2=\hat{\omega}_2}} \approx (\hat{b}_1\hat{b}_2 + \hat{b}_3\hat{b}_4)N^3 \left(\frac{\delta^4\sin(\delta)+2\delta^3\cos(\delta)-3\delta^2\sin(\delta)+\sin^3(\delta)}{16\delta^3[\sin(\delta)-\delta][\sin(\delta)+\delta]} \right).$$

If the true frequencies $\hat{\omega}_1$ and $\hat{\omega}_2$ are separated by two steps in the discrete Fourier transform, $\delta=2\pi$, we may reasonably ignore all but the δ^4 term to obtain

$$\frac{\partial^2 \overline{h^2}}{\partial \omega_1^2} \Bigg|_{\substack{\omega_1=\hat{\omega}_1 \\ \omega_2=\hat{\omega}_2}} \approx -\frac{(\hat{b}_1^2+\hat{b}_3^2)N^3}{48}$$

$$\frac{\partial^2 \overline{h^2}}{\partial \omega_2^2} \Bigg|_{\substack{\omega_1=\hat{\omega}_1 \\ \omega_2=\hat{\omega}_2}} \approx -\frac{(\hat{b}_2^2+\hat{b}_4^2)N^3}{48}$$

$$\left.\frac{\partial^2 \overline{h^2}}{\partial\omega_1\partial\omega_2}\right|_{\substack{\omega_1=\hat{\omega}_1 \\ \omega_2=\hat{\omega}_2}} \approx -\frac{(\hat{b}_1\hat{b}_2+\hat{b}_3\hat{b}_4)N^3\sin(\delta)}{16\delta} \ .$$

The accuracy estimates reduce to equation (60) when the frequencies are will separated. When the frequencies have approximately the same amplitudes and δ is order of 2π (the frequencies are separated by two steps in the discrete Fourier transform) the interaction term is down by four and one expects the estimates to be nearly the same as those for a single frequency. Probability theory indicates that two frequencies which are as close together as two steps in a discrete Fourier transform do not interfere with each other in any significant way.

To better understand the maximum theoretical accuracy with which two frequencies can be estimated we have prepared Table 1. To make these estimates comparable to those obtained in Section II we have again assumed there $N = 1000$ data points and $\sigma = 1$. There are three regions of interest: when the frequency separation is small compared to a single step in the discrete Fourier transform; when the separation is of order one step; and when the separation is large. Additionally we would like to understand the behavior when the signals are of the same amplitude, when one signal is slightly larger than the other, and when one signal is much larger than the other. When we prepared this table we used equation (27), not the "student t-distribution". In order to obtain the "best" theoretical estimates we did not include any noise in the data [just as noise was not included in (60)]. Had we used the "student t-distribution" the accuracy estimates would have been much much better. The estimates obtained are the "best" in the sense that in a real data set with $\sigma = 1$, containing $N = 1000$ data points the accuracy estimates one obtains will be, nearly always, slightly worse than those contained in Table 1.

TABLE 1			
Amplitudes/Description	$(\hat{f}_2-\hat{f}_1)=0.07$ Hz	$(\hat{f}_2-\hat{f}_1)=0.3$ Hz	$(\hat{f}_2-\hat{f}_1)=5.1$ Hz
The square magnitude of signal 2 is equal to signal 1	$(f_1)_{est}=\hat{f}_1\pm0.091$ $(f_2)_{est}=\hat{f}_2\pm0.091$	$(f_1)_{est}=\hat{f}_1\pm0.027$ $(f_2)_{est}=\hat{f}_2\pm0.027$	$(f_1)_{est}=\hat{f}_1\pm0.025$ $(f_2)_{est}=\hat{f}_2\pm0.025$
The square magnitude of signal 2 is four times larger than signal 1	$(f_1)_{est}=\hat{f}_1\pm0.091$ $(f_2)_{est}=\hat{f}_2\pm0.088$	$(f_1)_{est}=\hat{f}_1\pm0.027$ $(f_2)_{est}=\hat{f}_2\pm0.013$	$(f_1)_{est}=\hat{f}_1\pm0.025$ $(f_2)_{est}=\hat{f}_2\pm0.012$
The square magnitude of signal 2 is 128 times larger than signal 1	$(f_1)_{est}=\hat{f}_1\pm0.091$ $(f_2)_{est}=\hat{f}_2\pm0.034$	$(f_1)_{est}=\hat{f}_1\pm0.025$ $(f_2)_{est}=\hat{f}_2\pm0.0024$	$(f_1)_{est}=\hat{f}_1\pm0.025$ $(f_2)_{est}=\hat{f}_2\pm0.0022$

The three values of $(\omega_1-\omega_2)$ examined correspond to $\delta = 1/4$, $\delta = 4$, and $\delta = 16$: roughly these correspond to frequency separations of 1/12, 1, and 5 steps in the discrete Fourier transform. We held the squared magnitude of signal one constant at one and the second is either 1, 4 or 128 times larger.

When the separation frequency is 0.07 Hz the frequencies are indistinguishable. The smaller component cannot be estimated accurately. As the magnitude of the second signal increases the estimated accuracy of the second signal becomes better as one's intuition would suppose it should (the signal looks more and more like one frequency).

When the separation frequency is 0.3 Hz or about one step in the discrete Fourier transform, the accuracy estimates indicate that the two frequencies are well resolved. By this we mean one of the frequencies would have to be moved by eleven standard deviations before it would be confounded with the other (two parameters are said to be confounded when probability theory estimates their values to be the same). This is true for all the amplitudes in the table; it does however, improve with increasing amplitude. According to probability theory, when two frequencies are as close together as one step in the discrete Fourier transform those frequencies are clearly resolvable.

When the separation frequency is 5.1Hz, the accuracy estimates clearly determine both frequencies. Additionally, the accuracy estimates for the smaller frequency are essentially 0.025Hz which is the same as the estimate for a single harmonic frequency that we found previously (10). Examining Table 1 more carefully, we see that when the frequencies are separated by even a single step in the discrete Fourier transform, the accuracy estimates are essentially those for the single harmonic frequencies. The ability to estimate two close frequencies accurately is essentially independent of the separation frequency, as long as it is greater than or approximately equal to one step in the discrete Fourier transform.

To illustrate the two frequency probability density (59) we prepared a simple example, Fig. 4. This example was prepared from the following equation

$$d_i = \cos(0.3i + 1) + \cos(0.307i + 2) + e_i$$

where e_i has variance one and the index runs over the symmetric time interval $(-255.5 \leq i \leq 255.5)$ by unit steps. This time series Fig. 4(A) has two simple harmonic frequencies. The two frequencies are separated by approximately one step in the discrete Fourier transform, $|\omega_1 - \omega_2| \approx 2\pi/512$.

From looking at the raw time series one might just barely guess that there is more going on than a simple harmonic frequency plus noise, because the oscillation amplitude seems to vary slightly. If we were to guess that there are two close frequencies, then by examining the data one can guess that the difference between these two frequencies is not more that one cycle over the entire time interval. If the frequencies were separated by more than this we would expect to see beats in the data. If there are two frequencies, the second frequency must be within 0.012 of the first in dimensionless units. This is in the region were the frequency estimates are almost but not quite confounded.

Now Fig. 4(B) the periodogram (continuous curve) and the discrete Fourier transform (open circles) show only a single peak. The single frequency model has estimated a frequency which is essentially the average of the two. The two frequency posterior probability density Fig. 4(C) show two well resolved, symmetrical maxima. Thus the inclusion of just this one simple additional fact has greatly enhanced our ability to detect the two signals.

From the contours in Fig. 4(C) it appears that the upper frequency is being determined more accurately than the lower one. However, to be sure of this we should integrate out one frequency to see the marginal posterior distribution of the other.

Now that we know the data contain two partially resolved frequencies, we could proceed to obtain data over a longer time span and resolve the frequencies. Regardless, it is now clear that what one can detect clearly depends on what question one asks.

D. Multiple nonstationary frequencies estimation.

The general solution to this problem is given by the "student t-distribution". When the frequencies are harmonic (i.e. no decay or chirp) and well separated ($|\omega_j - \omega_k| > 2\pi/N$) the problem separates into multiple single frequency problems. When more than one frequency is close ($|\omega_j - \omega_k| \approx 2\pi/N$) we must use a more general model (around the close frequencies).

To understand this problem completely, one must look at the case when there are two nonstationary frequencies present. We already know the answer to this question: the Fourier transform would not work for estimating multiple nonstationary frequencies if the estimation problem did not separate. The details for this problem are just a straightforward generalization to the two frequency problem and we have not included them here.

From what we have learned and from what was shown by Jaynes[2] a procedure for estimating multiple frequencies can now be given. First, compute the log of the posterior probability using a model with a single harmonic frequency plus a constant and look for peaks. If there are r well resolved peaks above the noise level, it is a good bet there are at least r frequencies present. Second, use a single frequency model (with decay if necessary)

Figure 4.

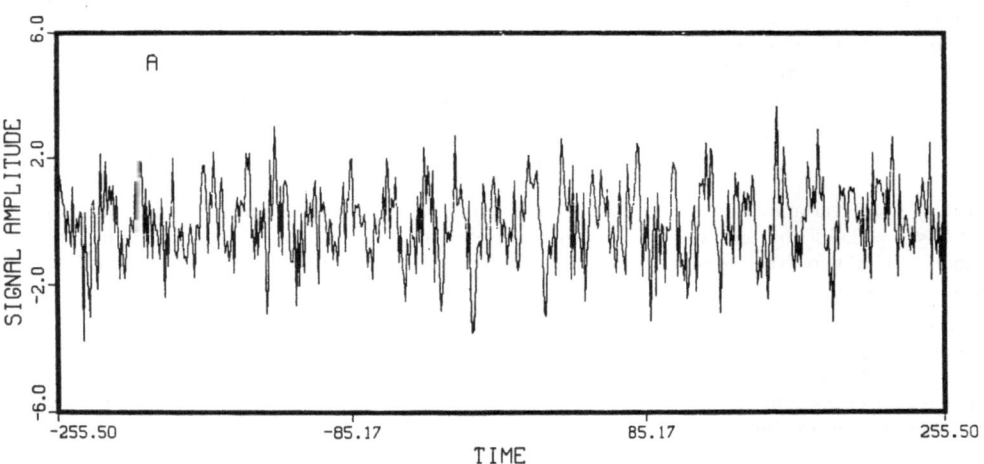

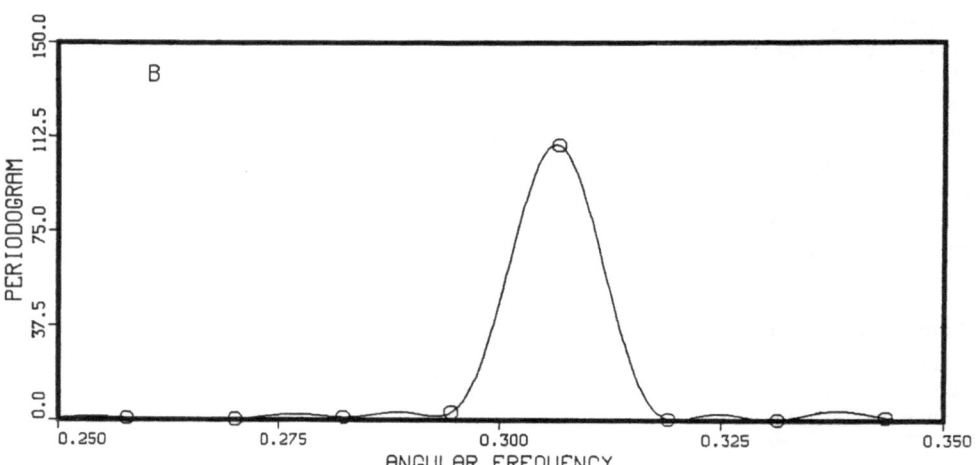

These data, Fig. 4(A) contains two frequencies. They are separated from each other by approximately a single step in the discrete Fourier transform. The periodogram, Fig. 4(B) shows only a single peak located between the two frequencies.

Figure 4. (continued)

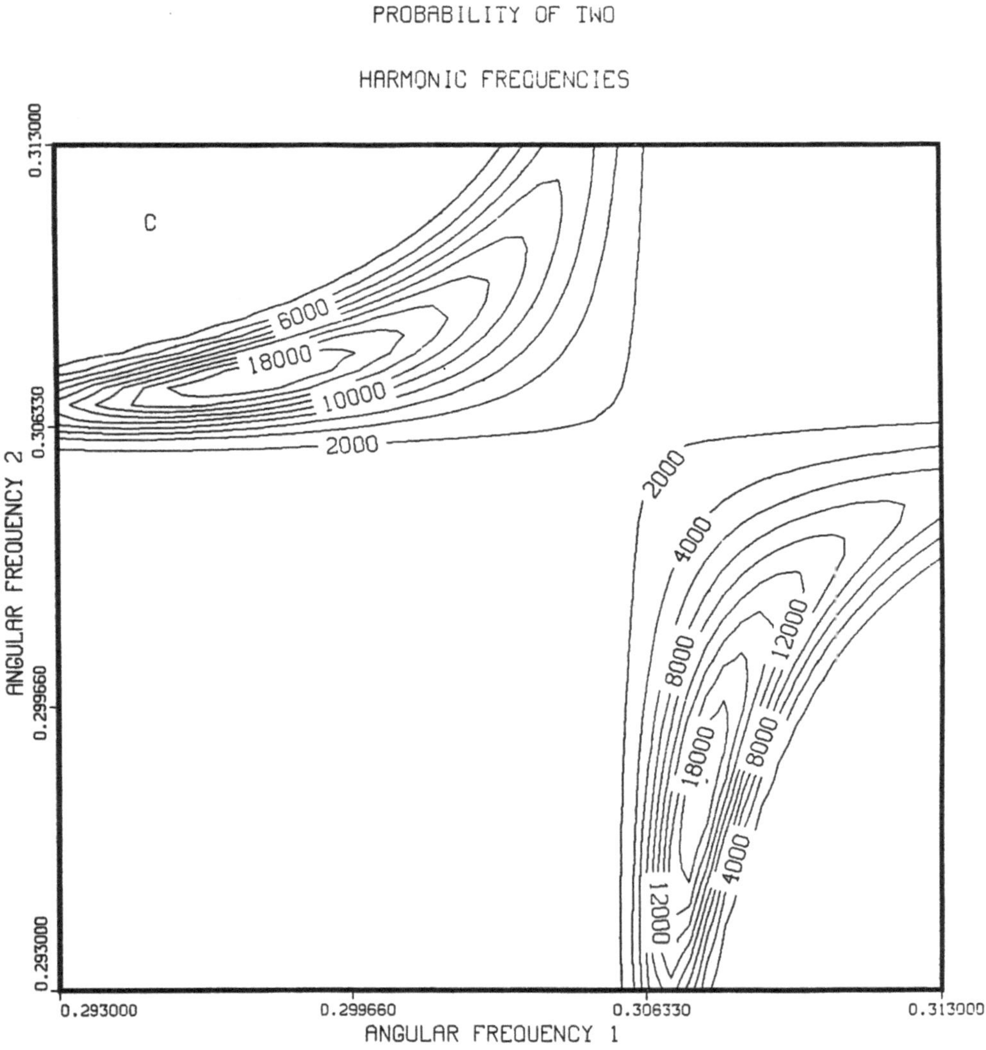

PROBABILITY OF TWO

HARMONIC FREQUENCIES

The two frequency probability density, Fig. 4(C) clearly indicates the presence of two frequencies. The odds ratio for these two (without including the prior) prefers the two frequency model by more than 10^{22} to one.

and locate the maximum of each peak in the Fourier transform. Third, use a two frequency model to examine each peak. The initial frequency estimates should be slightly above and below (about 1/4 step in the discrete Fourier transform) the location of the peak. If the peak is a single frequency, the two frequency model will confound it, because it cannot fit the data any better than does a one frequency model. If the peak is due to two resolvable frequencies it will find the second frequency. Remove any confounded parameters. Fourth, if one desires to know the "best" estimates of the parameters, then use the estimated values, from the preceding steps, as initial guesses in the general problem. Locate the maximum of the multiple frequency posterior probability distribution. This will improve the estimates by removing the interference effects between them. Then determine the accuracy of the estimates.

To obtain the accuracy estimates we must compute both $<\omega_j>$ and $<\omega_j^2>$ where ω_j is one member of the set of $\{\omega\}$ parameters. It might be a frequency, decay, chirp or any other parameter in the set. Then the estimates are given by

$$(\omega_j)_{est} = \hat{\omega}_j \pm \sigma_j$$

where

$$\sigma_j = \sqrt{<\omega_j^2> - <\omega_j>^2}$$

and $\hat{\omega}_j$ is the location of the maximum of the "student t-distribution" for parameter ω_j. But if the number of parameters in this set is large there is virtually no hope of performing the integrals represented by $<\omega_j>$ and $<\omega_j^2>$, either numerically or analytically. We will be forced to use an approximate result for σ_j.

We can approximate (27) by a Gaussian if we replace σ^2 in (27) by its expectation value (41). We can then Taylor expand to obtain a suitable Gaussian approximation. Define the matrix

$$H_{jk} \equiv -\frac{m}{4}\frac{\partial^2}{\partial\omega_j\partial\omega_k}\frac{\overline{h^2}}{<\sigma^2>}$$

then $P(\{\omega\}|D,\sigma,I)$ (27) is approximately given by

$$P(\{\omega\}|D,\sigma,I) = \frac{1}{z} \times \exp\left[\sum_{jk=1}^{r} H_{jk}\Delta_j\Delta_k\right]$$

where

$$\Delta_j \equiv \omega_j - \hat{\omega}_j$$

are just the Taylor expansion variables and z is a normalization constant.

We can transform the variables to an orthogonal set and then perform the r integrals just as we did with the amplitudes in Section III. These new variables are obtained from the eigenvalues and eigenvectors of H_{jk}. Let u_{jk} denote the k'th component of the j'th eigenvector of H_{jk} and let v_j be the eigenvalue. Then the orthogonal variables are given by

$$s_j = \sum_{k=1}^{r}\Delta_k u_{kj} \qquad \Delta_j = \sum_{k=1}^{r}s_k u_{jk} \; .$$

Making this change of variables we have

$$P(\{s\}|D,\sigma,I) \approx \prod_{k=1}^{r}\left(\frac{v_k}{\pi}\right)^{1/2}\exp\left[-\sum_{j=1}^{r}s_j^2\right] \tag{64}$$

From (64) we can compute $<s_j>$ and $<s_j^2>$. The Jacobian is just the determinant $|u_{jk}|$, but this is one since the transformation matrix u_{jk} is orthogonal. Of course $<s_j>$ is zero and the expectation value $<s_j s_k>$ is given by

$$<s_j s_k> = \frac{\delta_{jk}}{2\nu_k}$$

We can now work backward to obtain the standard deviation for the ω_j as

$$<\omega_j^2> - <\omega_j>^2 = \sum_{k=1}^{r} \frac{u_{kj}^2}{2\nu_k} \equiv \sigma_j^2 .$$

Then the estimated ω_j parameters are

$$\left(\omega_j\right)_{est} = \hat{\omega}_j \pm \sigma_j \qquad (65)$$

where $\hat{\omega}_j$ is the location of the maximum of the probability distribution as a function of the parameter ω_j.

For an arbitrary model the matrix H_{jk} cannot be calculated analytically; however, it can be evaluated numerically using the computer code given in Appendix A. We use a general searching routine to find the maximum of the probability distribution and then calculate this matrix numerically. The log of the "student t-distribution" is so sharply peaked that gradient searching routines do not work well. We use a "pattern" search routine described by Hooke and Jeeves.[12]

VI. EXAMPLES: APPLICATIONS TO REAL DATA.

The this section is devoted to applications. We will apply these procedures to a number of time series including NMR signals, economic time series, and Wolf's relative sunspot numbers. We do this in a effort to show how these procedures can be used to obtain optimal parameter estimates and to show the power and versatility of these methods.

A. NMR time series.

NMR is one of the best examples of how the introduction of modern computing machines has revolutionized a branch of science. With the aid of computers more data can be taken and summarized into a useful form faster than has ever been possible before. The standard way to analyze an NMR experiment is to obtain a quadrature data set, with two separate measurements, 90° out of phase with each other, and to do a complex Fourier transform on the data.[13] The global phase of the discrete complex transform is adjusted until the real part (called an absorption spectrum) is as symmetric as possible. The frequencies and decay rates are then estimated from the absorption spectrum. There are, of course, good physical reasons why the absorption spectrum of the "true signal" is important to physicists. However, as we have emphasized repeatedly since Section II, the discrete Fourier transform is an optimal frequency estimator only when a single simple harmonic frequency is present.

We will apply the procedures developed in the previous sections to a time series from a real NMR experiment, and contrast our analysis to the one done using the absorption spectrum. The NMR data used are of a free-induction decay,[14] Fig. 5. The sample contained a mixture of 63% liquid Hydrogen-Deuterium (HD) and Deuterium (D_2) at 20.2°K. The sample was excited with a 55MHz pulse, and then detected using a standard mixer-modulation technique. The resulting signal is in the audio range where it has several oscillations at about 100Hz. The data were sampled at $\Delta T = 0.0005$ seconds, and $N = 2048$ samples were taken for each channel. The data therefore, span a time interval of about one second. As was discussed earlier we are using dimensionless units. The relation to physical units are

$$f = \frac{\omega}{2\pi\Delta T} \text{ Hz} \quad \text{Period} = \frac{2\pi\Delta T}{\omega} \text{ Seconds}$$

where f is the frequency in Hertz, ω is the unitless frequency in radians per unit step, and

Figure 5.

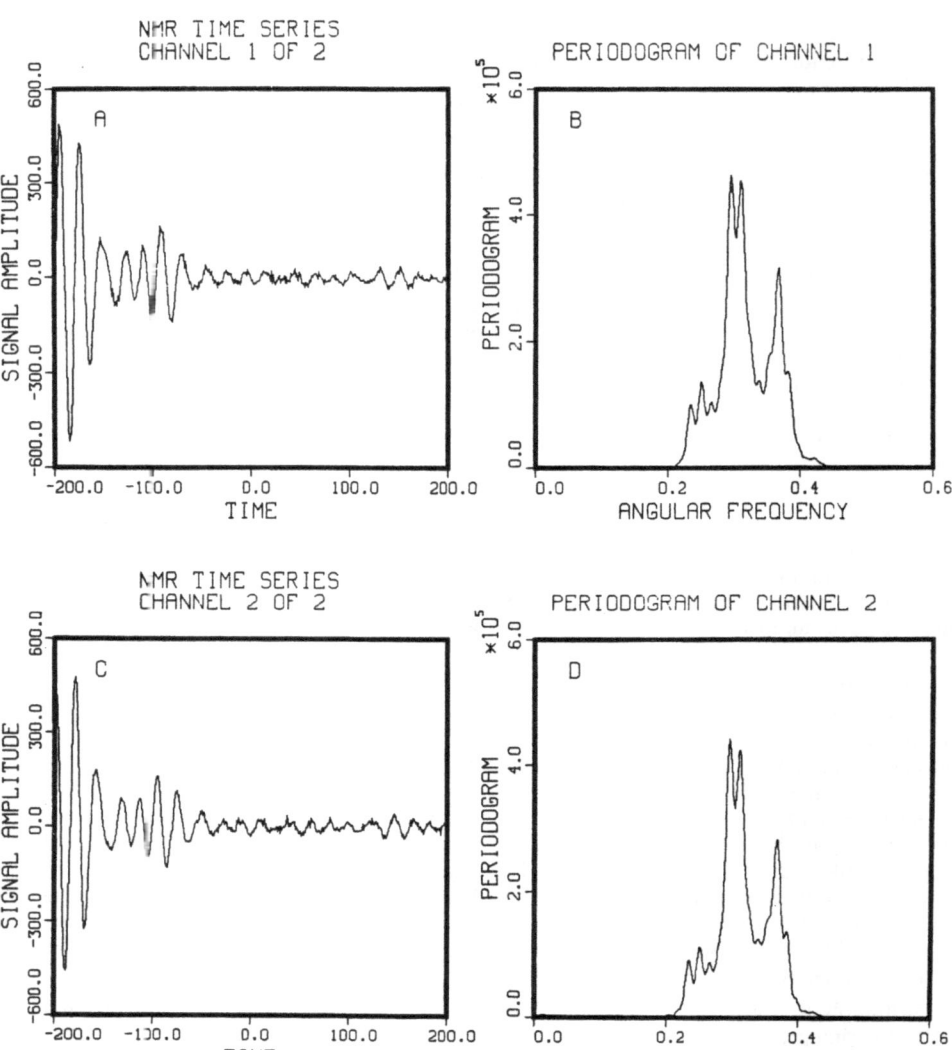

Figure 5(A) and 5(C) are channel one and two of an NMR experiment. The data are a free-induction decay for a sample containing a mixture of D_2 and HD in a liquid phase. Theory indicates there should be three frequencies in these data: A D_2 singlet, and an HD doublet with a 43Hz separation. The singlet should be approximately in the center of the doublet. There were 2048 data points in each channel, the data were sampled at 0.0005 seconds. In the discrete Fourier transform, Fig. 5(B) and 5(D), the singlet appears to be split into a doublet. This is caused by a non-Lorentzian decay. The envelope for the decay actually goes negative producing the double peak.

ΔT is the sampling time.

In these data there are a number of effects which we would like to investigate. First, the indirect J coupling[15] in the HD produces a doublet with a splitting of about 43Hz. The D_2 in the sample is also excited, its resonance is approximately in the middle of the HD doublet. One of the things we would like to determine is the shift of this frequency relative to the center of the HD doublet. In addition to the three frequencies there are two different characteristic decay times; the decay rate of the HD doublet is grossly different from that of D_2.[15] However, an inhomogeneous magnetic field could mask the true decay: the decay could be magnet limited. We would like to know how strongly the inhomogeneous magnetic field has affected the decay.

The analysis we did in Section III, although general, did not use a notation appropriate to two channels. We will need to generalize the notation briefly; this is very straightforward and we will only outline it here. There are two different measurements of this signal, (assumed to be independent) and we designate these measurements as $d_1(t_i)$ and $d_2(t_i)$. The model functions will be abbreviated as $f_1(t)$ and $f_2(t)$ with the understanding that each measurement of the signal has different amplitudes, and noise variance, but the same $\{\omega\}$ parameters.

We can write the likelihood (15) immediately to obtain

$$L(f_1, f_2) \propto (\sigma_1 \sigma_2)^{-N} \exp\left\{ -\frac{1}{2\sigma_1^2} \sum_{i=1}^{N} [d_1(t_i) - f_1(t_i)]^2 - \frac{1}{2\sigma_2^2} \sum_{i=1}^{N} [d_2(t_i) - f_2(t_i)]^2 \right\}$$

Because the amplitudes and noise variance are assumed different in each channel we may remove these using the same procedure developed in Section III. After removing the nuisance parameters the marginal posterior probability of the $\{\omega\}$ parameters is just the product of the "student t-distributions" (28) for each channel separately:

$$P(\{\omega\} | DI) \propto \left[1 - \frac{m \overline{h^2}_1}{N \overline{d^2}_1}\right]^{\frac{m-N}{2}} \left[1 - \frac{m \overline{h^2}_2}{N \overline{d^2}_2}\right]^{\frac{m-N}{2}} \tag{66}$$

where the subscripts refer to the channel number. As explained previously, (66) in effect estimates the noise level independently in the two channels.

A procedure for dealing with the multiple frequency problem was outlined in Section V, and we will apply that procedure here. The first step in any harmonic analysis is to plot the data and the log of the probability of a single harmonic frequency. If there is only one channel present, this is essentially the periodogram of the data, Fig. 5(B) and Fig. 5(D). When more than one channel is present the log probability of a single harmonic frequency is essentially the sum of the periodograms for each channel weighted by the appropriate variances. If the variances are unknown, then the appropriate statistic is the log of (66), Fig. 6.

Now as was noted in Section V, if the frequencies are well separated, a peak in the periodogram above the noise level is evidence of a frequency near that peak. From examining Fig. 6 we see there are a number of peaks near 0.3. There are many more peaks than theory indicates there should be. Is this evidence of more going on than theory predicts?

To answer this question we proceed to the next phase of the analysis and apply a two frequency model to each of these peaks. We know that these frequencies have some type of decay structure, we include a decay by adding an exponential factor. For this preliminary analysis we assume the same decay rate for all the frequencies. The model used in this investigation was

$$f_1(t) = [B_1 \cos(\omega_1 t) + B_2 \sin(\omega_1 t) + B_3 \cos(\omega_2 t) + B_4 \sin(\omega_2 t)] e^{-\alpha t} .$$

Figure 6.

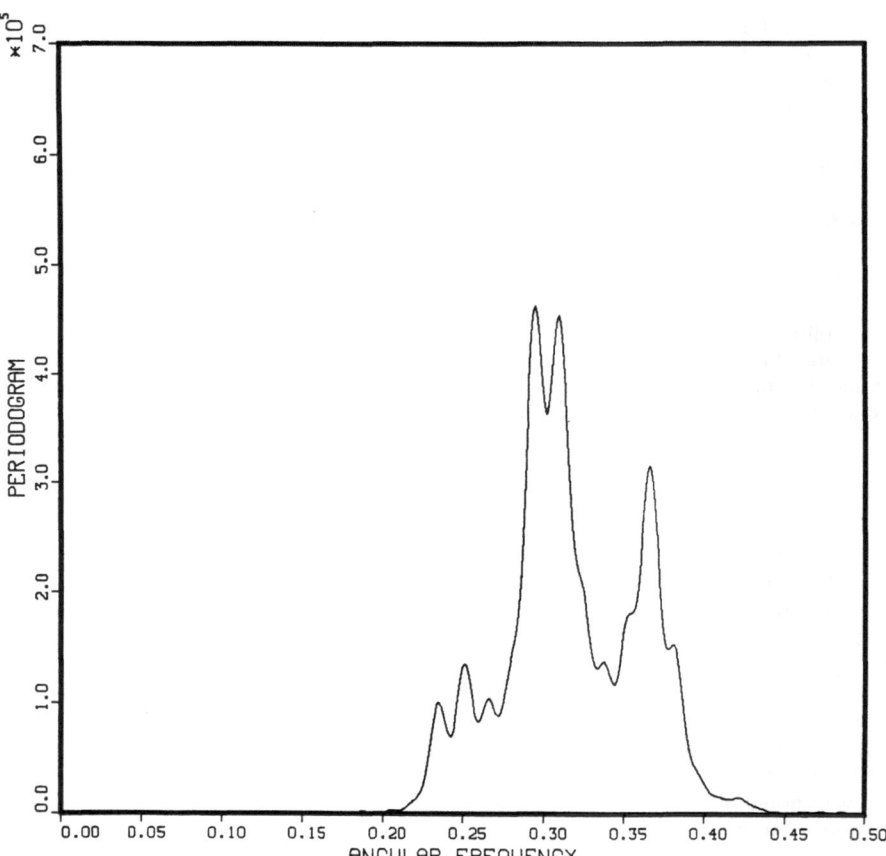

COMBINED PERIODOGRAM

FOR CHANNEL 1 AND 2

When more than one channel is present the periodogram is not the proper statistic to be examining for indications of a simple harmonic frequency. The analysis for multiple channels ndicates the proper statistic is essentially the sum of the periodograms for each channel weighted by the mean square variance of the data. A phase reversal in this data produces the splitting in the central peak.

After searching each of these peaks we found there is one center frequency at $\omega \approx 0.3$ and two others at $\omega \approx 0.25$ and $\omega \approx 0.35$. On the periodogram, therefore, the two highest peaks are not indicative of two frequencies but of a single frequency located at the minimum between them. Here we have the opposite effect from what we saw in Section IV; there we had only one peak in the periodogram, but finer analysis showed that there were two close frequencies present. Here, we have two peaks in the periodogram, but finer analysis shows only one frequency to be present. These examples just illustrate one of the major results of this work: If one asks a question about a single harmonic frequency, when the data have evidence of multiple complex phenomena, one can get answers which are misleading or simply incorrect in the light of more realistic models. Peaks in the Fourier transform are not always an indication of the frequencies present.

We investigated this splitting in the periodogram further. It is caused by a phase reversal in the signal. That is, if the signal is a simple cosine times a decay function $D(t)$, then the splitting can appear when $D(t)$ becomes negative. This type of feature can be present in an NMR signal because due to magnetic field inhomogeneity, the "line" may be a complex superposition of many small signals which have slightly different frequencies.

The next step in the analysis is to construct a plausible model for the data and apply it. As was demonstrated earlier, the exact decay model is not needed to obtain good estimates of a frequency. What is needed is a decay model which is reasonable for the phenomenon being observed: one which decays down to the noise level on the same time scale as the "true" decay. We simply assume the decay is magnet limited and that the decay is Lorentzian. The model we use has three frequencies and two decay rates,

$$f_1(t) = [B_1 \cos(\omega_1 t) + B_2 \sin(\omega_1 t)] \exp[-\alpha_1 t]$$
$$+ [B_3 \cos(\omega_2 t) + B_4 \sin(\omega_2 t)] \exp[-\alpha_2 t]$$
$$+ [B_5 \cos(\omega_3 t) + B_6 \sin(\omega_3 t)] \exp[-\alpha_1 t]$$

and similarly for channel 2

$$f_2(t) = [B_7 \cos(\omega_1 t) + B_8 \sin(\omega_1 t)] \exp[-\alpha_1 t]$$
$$+ [B_9 \cos(\omega_2 t) + B_{10} \sin(\omega_2 t)] \exp[-\alpha_2 t]$$
$$+ [B_{11} \cos(\omega_3 t) + B_{12} \sin(\omega_3 t)] \exp[-\alpha_1 t] .$$

There are five $\{\omega\}$ parameters, 12 amplitudes, and two noise variances. Probability theory has eliminated 14 of the nineteen parameters. We are primarily interested in the three frequencies; however, the decay rates will tell us just how strongly the inhomogeneous magnetic field is affecting the decay rates.

The computer code in Appendix A was used to evaluate the "student t-distribution" (28) for each channel, and these were multiplied to obtain (66). We searched in the five dimensional parameter space until we located the maximum of the distribution by the "pattern" searching procedure noted before. We then used the procedure given in Section V, equations (64-65), to estimate the standard deviation of the parameters. The derivatives which appear in this procedure were evaluated numerically. The results of this calculation are:

Parameter	estimate ± standard deviation
ω_1	74.51 ± 0.03 Hz
ω_2	96.68 ± 0.02 Hz
ω_3	117.38 ± 0.02 Hz
α_1	4.02 ± 0.02 Hz
α_2	5.07 ± 0.02 Hz .

We also estimated the signal-to-noise ratio (42) for each channel:

$$\frac{\text{Signal}}{\text{Noise}} \text{ channel } 1 = 14.5$$

$$\frac{\text{Signal}}{\text{Noise}} \text{ channel } 2 = 14.2$$

and the estimated variance (40) with $s = 1$:

$$<\sigma> \text{ channel } 1 = 10.9$$

$$<\sigma> \text{ channel } 2 = 10.7 \ .$$

The actual frequencies are of little importance in this experiment (the values are controlled by how close to 55Mhz a local oscillator is set). What is important is the relative separation of the three frequencies. The separation for the HD doublet $(\omega_3-\omega_1)$ is 42.87 ± 0.04Hz and the D_2 frequency is displaced from the center by $\Delta = \omega_2-(\omega_3+\omega_1)/2 = 0.74\pm0.04$Hz. The separation frequency is in excellent agreement with previous measurements and with theory.[15, 16]

The HD and D_2 components of the signal are known to have very different decay rates,[15] yet the values indicated by probability theory are nearly the same. We conclude that the inhomogeneous magnetic field has significantly affected the decay rates. The decay is substantially magnet limited.

We can compare these estimates directly to the absorption spectrum, Fig. 7(A). The absorption spectrum resolves the three frequencies. However, they are very close together. The reason the analysis of this experiment is so difficult with the absorption spectrum is that the full-width at half maximum for the D_2 peak, Fig. 7(A), is 16Hz. Figure 7(B) gives the estimates from (66). We have plotted these estimates as three normalized Gaussians each centered at the estimated frequency and having the same standard deviation as the estimated frequency. Clearly the resolution of these frequencies is much improved. With separately normalized distributions, the heights in Fig. 7(B) are indications of the accuracy of the three estimates, not of the power carried by the signal.

B. Economic data: Corn crop yields.

Economic data are hard to analyze, in part because the data are frequently contaminated by large spurious effects, and the time series are often very short. Here we will examine one example of economic data to demonstrate how to remove some unknown and spurious effects. In particular, we will analyze one hundred year's worth of the corn crop data from three states (Kansas, South Dakota, and Nebraska), Fig. 8(A) through Fig. 8(C).[17] We would like to know if there is any indication of periodic behavior in these data.

These data have been analyzed before. Currie[18] used a high pass filter and then applied the Burg algorithm[19] to the filtered data. Currie finds one frequency near 20 years which is attributed to the lunar 18.6 year cycle, and another at 11 years, which is attributed to the solar cycle.

There are three steps in Currie's analysis that are troublesome. First, the Burg algorithm is not optimal in the presence of noise (although it is for the problem it was formulated to solve). The fact that it continues to work means that the procedure is reasonably robust; that does not change the fact that it is fundamentally not appropriate to this problem.[19] Second, one has doubts about the filter: could it suppress the effect one is looking for or introduce other spurious effects? Third, to apply the Burg algorithm when the data consist of the actual values of a time series, the autoregression order (maximum lag to be used) must be chosen and there is no theoretical principle to determine this choice. We do not mean to imply that Currie's result is incorrect; only that it is provisional. We would like to apply probability theory as developed in this paper to check these results.

Figure 7.

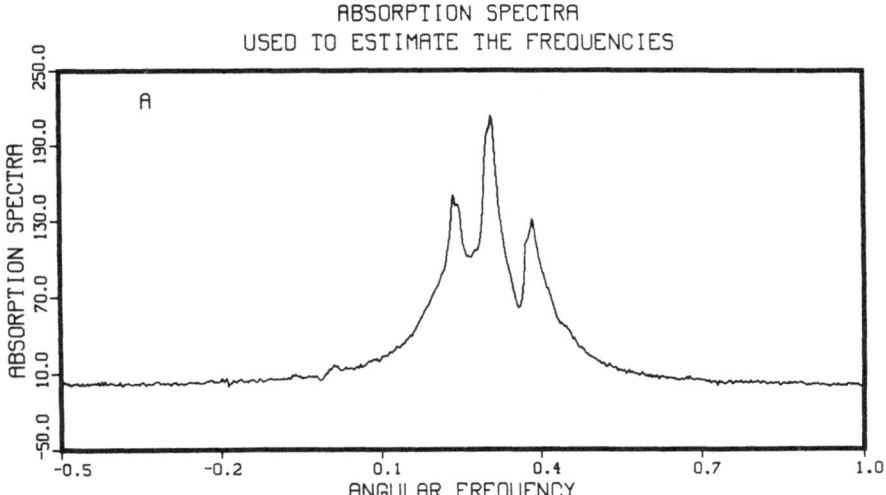

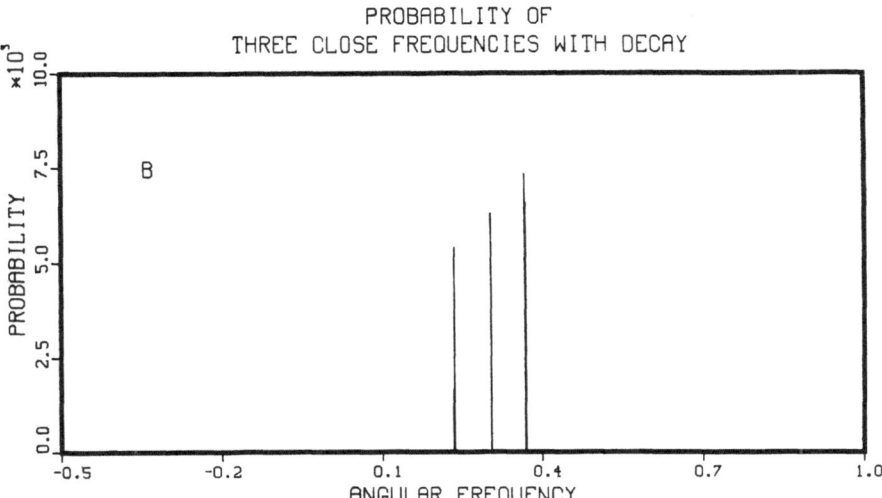

The absorption spectrum, Fig. 7(A) can be obtained from two channels by assuming the channels have the same amplitude and are 90° out of phase. A complex Fourier transform is calculated using one channel as the real and the other as the imaginary part of the signal. The global phase of the complex transform is adjusted until the real part has the largest possible area. The real part is the absorption spectrum. Of course the channels do not have the same amplitudes and are not exactly 90° out of phase. It requires an extensive procedure to put these two channels into a usable form. Using the full-width at half maximum of the absorption spectra to determine the accuracy estimate and converting to physical units one may determine the frequencies to about ±15Hz. The probability analysis, Fig. 7(B) used a three frequency model with two decay rates. The estimated accuracy is ±0.02Hz.

Figure 8.

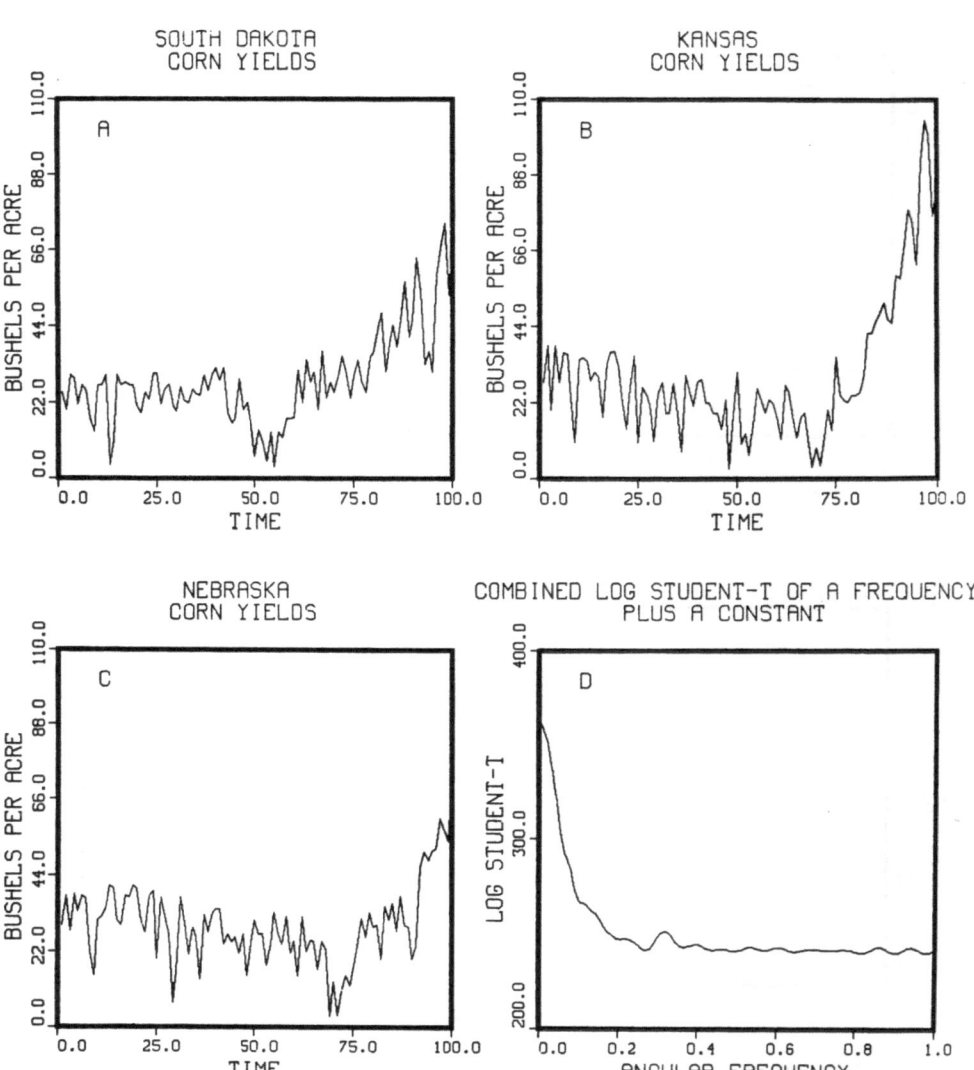

The three data sets analyzed were corn yields in bushels per acre for: South Dakota Fig. 8(A), Kansas Fig. 8(B), and Nebraska Fig. 8(C). The log probability of a single common frequency plus a constant is plotted in Fig. 8(D). The question we would like to answer is "Is that small bump located a approximately 0.3 corresponding to a 20 year period a real indication of a frequency or is it an artifact of the trend?" The sharp up turn in the yields occurs at about 1940 and is due to improved varieties, irrigation, etc.

The first step in a harmonic analysis is simply to plot the data, Fig. 8(A) through Fig. 8(C) and the log of the posterior probability of a single harmonic frequency. In the previous example we generalized the analysis for two measurements. The generalization to an arbitrary number of measurements is just a repeat of the previous arguments and we give the result here for any number of measurements

$$P(\{\omega\}|DI) \propto \prod_{j=1}^{r} \left\{ \left[1 - \frac{m_j \overline{h^2}_j}{N_j \overline{d^2}_j} \right]^{\frac{m_j - N_j}{2}} \right\} . \tag{67}$$

The subscripts refer to the j'th measurement, and each of the models have m_j amplitudes, and each data set contains N_j data values. Additionally it was assumed that the noise variance σ_j was unknown and possibly different for each measurement. The "student t-distributions" (28) for each measurement should be computed separately, thus estimating and eliminating the effects particular to one measurement, and then multiplied to obtain the posterior probability, for the common effects (67). As discussed earlier, this procedure leads to conservative estimates; if we incorporated more prior information (for example, if it were known that the σ_j are all equal) we would obtain slightly better results.

For this harmonic analysis we take the model to be a single sinusoid which oscillates about a constant. The model for one measurement may be written

$$f_j(t) = B_{j1} + B_{j2}\sin(\omega t) + B_{j3}\cos(\omega t) . \tag{68}$$

We allow B_{j1}, B_{j2}, and B_{j3} to be different for each measurement; thus there are a total of nine amplitudes, one frequency, and three noise variances. To compute the posterior probability for each measurement, we used the computer code in Appendix A. The log of each "student t-distribution" (28) was computed, and added to obtain the log of the posterior probability of a single common harmonic frequency, Fig. 8(D).

What we would like to know is, "Are those small bumps in Fig. 8(D) indications of periodic behavior, or are they artifacts of the noise or trend?" To attempt to answer this, consider the following model function

$$f_j(t) = T_j(t) + B_{j,1}\cos(\omega t) + B_{j,2}\sin(\omega t)$$

where we have augmented the standard frequency model by a trend $T_j(t)$. The only parameter of interest is the frequency ω. The trend $T_j(t)$ is a nuisance function. To eliminate the nuisance function $T_j(t)$ we expand the trend in orthonormal polynomials $L_j(t)$. These orthonormal polynomials could be any complete set. We use the Legendre polynomials with an appropriate scaling of the independent variable to make them orthonormal on the region $(-49.5 \leq t \leq 49.5)$. This is the range of values used for the time index in the sine and cosine terms. After expanding the trend, the model function for the j'th measurement can be written

$$f_j(t) = \sum_{k=0}^{r} B_{j,k+1}L_k(t) + B_{j,r+2}\cos(\omega t) + B_{j,r+3}\sin(\omega t) .$$

Notice, that if $r = 0$ the problem is reduced to the previous problem (68). The cosine and sine model functions have been renumbered to remain consistent with the notation used earlier.

The expansion order r must be set to some appropriate value. From looking at these data one sees that it will take at least a second order expansion to remove the trend. The actual expansion order for the trend is unknown. However, it will turn out that the estimated frequencies are insensitive to the expansion order, as long as the expansion is sufficient to represent the trend without representing the signal of interest. Of course, different orders would have very different implications about other questions than the ones we are asking; for example, predicting the future trend. That is an altogether more difficult problem than the one we are solving.

The effects of increasing the expansion order r can be demonstrated by plotting the posterior probability for several expansion orders; see Fig. 9(A) through Fig. 9(H). For expansion orders zero, Fig. 9(A) through expansion order 2, Fig. 9(C) the trend has not been removed: the posterior probability continues to pick out the low frequency trend. When a third order trend is used, Fig. 9(D) a sudden change in the behavior is seen. The frequency near $\omega \approx 0.31$ suddenly shows up, along with a spurious low frequency effect due to the trend. In expansion orders four through seven, Fig. 9(E) through Fig. 9(H) the trend has been effectively removed and the posterior probability indicates there is a frequency near 0.31 corresponding to a 20.4 year period.

The amount of variability in the frequency estimates as a function of the expansion order will show how strongly the trend expansion is affecting the estimated frequency. The frequency estimates for the fourth through seventh order expansions are

$$(f_4)_{est} = 20.60 \pm 0.08 \text{ years}$$

$$(f_5)_{est} = 20.47 \pm 0.09 \text{ years}$$

$$(f_6)_{est} = 20.20 \pm 0.07 \text{ years}$$

$$(f_7)_{est} = 20.47 \pm 0.09 \text{ years}.$$

Here the estimated errors represent one standard deviation of the posterior distribution. Generally it is considered good policy to claim an accuracy corresponding to two standard deviations. Thus, given the spread in the estimates it appears there is indeed evidence for a frequency of a period 20.4 ± 0.2 years.

Now that the effects of removing a trend are better understood, we can proceed to a two frequency model plus a trend to see if we can verify Currie's two frequency results. Figure 10 is a plot of the log of this probability distribution after removing a fifth order trend. The behavior of this plot is the type one would expect when a two frequency model is applied to a data set that contains only one frequency. From this we cannot verify Currie's results. That is, for the three states taken as a whole these data show evidence for a oscillation near 20.4 years as he reports, but we do not find evidence for an 11 year cycle. This does not say that Currie's result is incorrect; he incorporated much more data into his calculation and to check it we would need to include data from at least a dozen more states. While this is a worthy project it is beyond the scope of this simple demonstration.

C. Another NMR example.

Now that the tools have been developed we can demonstrate how one can incorporate partial information about a model. In the corn crop example the trend was unknown, so it was expanded in orthonormal polynomials and integrated out of the problem, while we included what partial information we had in the form of the sine and cosine terms. In this NMR example let us assume that the decay function is of interest to us. We would like to determine this function as accurately as possible.

The data we used, Fig. 11(A), in this example are one channel of a pure D_2 spectrum.[14] Figure 11(B), contains the periodogram for these data. For this demonstration we will use the first $N = 512$ data points because it contains most of the signal.

For D_2, theory indicates there is a single frequency with decay.[15] Now we expect the signal should have the form

$$f(t) = \left\{ B_1 \sin(\omega t) + B_2 \cos(\omega t) \right\} D(t)$$

Figure 9.

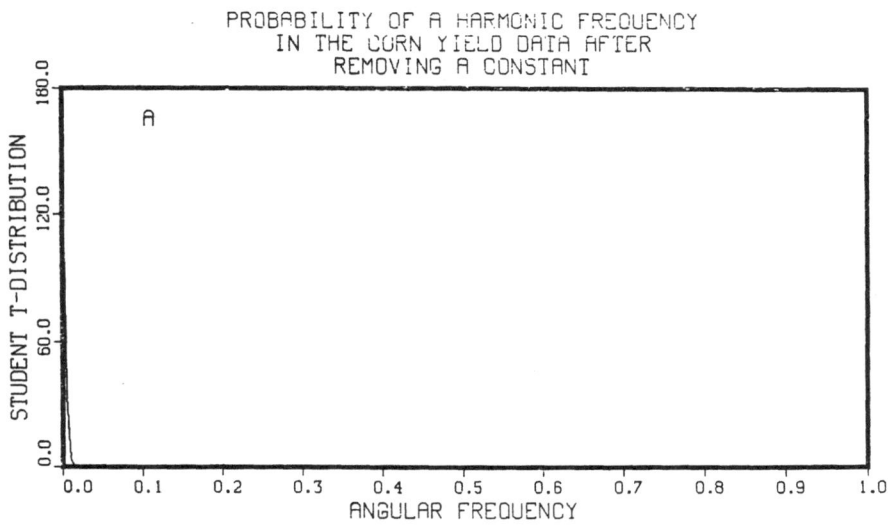

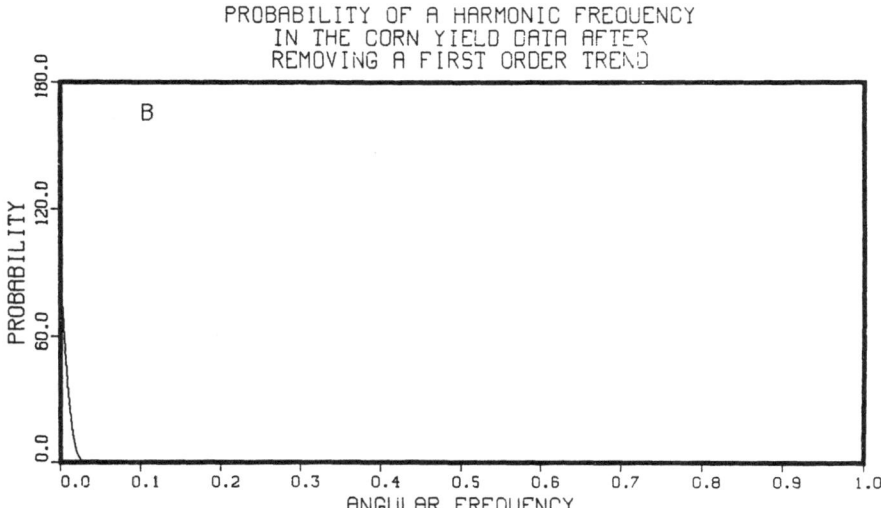

The probability of a single harmonic frequency plus a constant has a large sharp peak at zero corresponding to the trend in the data, Fig. 9(A). When a linear trend is included in the analysis, the peak lowers but the estimated frequency still corresponds to the low frequency trend, Fig. 9(B).

Figure 9. (continued)

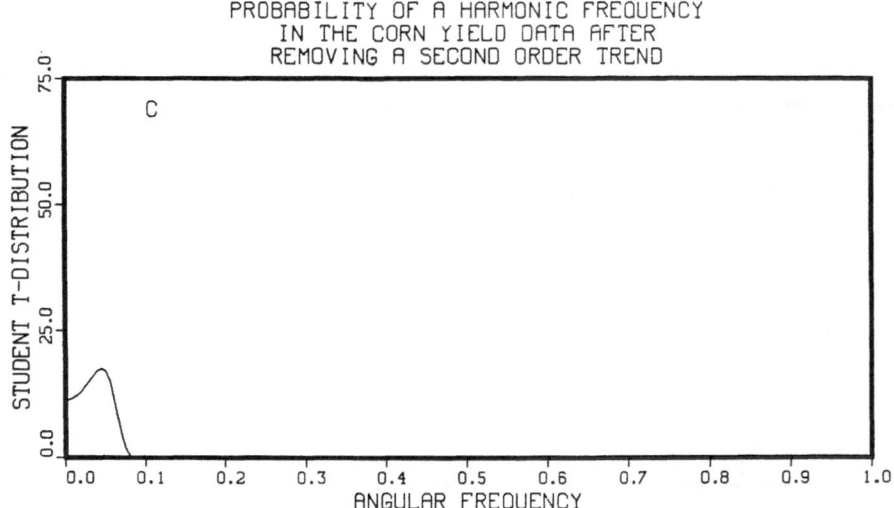

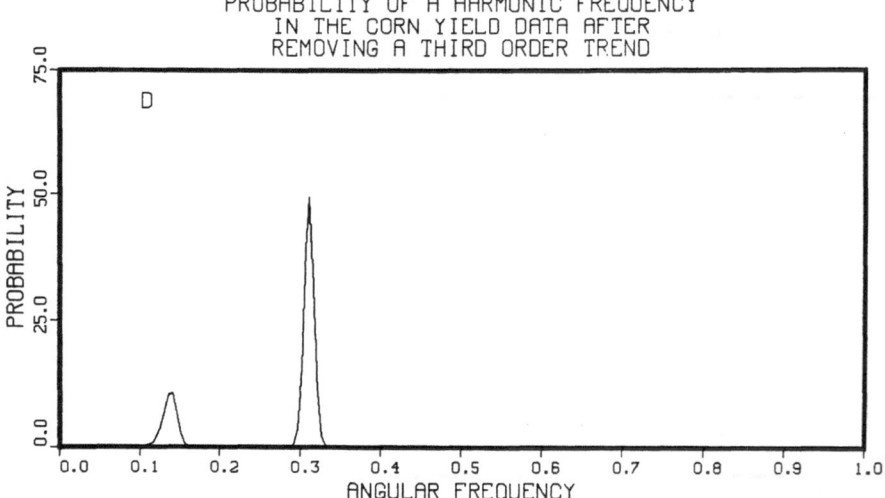

The probability of a single harmonic frequency plus a second order trend, Fig. 9(C) continues to pick out the low frequency trend. However, the level and spread of the probability is such that the trend has almost been removed. When the probability of a single harmonic frequency plus a third order trend is computed, the probability density suddenly changes behavior. The frequency near 0.3 is now the dominant feature, Fig. 9(D). The trend has not been completely removed; a small artifact persists at low frequencies.

Figure 9. (continued)

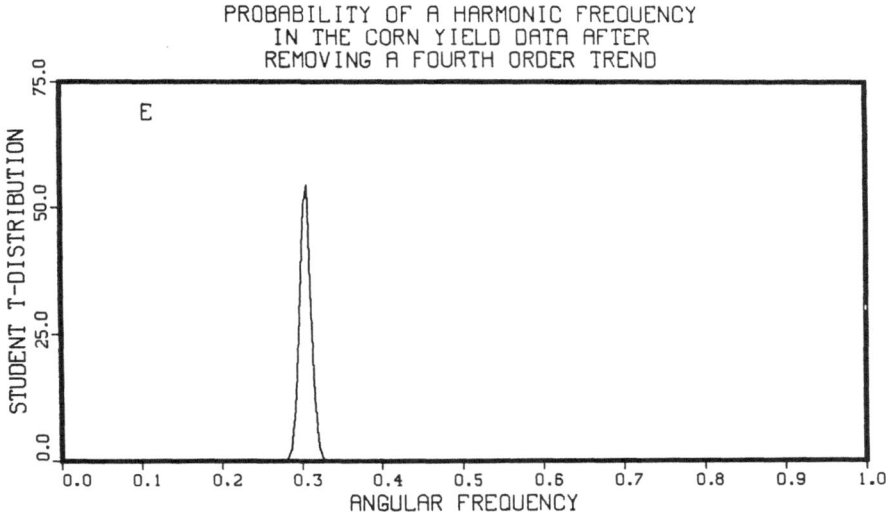

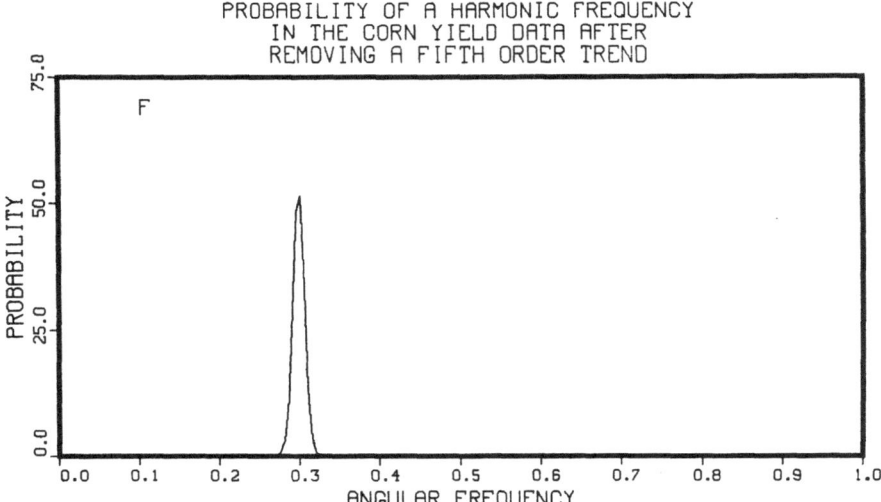

When the probability of a fourth order trend plus a harmonic frequency is computed the trend is now completely gone and only the frequency at 20 years remains, Fig. 9(E). When the expansion order is increased in Fig. 9(F) the frequency estimate is not essentially changed.

Figure 9. (continued)

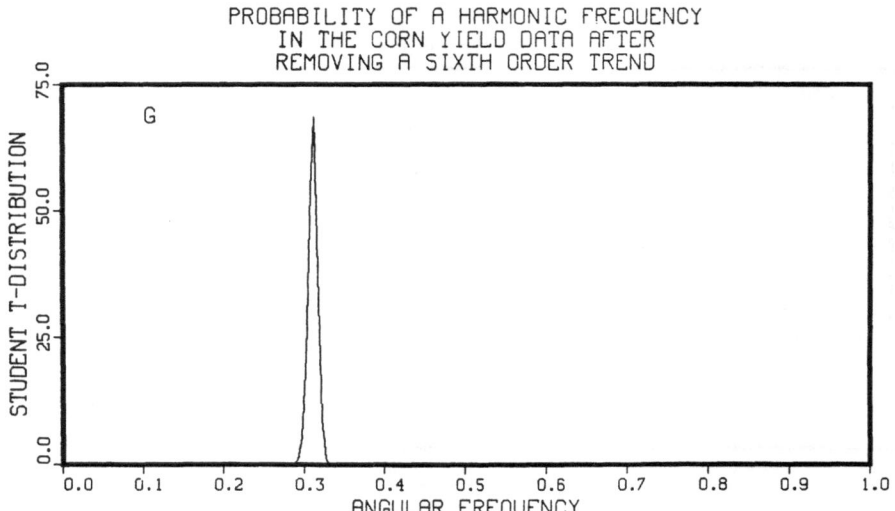

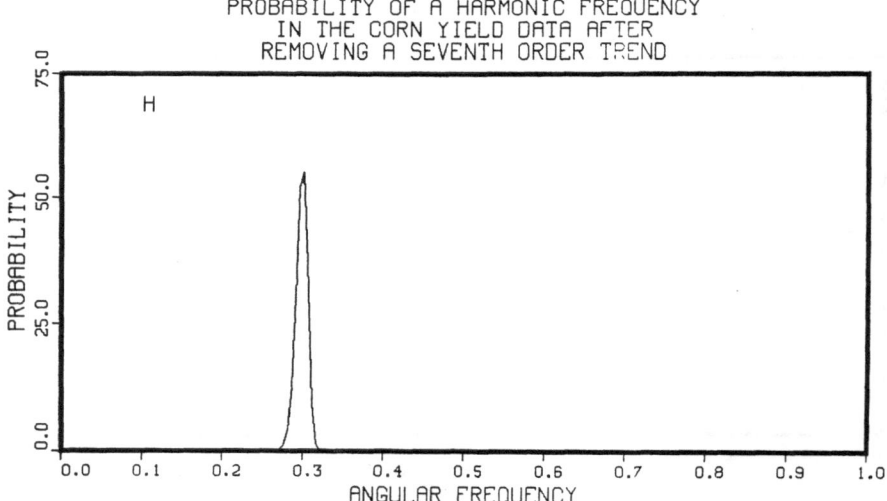

Increasing the expansion order further does not significantly affect the estimated frequency, Fig. 9(G) and Fig. 9(H). If the expansion order is increased sufficiently, the expansion will begin to remove the harmonic oscillation; the posterior probability density will gradually decrease in height.

Figure 10.

LOG PROBABILITY OF TWO HARMONIC FREQUENCIES

AFTER REMOVING A FIFTH ORDER TREND

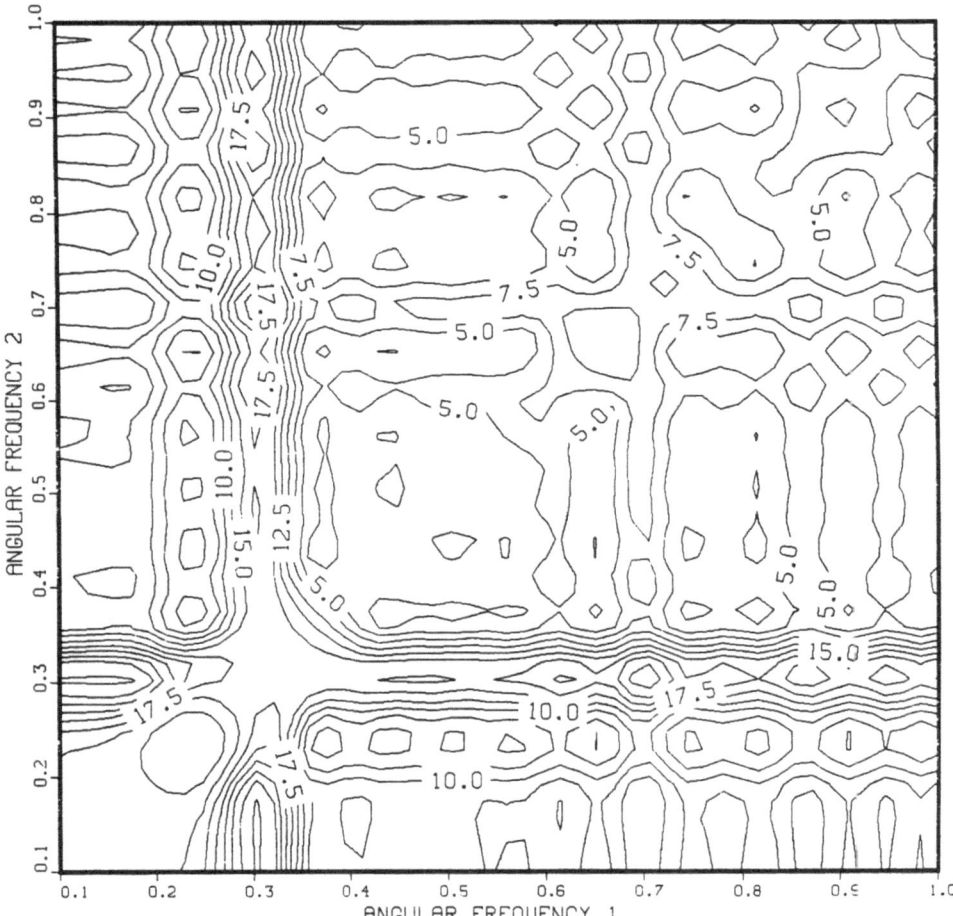

This is the $\log_e$ of the probability of two common harmonic frequencies in the crop yield data with a fifth order trend. This type of structure is what one expects from the sufficient statistic when there is only one frequency present. Notice the maximum is located roughly along a vertical and horizontal line at 0.3. The slight increase in the sufficient statistic below 0.2 just indicates we have not completely removed the trend.

Figure 11.

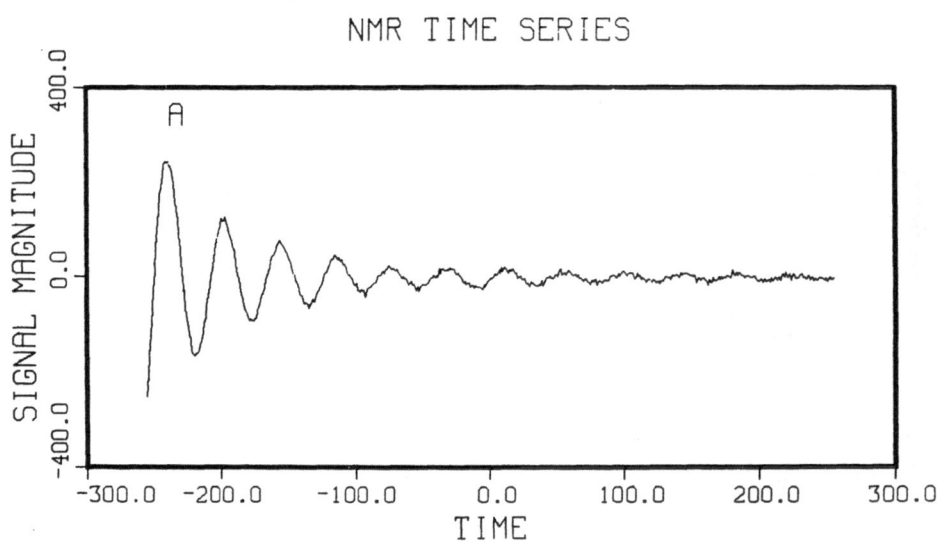

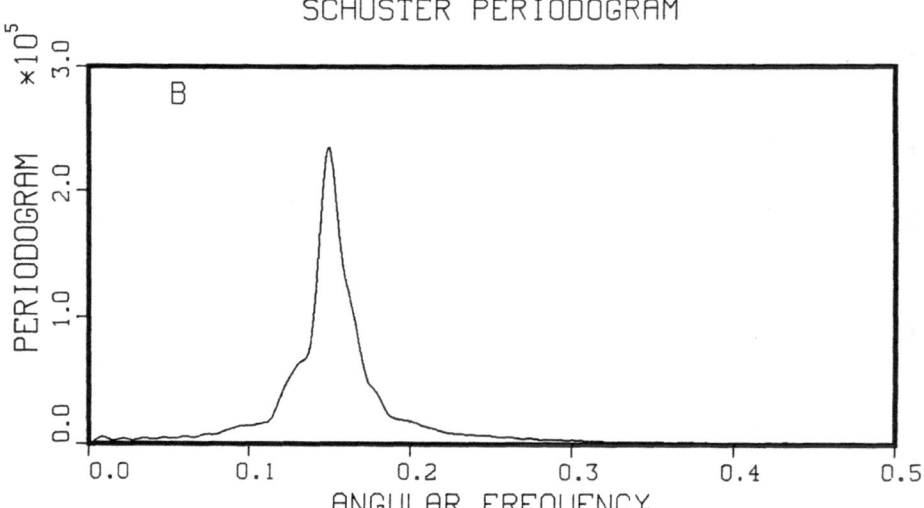

This NMR data, Fig. 11(A) is a free-induction decay for a D_2 sample. The sample was excited using a 55MHz pulse and the signal detected using a mixer-demodulator. We used 512 data samples to compute the periodogram, Fig. 11(B). We would like to use probability theory to obtain an estimate of the decay function while incorporating what little we know about the oscillations.

where $D(t)$ is the decay function, and the sine and cosine effectively express what partial information we have about the signal. We will expand the decay function $D(t)$ to obtain

$$f(t) = \left\{ B_1\sin(\omega t) + B_2\cos(\omega t) \right\} \sum_{j=0}^{r} D_j L_j(t)$$

where D_j are the expansion coefficients for the decay function, B_1 and B_2 are effectively the magnitude and phase of the sinusoidal oscillations, and L_j are the Legendre polynomials with the appropriate change of variables. This model can be rewritten as

$$f(t) = \sum_{j=0}^{r} D_j B_1 \left\{ L_j(t)[\sin(\omega t) + \frac{B_2}{B_1}\cos(\omega t)] \right\}.$$

There is an indeterminacy in the overall scale. That is, the amplitude of the sinusoid and the amplitude of the decay $D(t)$ cannot both be determined. One of them is necessarily arbitrary. We chose the amplitude of the sine term to be one because it effectively eliminates one $\{\omega\}$ parameter from the problem. We have a choice, in this problem, on which parameters are to be removed by integration. We chose to eliminate $\{D_j B_1\}$ because there are more of them, even though they are really the parameters of interest.

When we eliminate a parameter from the problem, it does not mean that it cannot be estimated. In fact, we can always calculate these $\{D_j B_1\}$ parameters from the linear relations between models (19). For this problem it is computationally simpler to search for the maximum of the probability distribution as a function of frequency ω and the ratio B_1/B_2, and then use equation (19) to compute the expansion coefficients. If we choose to eliminate the amplitudes of the sine and cosine terms then we must search for the maximum of the probability distribution as a function of the expansion parameters; there could be a large number of these.

We must again set the expansion order r; here we have plenty of data so in principle we could take r to be large. However, unless the decay is rapidly varying we would expect a moderate expansion of perhaps 5th to 10th order to be more than adequate. In the examples given here we set the expansion order to 10. We solved the problem also with the expansion order set to 5, and the results were effectively identical to the 10'th order expansion.

To solve this problem we again used the computer code in Appendix A, and the "pattern" search routine discussed earlier. We located the maximum of the two dimensional "student t-distribution" (28) and used the procedure given in Section V, equations (64-65), to estimate the standard deviation of the parameters. We find these to be

$$(\omega)_{est} = 0.14976 \pm 10^{-5}$$

$$\left(\frac{B_2}{B_1} \right)_{est} = -.475 \pm 2.7 \times 10^{-3} .$$

The variance of these data was $\overline{d^2} = 2902$, the estimated noise variance $<\sigma^2>_{est} \approx 27.1$, and the signal-to-noise ratio was 23.3.

After locating the maximum of the probability density we used the linear relations (19) between the orthonormal model and the nonorthonormal model to compute the expansion coefficients. As noted earlier there is an arbitrary choice in the scale (magnitude) of the decay function. We set the scale by requiring the decay function and the reconstructed model function to touch at one point near the global maximum. We have plotted the data and the estimated decay function, Fig. 12(A). In Fig. 12(B) we have a close up of the data, the decay function, and the reconstructed signal.

It is apparent from this plot that the decay is not Lorentzian. The decay function drops rapidly and then begins to oscillate. This is a real effect and is not an artifact of the procedure we are using. There are two possible interpretations: there could be a second small

Figure 12.

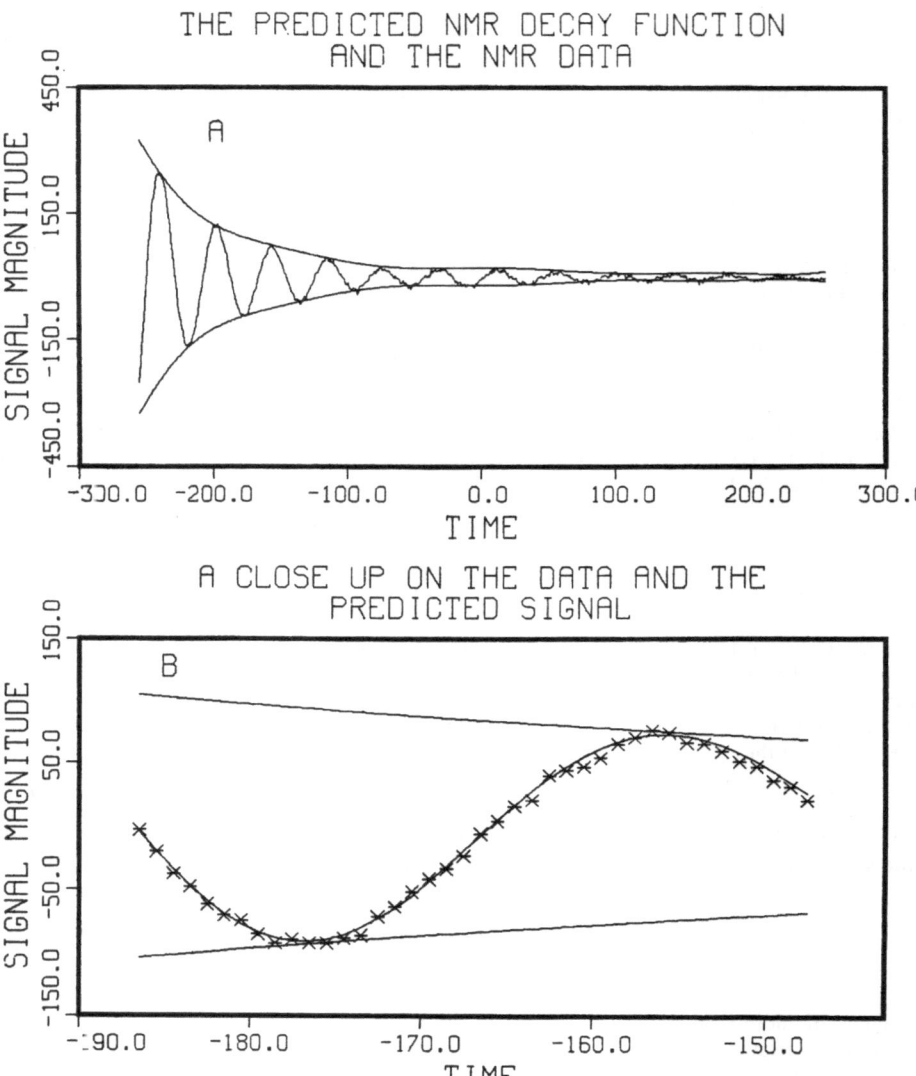

The decay function in Fig. 12(A) comes down smoothly and then begins to oscillate. This effect is a real effect and is not an artifact of the analysis. This type of behavior is characteristic of an inhomogeneous magnetic field. In Fig. 12(B) we have plotted a blow up of the data, the predicted signal, and the decay function.

signal which is beating against the primary signal, or the inhomogeneous magnetic field could be causing it. The most likely cause is the inhomogeneous magnetic field, because one can literally change a dial setting on the equipment and get the decay envelope to change shape.[14] In problems with multiple signals, or even with this D_2 signal, when the magnetic field is particularly inhomogeneous the decay function can show much stronger oscillations and even become negative.

D. Wolf's relative sunspot numbers.

In 1848 Rudolf Wolf introduced the relative sunspot numbers as a measure of solar activity. These numbers, defined earlier, are available as yearly averages since 1700, Fig. 1(A). The importance of these numbers is primarily due to the fact that they are the longest available quantitative index of the sun's internal activity. The most prominent feature in these numbers is the 11.04 year cycle discussed earlier. In addition to this cycle a number of others have been reported including 180, 90, 45, and a 22 year cycle as well as a number of others.[20,21] We will apply probability theory to these numbers to see what can be learned. We must stress that in what follows we have no idea what the "true" model is, but can only examine a number of different possibilities. We begin by asking "What is the evidence for multiple harmonic frequencies in these data?"

These numbers have been analyzed before by many writters. We will contrast our results to those obtained recently by Sonnet[21] and Bracewell.[22] The analysis done by Sonnet concentrated on determining the spectrum of the relative sunspot numbers. He used the Burg[19] algorithm. This routine is extremely sensitive to the frequencies. In addition to finding the frequencies, this routine will sometimes shift the location of the predicted frequency, and it estimates a spectral density (a power normalized probability distribution), not the power carried in a line. Consequently, no accurate determination of the power carried by these lines has been done. We will use probability theory to estimate the frequencies, their accuracy, the amplitudes, the phases, as well as the power carried by each line.

Again, we plot the log of the probability of a single harmonic frequency plus a constant, Fig. 13(A). We include the constant and allow probability theory to remove it the correct way, instead of subtracting the average from the data as was done in Section II. We do this to see if this theoretically correct way of eliminating a constant will make any difference in the predicted frequencies. We plot the log of the marginal posterior probability (28) using

$$f(t) = B_1 + B_2\cos\omega t + B_3\sin\omega t$$

as the model. The periodogram, Fig. 13(B), is a sufficient statistic for harmonic frequencies if and only if the time series has zero mean. Under these conditions the periodogram must go to zero at $\omega = 0$. In the periodogram, Fig. 13(B), that small peak near zero is a spurious effect due to subtracting the average value from the data. Probability analysis using a simple harmonic frequency plus a constant does not show any evidence for this period, Fig. 13(A).

Now we examined each of the peaks in Fig. 13(A) with a two frequency model plus a constant to determine if there is any evidence for doublets. There are two, one located near 0.11 and another one near 0.72. Additionally, we examined the low frequency region very closely to see if there was evidence for a very low frequency and found none. The fact that we could not find it does not prove conclusively that the period is not there. We had to search for the peak in a high dimensional space. The peak is small, and the search routine could step over it.

We had to decide which peaks to include in the model; we simply took the 13 largest. We choose 13 because we wanted at least a 12 frequency model to be able to compare to Sonnet's model. We then applied a multiple frequency model using

$$f(t) = B_1 + \sum_{j=1}^{13}\left\{B_{j+1}\cos(\omega_j t) + B_{2j+1}\sin(\omega_j t)\right\}.$$

Figure 13.

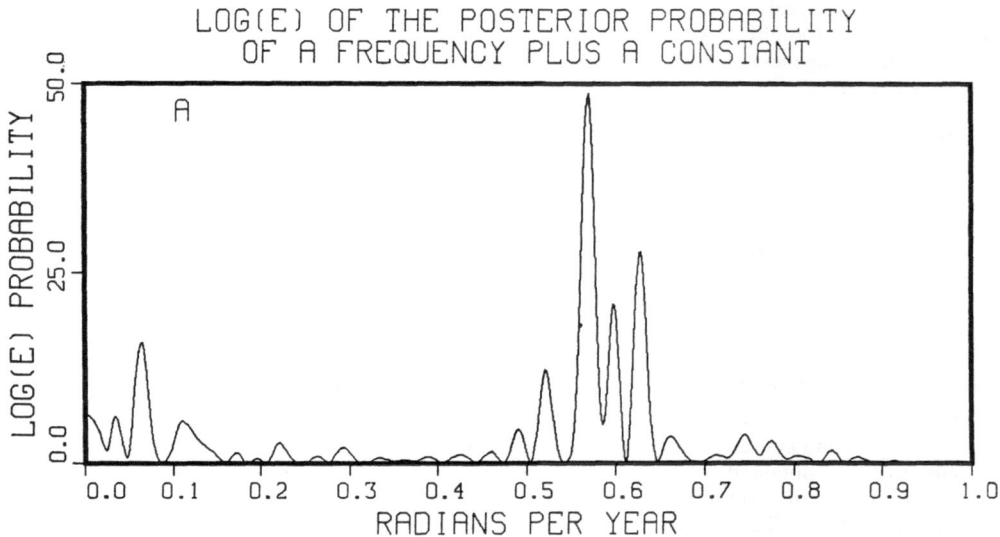

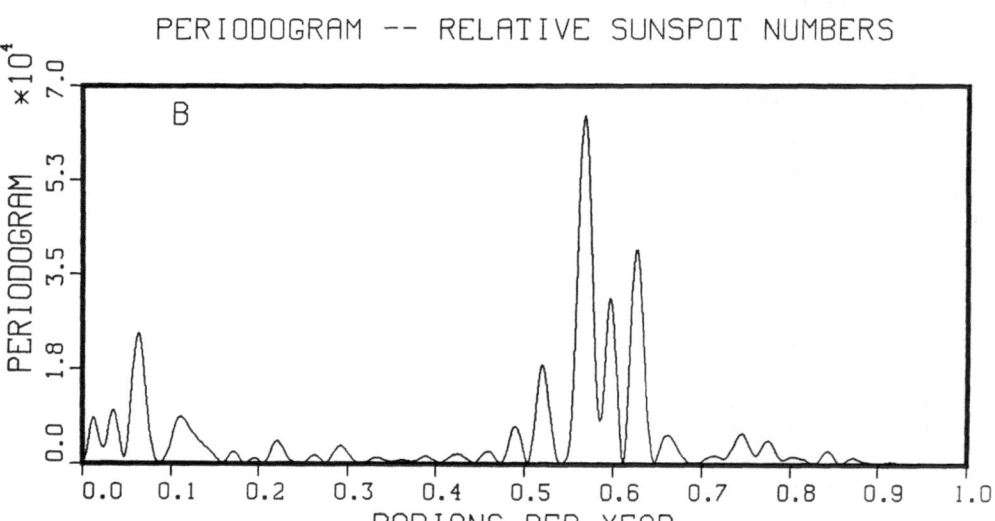

The $\log_e$ of the "student t-distribution" Fig. 13(A), and the periodogram 13(B) are almost identical. There are two major differences. The periodogram is related to the posterior probability when σ^2 is known. If the σ^2 is unknown then the scale does not tell one anything about the evidence for a frequency. The $\log_e$ of the "student t-distribution" does not have this problem. If the peak is higher by 2 then the probability is higher by e^2. The other major difference is at the origin, for a data set with zero mean the periodogram must go to zero. This will create a small peak near the zero. The $\log_e$ of the "student t-distribution" goes to zero only if there is no evidence of a constant. Thus, the "student t-distribution" does not indicate the presence of a spurious peak near zero.

We computed the probability of the frequencies $\{\omega_1, \cdots \omega_{13}\}$ using the computer code given in Appendix A. The pattern search routine discussed earlier was used to locate the maximum of this 13 dimensional space to five significant digits. Two of the 13 frequencies converged to the same numerical value, indicating that what we thought was two frequencies in Fig. 13(A) was in fact only one frequency. We removed one of these frequencies to obtain a 12 frequency model, and repeated the search using the previous values as our initial estimates. We computed the standard deviation using the procedure developed in Section V, equations (64-65). Last, we used the linear relations between the models (19) to compute the nonorthonormal amplitudes as well as their second moments. These are summarized as follows

$<\hat{f}>_{est}$	$<B_1>$	$<B_2>$	$<B_1^2 + B_2^2>$
95.29±0.62 years	7.522	17.624	371.01
58.94±0.38 years	11.165	5.247	157.91
51.07±0.18 years	1.499	-8.789	84.73
28.16±0.07 years	6.146	-3.522	54.07
13.03±0.01 years	6.728	-1.637	51.08
11.86±0.02 years	-15.613	-0.840	247.52
10.99±0.02 years	-37.569	-10.329	1521.95
10.75±0.01 years	23.071	-4.526	555.29
9.97±0.01 years	13.509	-11.932	328.87
9.41±0.01 years	1.971	7.038	58.63
8.48±0.01 years	-9.222	4.655	109.81
8.12±0.01 years	1.654	7.254	60.12

With these 12 frequencies and one constant the estimated noise variance is $<\sigma^2>_{est} = 398$, and the signal-to-noise ratio is 15.2. The constant term had a value of 48.22.

We have plotted these 12 frequencies as normalized Gaussians 14(A) to get a better understanding of their determination. The best determined frequency is, of course, the 10.99±0.016 year cycle. When we performed this calculation using the single frequency model our estimate was 11.04±0.02 years; we have moved the estimated frequency over three standard deviations. This illustrates that the periodogram can give misleading estimates when there are multiple close frequencies. However, as long as they are reasonably well separated the estimates should not be off by very much.

We could not verify the 180 year period. We included this one in the original 13 frequencies. However, the pattern search consistently confounded it with the 95 year period. This could be due to poor searching procedures or it could indicate that the data do not show evidence for this frequency. Regardless, this frequency needs to be examined more closely. Additionally, there are a number of other small frequencies on the periodogram; we did not include these even though we suspect they are real frequencies.

We can plot an approximation to the power spectral density just by normalizing 14(A) to the appropriate power level, Fig. 14(B). The dotted line on this plot is the periodogram normalized to the highest value in the power spectral density. This plot brings home the fact that when the frequencies are close the periodogram is not even approximately the correct sufficient statistic for estimating a harmonic frequency. At least one of the predicted frequencies occurs right at a minimum of the periodogram. Also notice that the normalized power is more or less in fair agreement with the periodogram when the frequencies are well separated. That is because for a simple harmonic frequency the peak of the periodogram is indeed a good estimate of the energy carried in that line.

In addition to the power spectral density we can plot what this model thinks is the sunspot series -- less the noise, Fig. 15(A). We have repeated the plot of the sunspot numbers, Fig. 15(B) for comparison.

Figure 14.

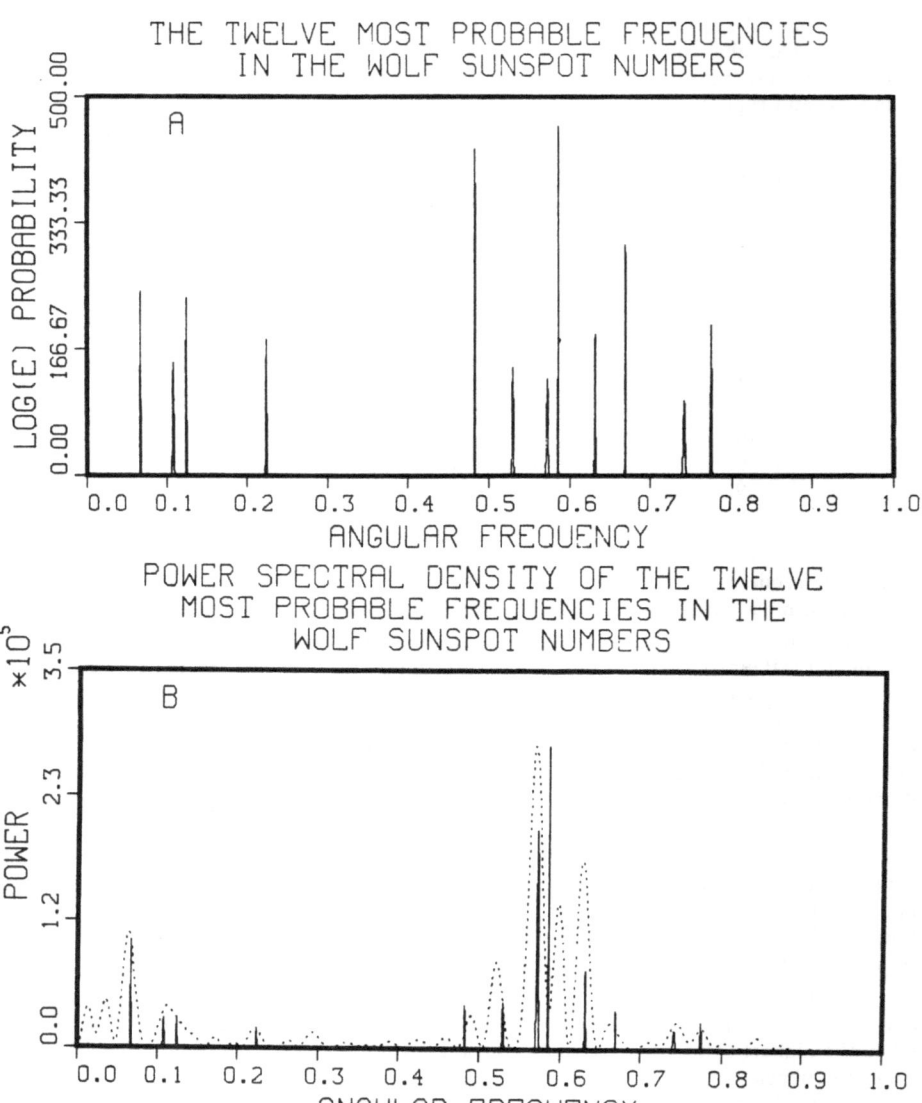

The posterior probability of twelve frequencies in the relative sunspot numbers Fig 14(A), has twelve well resolved peaks. With normalized Gaussians the height represents the accuracy of the estimates not the power carried them. Figure 14(B) has been normalized to the power. The peak value of the periodogram is an accurate estimate of the energy carried in a line so long as there is only one isolated frequency present. Notice that the periodogram does estimate the isolated peaks reasonably accurately. However, around the 11 year period ($\omega \approx 0.58$) the periodogram not only does not get the power correctly, at least one of the estimated frequencies is located at a minimum of the periodogram.

Figure 15.

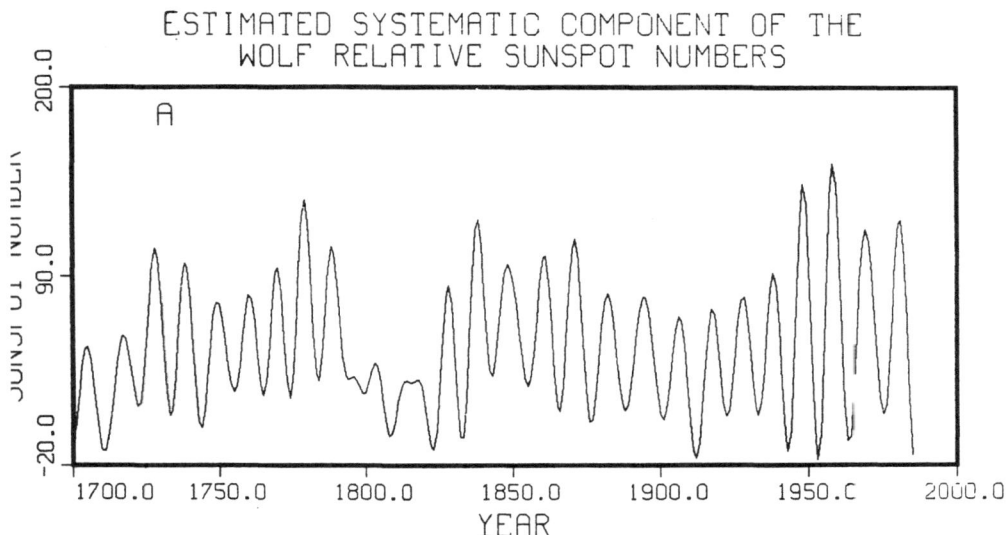

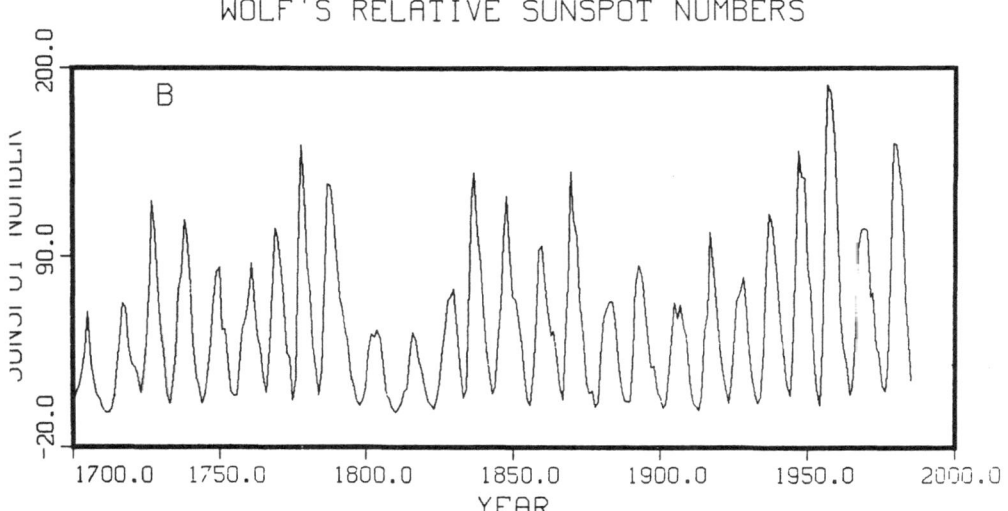

Not only can one obtain the estimated power carried by the signal, one can use the amplitudes to plot what probability theory has taken to be the signal with the noise removed. Of course a reconstruction of this nature is only as good as the model, Fig. 15(A). We have included the relative sunspot numbers. Fig. 15(B), for easy comparison. The predicted series can probably be made better by including some of the smaller cycles we ignored in this analysis.

This simple 12 frequency model reproduces most, but not all of the features of the sunspot numbers. There is still something missing from the model. In particular the data values drop uniformly to zero at the minima. This behavior is not repeated in the 12 frequency model. Also, the data have sharper peaks than troughs, while our sinusoidal model, of course, does not. This is, as has been noted before, evidence of some kind of "rectification" process. A better model could easily reproduce these effects.

We chose to examine 12 frequencies because that was the number of frequencies used in a model proposed by Sonnet.[21] He has attempted to explain these numbers in terms of harmonic frequencies: 180, 90, and 45 are examples of harmonically related frequencies. In 1982, C. P. Sonnet[21] published a small paper in which the sunspot number spectrum was be explained using

$$f(t) = [1 + \alpha\cos(\omega_m t)][\cos(\omega_c t) + \Delta]^2$$

as a model, where Sonnet's estimate of magnetic cycle ω_m is approximately 90 years, and his estimate of the solar cycle ω_c is 22 years. The rectification effect is present here.

This model is written in a deceptively simple form and a number of constants (phases and amplitudes) have been suppressed. We propose to apply probability theory using this model to determine ω_c and ω_m. To do this we first square the term in brackets and then use trigonometric identities to reduce this mode to a form where probability theory can readily estimate the amplitudes and phases:

$$\begin{aligned}
f(t) = {} & A_1 + A_2\cos([\omega_m]t) + A_3\sin([\omega_m]t) \\
& + A_4\cos([2\omega_m]t) + A_5\sin([2\omega_m]t) \\
& + A_6\cos([\omega_c-2\omega_m]t) + A_7\sin([\omega_c-2\omega_m]t) \\
& + A_8\cos([\omega_c-\omega_m]t) + A_9\sin([\omega_c-\omega_m]t) \\
& + A_{10}\cos([\omega_c]t) + A_{11}\sin([\omega_c]t) \\
& + A_{12}\cos([\omega_c+\omega_m]t) + A_{13}\sin([\omega_c+\omega_m]t) \\
& + A_{14}\cos([\omega_c+2\omega_m]t) + A_{15}\sin([\omega_c+2\omega_m]t) \\
& + A_{16}\cos([2\omega_c-2\omega_m]t) + A_{17}\sin([2\omega_c-2\omega_m]t) \\
& + A_{18}\cos([2\omega_c-\omega_m]t) + A_{19}\sin([2\omega_c-\omega_m]t) \\
& + A_{20}\cos([2\omega_c]t) + A_{21}\sin([2\omega_c]t) \\
& + A_{22}\cos([2\omega_c+\omega_m]t) + A_{23}\sin([2\omega_c+\omega_m]t) \\
& + A_{24}\cos([2\omega_c+2\omega_m]t) + A_{25}\sin([2\omega_c+2\omega_m]t) .
\end{aligned}$$

Now Sonnet specifies the amplitudes, but not the phases.[21] We will take a more general approach and not constrain these amplitudes. We will simply allow probability theory to pick the amplitudes which fit the data best. Thus any result we find will have the Sonnet frequencies but the amplitudes and phases will be chosen to fit the data "better" than the Sonnet model. After integrating out the amplitudes we have only two parameters to determine, ω_c and ω_m.

We located the maximum of the posterior probability density using he computer code in Appendix A, and using the pattern search routine. The "best" estimated value for ω_c (in years) is approximately 21.0 years, and ω_m approximately 643 years. The values for these parameters given by Sonnet are $\omega_c = 22$ years and $76 < \omega_m < 108$ years with a mean value of $\omega_m \approx 89$ years. Our probability analysis estimates the value of ω_c and ω_m to be substantially different from those given by Sonnet. The most indicative value is the estimated variance for this model $\sigma^2_{\text{Sonnet}} = 605$. This is worse than that predicted for the simple 12 frequency model by almost a factor of 1.5 and is comparable to the fit achieved by a five frequency model.

We have so far investigated two variations of harmonic analysis on the relative sunspot numbers. Let us proceed to investigate a more complex case to see if there might be more going on in the relative sunspot numbers than just simple periodic behavior. These data have been looked at from this standpoint at least once before. Bracewell[22] has analyzed these numbers to determine if they could have a time-dependent "instantaneous phase". The model used by Bracewell can be written as

$$f(t) = B_1 + \text{Re}\left[E(t)\exp\left(i\phi(t) + i\omega_{11}t\right)\right]$$

where B_1 is a constant term in the data, $E(t)$ is a time varying magnitude of the oscillation, $\phi(t)$ is the "instantaneous phase", and ω_{11} is the 11 year cycle.

This model does not incorporate any prior information into the problem. It is so general that any function can be written in this form. Nevertheless, the idea that the phase $\phi(t)$ could be varying slowly with time is interesting and worth investigation.

An "instantaneous phase" in the notation we have been using is a chirp. Let $\phi(t)$ stand for the phase of the signal, and ω its frequency. Then we may Taylor expand $\phi(t)$ around $t = 0$ to obtain

$$\omega t + \phi(t) \approx \phi_0 + \omega t + \frac{\phi''}{2}t^2 + \cdots$$

where we have assumed the first derative $\phi'(t)$ is zero. If this were not so then ω is not the frequency as presumed here. The Bracewell model can then be approximated as

$$f(t) = B_1 + E(t)[\cos(\omega t + \alpha t^2) + B_2\sin(\omega t + \alpha t^2)]$$

To second order, the Bracewell model is just a chirped frequency with a time varying envelope.

We can investigate the possibility of a chirped signal using

$$f(t) = B_1 + B_2\cos(\omega t + \alpha t^2) + B_3\sin(\omega t + \alpha t^2)$$

as the model, where α is the chirp rate, B_1 is a constant component, ω is the frequency of the oscillation, and B_2 and B_3 are effectively the amplitude and phase of the oscillation. This model is not a substitute for the Bracewell model. Instead this model is designed to allow us to investigate the possibility that the sunspot numbers contain evidence of a chirp, or "instantaneous phase" in the Bracewell terminology.

A plot of the log of the "student t-distribution" using this as a model is the statistic to look for chirp. However, we now have two parameters to plot, not one. We have constructed a contour plot around the 11 year cycle, Fig. 16. We expect this plot to have a peak near the location of a frequency. It will be centered at zero chirp rate if there is no evidence for chirp, and at some nonzero value when there is evidence for chirp. Notice, that along the line $\alpha = 0$ this "student t-distribution" is just the simple harmonic probability distribution studied earlier, Fig. 1(A). As with the Fourier transform if there are multiple well separated chirped frequencies (with small chirp rates) then we expect there to be multiple peaks in, Fig. 16.

There are a number of peaks; the single largest point on the plot is located off the $\alpha = 0$ axis. The data contain evidence for chirp. The low frequencies also show evidence for chirp. To the extent that the Bracewell "instantaneous phase" may be considered as a chirp we must agree with him; there is evidence in these data for chirp.

In light of this discussion, exactly what these numbers represent and exactly what is going on inside the sun to produce them must be reconsidered. Certainly we have not answered any real questions about what is going on; indeed that was not our intention.

Figure 16.

LOG(E) OF THE PROBABILITY OF ONE FREQUENCY

WITH CHIRP IN THE WOLF SUNSPOT NUMBERS

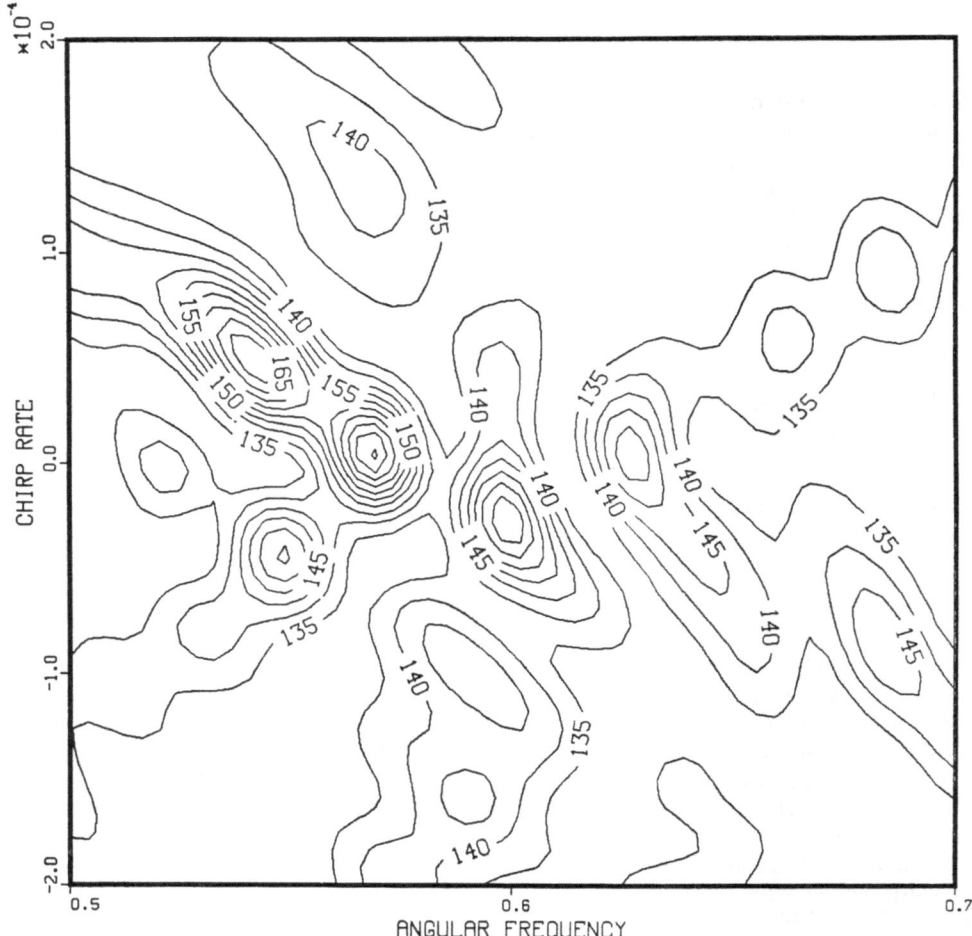

To check for chirp we take $f(t) = A_1 + A_2\cos(\omega t + \alpha t^2) + A_3\sin(\omega + \alpha t^2 t)$ as the model. After integrating out the nuisance parameters the posterior probability is a function of two variables: the frequency ω, and the chirp rate α. We then plotted the $\log_e$ of the posterior probability, Fig. 16(A). If there is evidence of chirp in these data we expect the peak to be displaced into the α direction. The single highest peak is located at a positive value of α: there is evidence of chirp.

Instead we have shown how use of probability theory for data analysis can facilitate future research by testing various hypotheses more sensitively than could the traditional intuitive *ad hoc* procedures.

VII. SUMMARY AND CONCLUSIONS.

In this analysis we have attempted to explore some of the aspects of Bayesian parameter estimation as they might apply to time series, even though the analysis as formulated is applicable to any data set, be it a time series or not.

A. Summary

We began this analysis in Section II, by applying probability theory to estimate the spectrum of a data set that, we postulated, contained only a single sinusoid plus noise. In Section III, we generalized these simple considerations to relatively complex models including the problem of estimating the spectrum of multiple nonstationary harmonic frequencies in the presence of noise. This led us to the "student t-distribution": the posterior probability of the $\{\omega\}$ parameters, whatever their meaning. In Section IV, we estimated the nuisance parameters and calculated, among other things, the power spectral density, and the noise variance. In Section V, we specialized to spectrum analysis and explored some of the implications of the "student t-distribution" for this problem. At the end of Section V, we developed a procedure for estimating the accuracy of the $\{\omega\}$ parameters. In Section VI, we applied these analyses to a number of real time series with the aim of exploring some of the techniques needed to apply these procedures. In particular, we demonstrated how to use them to estimate multiple nonstationary frequencies, and how to incorporate incomplete information into the estimation problem. In the sunspot example we did not know which model was appropriate, so we applied a number of different models with the intention of discovering as much about them as possible.

B. Conclusions

Perhaps the single biggest conclusion of this work is that what one can learn about a data set depends critically on what questions one asks. If one insists on doing Fourier transform analysis on a data set, then our analysis shows that one will always obtain answers of the form "What is the evidence of a single stationary harmonic frequency in these data?" This will be more than adequate if there are plenty of data and no evidence of complex phenomena. However, if the data show evidence for multiple frequencies or complex behaviors, the Fourier transform gives answers which can be misleading or incorrect in light of more realistic models.

APPENDIX A
A Computer Algorithm for Computing the Posterior
Probability (28) for an Arbitrary Set of Model Equations.

This subroutine was used to prepare all of the numerical analysis presented in this work. This is a general purpose implementation of the calculation that will work for any model functions and for any setting of the parameters, independent of the number of parameters and their values. In order to do this, the subroutine requires five pieces of input data and one work area. On return one receives $H_i(t_j)$, h_i, $\overline{h^2}$, $P(\{\omega\}|DI)$, $<\sigma>$, and $\hat{p}(\{\omega\})$. The parm list is as follows:

Parm	LABEL	i/o	Description/function
N	INO	input	The number of discrete time samples in the time series to be analyzed.
m	IFUN	input	This is the order of the matrix g_{jk} and is equal to the number of model functions.
d_j	DATA	input	The time series (length N): this is the data to be analyzed.
G_{ij}	GIJ	input	This matrix contains the j nonorthogonal model functions [dimensioned as GIJ(INO,IFUN)] and evaluated at t_i.
ZLOGE	ZLOGE	i/o	This is the $\log_e$ of the normalization constant. On the initial call to this routine this field should be initialized to zero. The subroutine never computes the "student t-distribution" when ZLOGE is zero: instead the $\log_e$ of the "student t-distribution" is computed. It is up to the user to locate a value of $\log_e[P(\{\omega\}|DI)]$ close to the maximum of the probability density. This log value should then be placed in ZLOGE to act as an upper bound on the normalization constant. With this value in place the subroutine will return the value of the probability; then, an integral over the probability density can be done to find the correct value of the normalization constant. For an example of this procedure see the driver routine in Appendix B.
$H_i(t_j)$	HIJ	output	These are orthonormal model functions (17) evaluated at the same time and parameter values as GIJ.
h_i	HI	output	These are projections of the data onto the orthonormal model functions (24) and (36).
$\overline{h^2}$	H2BAR	output	The sufficient statistic $\overline{h^2}$ (26) is always computed.
$P(\{\omega\}|DI)$	ST	output	The "student t-distribution" (28) is not computed when the normalization constant is zero. To insure this field is computed the normalization constant must be set to an appropriate value. The calling routine in Appendix B has an example of how to do this.

STLE STLE output This is the $\log_e$ of the "student t-distribution" (28). This field is always computed even when the normalization is zero.

$<\sigma>$ SIG output This is the expected value of the noise variance σ as a function of the $\{\omega\}$ parameters (40) with $s=1$.

$\hat{p}(\{\omega\})$ PHAT output This is the power spectral density (39) as a function of the $\{\omega\}$ parameters.

WORK scratch This work area must be dimensioned $5m^2$. The dimension in the subroutines was set high to avoid possible "call by value" problems in FORTRAN. On return WORK contains the eigenvectors and eigenvalues of the g_{jk} matrix. The eigenvector matrix occupies m^2 continuous storage locations. The m eigenvalues immediately follow the eigenvectors.

This subroutine makes use of a general purpose "canned" eigenvalue and eigenvector routine which has not been included. If one chooses to implement this program one must replace the call (clearly marked in the code) with a call to an equivalent routine. Both the eigenvalues and eigenvectors are used by the subroutine and it assumes the eigenvectors are normalized.

```
      SUBROUTINE PROB
    C (INO,IFUN,DATA,GIJ,ZLOGE,HIJ,HI,H2BAR,ST,STLOGE,SIGMA,PHAT,WORK)
      IMPLICIT REAL*08(A-H,O-Z)
      DIMENSION DATA(INO),HIJ(INO,IFUN),HI(IFUN),GIJ(INO,IFUN)
      DIMENSION WORK(IFUN,IFUN,20)
    C
    C
      CALL VECTOR(INO,IFUN,GIJ,HIJ,WORK)
    C
      H2=0D0
      DO 1600 J=1,IFUN
      H1=0D0
      DO 1500 L=1,INO
 1500 H1=H1 + DATA(L)*HIJ(L,J)
      HI(J)=H1
      H2=H2 + H1*H1
 1600 CONTINUE
      H2BAR=H2/IFUN
    C
      Y2=0D0
      DO 1000 I=1,INO
 1000 Y2=Y2 + DATA(I)*DATA(I)
      Y2=Y2/INO
    C
      QQ=1D0 - IFUN*H2BAR / INO / Y2
      STLOGE=DLOG(QQ) * ((IFUN - INO)/2D0)
    C
      AHOLD=STLOGE - ZLOGE
      ST    =0D0
      IF(DABS(ZLOGE).NE.0D0)ST=DEXP(AHOLD)
    C
      SIGMA=DSQRT( INO/(INO-IFUN-2) * (Y2 - IFUN*H2BAR/INO) )
    C
    C
      PHAT = IFUN*H2BAR * ST
    C
      RETURN
      END
```

```
      SUBROUTINE VECTOR(INO,IFUN,GIJ,HIJ,WORK)
      IMPLICIT REAL*8(A-H,O-Z)
      DIMENSION HIJ(INO,IFUN),GIJ(INO,IFUN),WORK(IFUN,IFUN,20)
C
      DO 1000 I=1,IFUN
      DO 1000 J=1,INO
 1000 HIJ(J,I)=GIJ(J,I)
C
      CALL ORTHO(INO,IFUN,HIJ,WORK)
C
      DO 5000 I=1,IFUN
      TOTAL=0D0
      DO 4500 J=1,INO
 4500 TOTAL=TOTAL + HIJ(J,I)**2
      ANORM=DSQRT(TOTAL)
      DO 4000 J=1,INO
 4000 HIJ(J,I)=HIJ(J,I)/ANORM
 5000 CONTINUE
      RETURN
      END

      SUBROUTINE ORTHO(INO,NMAX,AIJ,W)
      IMPLICIT REAL*8 (A-H,O-Z)
      REAL*8  AIJ(INO,NMAX),W(NMAX)
C
      IT=1
      IE=IT + NMAX*NMAX
      IM=IE + NMAX*NMAX
      IW=IM + NMAX*NMAX
      I2=IW + NMAX*NMAX
      CALL TRANS(INO,NMAX,AIJ,W(IM),W(IT),W(IE),W(IW),W(I2))
      RETURN
      END

      SUBROUTINE TRANS
     C (INO,NMAX,AIJ,METRIC,TRANSM,EIGV,WORK1,WORK2)
      IMPLICIT REAL*8 (A-H,O-Z)
      REAL*8  AIJ(INO,NMAX)
      REAL*8  METRIC(NMAX,NMAX),EIGV(NMAX)
      REAL*8  TRANSM(NMAX,NMAX),WORK1(NMAX),WORK2(NMAX)
C
      DO 2000 I=1,NMAX
      DO 2000 J=1,NMAX
      TOTAL=0D0
      DO 1000 K=1,INO
 1000 TOTAL=TOTAL + AIJ(K,I)*AIJ(K,J)
      METRIC(I,J)=TOTAL
 2000 CONTINUE
C=*****************************************************************
C-***  THIS CALL MUST BE REPLACED WITH THE CALL TO AN EIGENVALUE
C-***  AND EIGENVECTOR ROUTINE
      CALL EIGERS(NMAX,NMAX,METRIC,EIGV,1,TRANSM,WORK1,WORK2,IERR)
C=***  NMAX   IS THE ORDER OF THE MATRIX
C=***  METRIC IS THE MATRIX FOR WHICH THE EIGENVALUES AND VECTORS
C=***         ARE NEEDED
C=***  EIGV   MUST CONTAIN THE EIGENVALUES ON RETURN
C=***  TRANSM MUST CONTAIN THE EIGENVECTORS ON RETURN
C=***  WORK1  IS A WORK AREA USED BY MY ROUTINE AND MAY BE USED
C=***         BY YOUR ROUTINE.  ITS  DIMENSION IS NMAX
C****         IN THIS ROUTINE. HOWEVER IT MAY BE DIMENSIONED
C****         AS LARGE AS NMAX*NMAX WITHOUT AFFECTING ANYTHING.
C****  WORK2  IS A SECOND WORK AREA AND IS OF DIMENSION NMAX
C****         IN THIS ROUTINE, IT MAY ALSO BE DIMENSIONED AS
C****         LARGE AS NMAX*NMAX WITHOUT AFFECTING ANYTHING.
C=*****************************************************************
C
C     SET UP THE ORTHOGONAL VECTORS
      DO 5120 K=1,INO
      DO 3100 J=1,NMAX
 3100 WORK1(J)=AIJ(K,J)
      DO 5120 I=1,NMAX
      TOTAL=0D0
      DO 3512 J=1,NMAX
 3512 TOTAL=TOTAL + TRANSM(J,I)*WORK1(J)
 5120 AIJ(K,I)=TOTAL
      RETURN
      END
```

APPENDIX B
An Example of how to Use Subroutine PROB

The following program was designed and used to prepare one example (the single harmonic frequency with Lorentzian decay) in the text. The steps needed to create this example may be generally described as follows: we read in the data; got an initial estimate of the log normalization constant, then integrate over the probability distribution, update the normalization constant, and evaluate the normalized probability distribution over the desired range of parameter values.

There are two basic steps involved in using subroutine PROB: first one must evaluate the nonorthogonal model functions at the desired values of the parameters; then, the subroutine PROB must be called to evaluate the "student t-distribution" for these parameter settings.

There are three routines in this example: the main line routine performs the steps just described; SETGIJ will evaluate the nonorthogonal model functions at the desired time points for the desired parameter values; ALIKE evaluates the probability density at the parameter values requested by the Gaussian quadrature routine. We have not included the integration routine since such routines are easily available, or easily written if need be.

```
        IMPLICIT REAL*08(A-H,O-Z)
        DIMENSION DATA(512),HIJ(512,2),HI(2),GIJ(512,2),WA(2,2,20)
        COMMON    DATA,HIJ,HI,GIJ,WA,ZLE,INO
        EXTERNAL  ALIKE
C
C
        CALL PROB(INO,2,DATA,GIJ,ZLE,HIJ,HI,H2,ST,STLE,SIG,PHAT,WA)
C
C       INO             THE NUMBER OF DATA POINTS
C       2               THE NUMBER OF MODEL FUNCTIONS
C       DATA            THE TIME SERIES
C       GIJ             THE NON-ORTHONORMAL MODEL FUNCTIONS
C       ZLE             LOG BASE E OF THE NORMALIZATION CONSTANT
C       HIJ             THE ORTHONORMAL MODEL FUNCTIONS
C       HI              THE PROJECTIONS OF THE DATA ONTO HIJ
C       H2              THE H**2 BAR STATISTIC
C       ST              STUDENT T-DISTRIBUTION
C       STLE            LOG BASE E OF THE STUDENT T-DISTRIBUTION
C       SIG             THE VARIANCE OF THE DATA
C       PHAT            POWER SPECTRAL DENSITY
C       WA              A WORK AREA USED BY THE SUBROUTINE
C
C
        INO=512
        READ(8,1000)(DATA(I),I=1,INO)
 1000 FORMAT(1X,19A4)
C
        ID=50
        FLOW=0.295D0
        FHI =0.31D0
        DF  =(FHI - FLOW)/(ID-1)
        ALOW=-0.03D0
        AHI =-0.01D0
        DA  =(AHI - ALOW)/(ID-1)
C
C       THIS ROUTINE WILL SET THE NORMALIZATION CONSTANT
C
        ZLE=0D0
C
        CALL SETGIJ(INO,2,GIJ,0.3D0,-.02D0)
C
        CALL PROB(INO,2,DATA,GIJ,ZLE,HIJ,HI,H2,ST,STLE,SIG,PHAT,WA)
C
        ZLE=STLE
C
```

```
C       INTEGRATE THE STUDENT T-DISTRIBUTION AROUND THE MAXIMUM
C       THIS IS A 24POINT GAUSSIAN QUADRATURE ROUTINE
        AN2=XYINT(0.299D0,0.301D0,1,-.03D0,-.01D0,1,ALIKE)
C
        ZLE=ZLE + DLOG(AN2)
C
C       THIS LOOP EVALUATES THE NORMALIZED DISTRIBUTION
C       ON A 50 BY 50 GRID. THESE POINTS ARE USED IN THE
C       PLOT ROUTINES AND THEY WERE ALSO USED TO INTEGRATE
C       OUT THE DECAY OR THE FREQUENCY PARAMETERS
C
        TOTAL=0D0
        DO 2000 I=1,ID
        DO 2000 J=1,ID
C
        W=(I-1)*DF + FLOW
        ALPHA=(J-1)*DA + ALOW
C
        CALL SETGIJ(INO,2,GIJ,W,ALPHA)
C
C       EVALUATE THE PROBABILITY DENSITY AT THESE POINTS
        CALL PROB(INO,2,DATA,GIJ,ZLE,HIJ,HI,H2,ST,STLE,SIG,PHAT,WA)
C
        TOTAL=TOTAL + ESTP*DF*DA
C
        WRITE(7,3333)W,ALPHA,ST,SIG,PHAT
 2000   WRITE(6,3333)W,ALPHA,ST,SIG,PHAT
 3333   FORMAT(5D15.5)
C
        STOP
        END

        SUBROUTINE SETGIJ(INO,IVEC,GIJ,W,ALPHA)
        IMPLICIT REAL*08(A-H,O-Z)
        DIMENSION GIJ(INO,IVEC)
C
C       THIS ROUTINE WILL EVALUATE THE MODEL FUNCTIONS AT
C       FREQUENCY W AND DECAY RATE ALPHA
C
        ADELTA=0.5D0*INO + 0.5D0
C
        DO 1000 I=1,INO
C
        TIME=I - ADELTA
C
C       EVALUATE MODEL FUNCTION 1
        GIJ(I,1)=DCOS(W*TIME)*DEXP(ALPHA*TIME)
C
 1000   GIJ(I,2)=DSIN(W*TIME)*DEXP(ALPHA*TIME)
C
        RETURN
        END

        REAL FUNCTION ALIKE*8(W,ALPHA)
        IMPLICIT REAL*08(A-H,O-Z)
        DIMENSION DATA(512),HIJ(512,2),HI(2),GIJ(512,2),WA(2,2,20)
        COMMON    DATA,HIJ,HI,GIJ,WA,ZLE,INO
C
C       THIS ROUTINE IS USED BY THE INTEGRATION ROUTINE
C       IT RETURNS THE VALUE OF THE INTEGRAND AT THE REQUESTED
C       VALUES
C
        CALL SETGIJ(INO,2,GIJ,W,ALPHA)
C
        CALL PROB(INO,2,DATA,GIJ,ZLE,HIJ,HI,H2,ST,STLE,SIG,PHAT,WA)
C
        ALIKE=ST
C
        RETURN
        END
```

References

1. G. Larry Bretthrost, *Bayesian Spectrum Analysis and Parameter Estimation*, Ph.D. thesis, University Microfilms Inc., Washington University, St. Louis, MO, Aug. 1987.

2. E. T. Jaynes, "Bayesian Spectrum and Chirp Analysis," in *Proceedings of the Third Workshop on Maximum-Entropy and Bayesian Methods (1983)* , ed. C. Ray Simth, D. Reidel, Boston, 1987. (the third workshop was held in Laramie, Wyoming)

3. R. B. Blackman and J. W. Tukey, *The Measurement of Power Spectra*, Dover Publications, Inc., New York, 1959.

4. E. T. Jaynes, *Papers on Probability, Statistics and Statistical Physics*, D. Reidel, Boston, 1983.

5. A. Schuster, "The Periodogram an its Optical Analogy," *Proceedings of the Royal Society of London*, vol. 77, p. 136, 1905.

6. Lord Rayleigh, *Philosophical Magazine*, vol. (5) 8, p. 261, 1879.

7. J. W. Tukey, several conversations with E. T. Jaynes, in the period 1980-1983.

8. Sir Harold Jeffreys, *Theory of Probability*, Oxford University press, London, 1939. (Later editions, 1948, 1961)

9. E. T. Jaynes, "Prior Probabilities," in *Papers on Probability, Statistics and Statistical Physics*, ed. R. D. Rosenkrantz, D. Reidel, Boston, 1983.

10. E. T. Jaynes, "Marginalization and Prior Probabilities," in *Papers on Probability, Statistics and Statistical Physics*, ed. R. D. Rosenkrantz, D. Reidel, Boston, 1983.

11. M. Waldmeier, in *The Sunspot Activity in the Years 1610-1960*, Schulthes, Zurich, 1961.

12. Robert Hooke and T. A. Jeeves, "'Direct Search' Solution of Numerical and Statistical Problems," *J. Assoc. Comp. Mach.*, p. 212, March 1962.

13. Derek Shaw, in *Fourier Transform NMR Spectroscopy*, Elsevier Scientific Pub. Co., New York , 1976.

14. Joseph W. Ganem and R. E. Norberg, *Private Communication*, 1987.

15. A. Abragam, in *Principles of Nuclear Magnetism*, p. 187, Oxford Science Publications, London, 1961. reprint (1985)

16. Richard James Beckett, *The Temperature and Density Dependence of Nuclear Spin-Spin Interactions in Hydrogen-Deuteride Gas and Fluid.*, Rutgers University Ph.D. Thesis, New Brunswick, New Jersey, 1979. (unpublished)

17. Robert Guinn Currie, *Private Communication*, 1985.

18. Robert Guinn Currie and Sultan Hameed, "Climatically Induced Cyclic Variations in United States Corn Yield and Possible Economic Implications," *presented at the Canadian Hydrology Symposium*, Regina, Saskatchewan, June 3, 1986.

19. John Parker Burg , *Maximum Entropy Spectral Analysis, Ph.D. Dissertation*, (University Microfilms No. 75-25), Stanford University, 1975.

20. T. J. Cohen and P. R. Lintz, "Long Term Periodicties in the Sunspot Cycle," *Nature*, vol. 250, p. 398, 1974.

21. C. P. Sonnet, "Sunspot Time Series: Spectrum From Square Law Modulation of the Half Cycle," *Geophysical Research Letters*, vol. 9 NO 12, pp. 1313-1316, 1982.

22. R. N. Bracewell, "Simulating the Sunspot Cycle," *Nature*, vol. 323, p. 516, Oct. 9, 1986.

DETECTION OF EXTRA-SOLAR SYSTEM PLANETS

E.T. Jaynes
Arthur Holly Compton Laboratory of Physics
Washington University, St. Louis, MO 63130

Abstract: Stimulated by a proposal of Wm. Hayden Smith and co-workers to detect planets on nearby stars by high resolution imaging, we speculate on the appropriate data analysis method, making use of probability theory to perform the optimal deconvolution of the point-spread function.

In this preliminary study, we seek to understand what probability theory has to say about the fundamental problem, by analyzing a simple one-dimensional version. The necessary theoretical principles are developed in the thesis of G.L. Bretthorst (Washington University, May 1987).

Our main message is this: once one is committed to using a computer to analyze the data, the high-resolution imaging problem is completely changed. What one has tried to do in the past by fancy optical and mechanical engineering (apodizing, image stabilizing) can be done better, and at a small fraction of the cost, by the computer.

Contents

One-Dimensional "Baby" Version of the Problem

Our optical system has a point-spread function $(\sin x/x)^2$. A "star" whose image should be a sharp point at position $x = a$ then produces a smeared image proportional to

$$G_1(x) = \frac{\sin^2(x - a)}{(x - a)^2} \tag{1}$$

and a "planet" at $x = b$ gives an offset diffraction pattern

$$G_2(x) = \frac{\sin^2(x - b)}{(x - b)^2}. \tag{2}$$

G. J. Erickson and C. R. Smith (eds.),
Maximum-Entropy and Bayesian Methods in Science and Engineering (Vol. 1), 147–160.
© *1988 by Kluwer Academic Publishers.*

These are the "model functions" to be built into the computer program. The star
and planet have brightnesses A_1, A_2 and so they produce jointly a smeared image

$$f(x) = A_1\, G_1(x) + A_2\, G_2(x) \tag{3}$$

This is observed at the positions $(x_1 \ldots x_N)$, getting a data set $D = (d_1 \ldots d_N)$:

$$d_i = f(x_i) + e_i, \qquad 1 \le i \le N \tag{4}$$

where the e's are white (*i.e.* uncorrelated) Gaussian noise measurement errors.
Don't worry about the assumption of whiteness; it will turn out that this assumption
can be removed almost trivially at the end, as explained below, so let's keep the
problem simple for now.

We shall assume that the noise level is not known in advance, so the computer
must estimate it from the data, and use it to determine the accuracy of the other
estimates. If the true noise level is known, this extra information can also be given
to the computer and it will enable us to improve the accuracy of the estimates, but
only slightly.

And for this first look at the problem, it doesn't matter whether the noise e is
thought of as measurement noise in the apparatus, atmospheric turbulence noise,
or "photon noise" in the phenomenon; the computer is going to estimate the total
noise in the data and make proper allowance for it, whatever its source. Later we
will see how to teach the computer to distinguish between atmospheric distortion
and true noise.

Also, don't worry about the "assumption" of Gaussianity; this will turn out to
be not really an assumption at all, but rather the most conservative assignment we
could make. If we have any additional information about the noise, which would
lead us to assign a nonGaussian probability distribution, this can be built into the
computer program and will enable one to get slightly better results than we will
obtain below. However, they will not be much better unless that information leads
to a wildly nonGaussian noise probability distribution, with a sharp upper bound
cutoff.

In short, all the assumptions we are making now are conservative, and removing
them will enable us to do still better (at the cost of more computation; but today
computing power is plentiful and cheap). The real power of computers to perform
sophisticated data analysis is only now beginning to be realized.

The computer's job is now: given the model functions (1), (2) and the data D,
tell us whether there is evidence for existence of a planet, and if so, give us the best
estimates of the separation $r = b - a$ and the relative brightness of planet and star;
and indicate the accuracy of those estimates.

The principles of probability theory, explained in Larry Bretthorst's thesis, determine the data analysis procedure to answer these questions in a way that is optimal (*i.e.*, makes full use of all the relevant information in the data) but is also conservative (*i.e.*, does not mislead us as to the accuracy of its conclusions). The reader is assumed familiar with the general formalism and notation used in the Bretthorst thesis, and we merely apply them to the present problem.

First, the computer is given the model functions $G(x)$, either analytically or measured at the observation points. Then the interaction matrix is

$$g_{jk} = \sum_{i=1}^{N} G_j(x_i)G_k(x_i) \tag{5}$$

and this matrix will be calculated by the computer program. For now, approximate it by an integral:

$$g_{12} = \int_{-\infty}^{\infty} G_1(x)G_2(x)dx = 4\pi \left[\frac{2r - \sin 2r}{(2r)^3}\right]. \tag{6}$$

It depends only on $r = b - a$, the planetary separation. As $r \to 0$,

$$g_{12} \to g_{11} = \frac{2\pi}{3} \tag{7}$$

[Check: verify $\int (\sin x/x)^4 dx = 2\pi/3$ directly.] Therefore, define

$$g_{12} = \frac{2\pi}{3}u(r) \tag{8}$$

where

$$u(r) = 6\left[\frac{2r - \sin 2r}{(2r)^3}\right] \tag{9}$$

is the overlap function, normalized to $u(0) = 1$. Note, for later purposes, that as $r \to 0$, $u(r)$ is given asymptotically by

$$u(r) \sim 1 - \frac{1}{5}r^2 + \frac{2}{105}r^4 + \dots \tag{10}$$

while as $r \to \infty$,

$$u(r) \sim \frac{3}{2}r^{-2} + \dots \tag{11}$$

The matrix g now becomes

$$g = \frac{2\pi}{3}\begin{bmatrix} 1 & u(r) \\ u(r) & 1 \end{bmatrix}. \tag{12}$$

Since the diagonal elements are equal, a fixed transformation will diagonalize this for any value of r; that is, the orthonormal model functions always have the form

$$H(x) = C[G_1(x) \pm G_2(x)]$$

so, supplying the proper normalization factors, we have

$$H_1(x; a, b) = \left[\frac{3}{4\pi(1 + u)}\right]^{1/2} [G_1(x) + G_2(x)] \qquad (13a)$$

$$H_2(x; a, b) = \left[\frac{3}{4\pi(1 - u)}\right]^{1/2} [G_1(x) - G_2(x)] \qquad (13b)$$

which somehow remind one of molecular orbitals. As we see from (1), (2), and (9), they contain the quantities of interest (a, b) as parameters. Once the computer is set to calculate these functions for all (x, a, b), we are ready to analyze any number of data sets (*i.e.*, any number of stars) with it.

The Computation Algorithm

Given a data set $(d_1 \ldots d_N)$, calculate the projections of the data onto the $H(x)$ functions:

$$h_j(a, b) = \sum_i d_i H_j(x_i), \qquad j = 1, 2 \qquad (14)$$

then a jointly sufficient statistic, which contains all the information the data have to give us about the unknown parameters (a,b) is simply

$$\sum h^2 = 2\overline{h^2}(a, b) = h_1^2 + h_2^2. \qquad (15)$$

If the computer is to estimate the noise level from the data and use the student t-distribution, it must also calculate

$$\sum d^2 = N\overline{d^2} = \sum_i d_i^2. \qquad (16)$$

Then the joint posterior probability density function for the parameters (a, b), as derived in the Bretthorst thesis, is proportional to

$$p(a, b \mid D, I) \sim \left[\frac{1}{\sum d^2 - \sum h^2}\right]^{(N-m)/2} \qquad (17)$$

where m (=2 in the present case) is the number of model functions that we are fitting to the data. But the only parameter of interest is $r = b - a$, so go to the mean and relative coordinates

$$R = (a + b)/2; \qquad r = b - a. \qquad (18)$$

The Jacobian of the transformation (18) is one, so the joint posterior probability density for R and r is the same quantity:

$$p(R, r \mid D, I) = p(a, b \mid D, I). \tag{19}$$

The final step is to integrate out the uninteresting parameter R, getting a function of r :

$$p(r) = p(r \mid D, I) = \int dR \; p(R, r \mid D, I) \tag{20}$$

which tells us everything the data have to say about the planetary separation, independently of the brightness and absolute positions of star and planet.

If the data contain evidence for a planet, then $p(r)$ will exhibit a peak at a value of r that represents the "best" estimate of its distance, and the width of the peak will indicate the probable error of that estimate.

The computer program can also give us its best estimate of the relative brightness of the star and planet, and the accuracy of that estimate. The way to do this is explained in the Bretthorst thesis; it is a small further detail available at essentially zero additional computation cost (the estimated brightnesses are just linear combinations of the coefficients h_i already calculated).

If there is no evidence in the data for a planet, then the computer will not be able to fit the data to a model function with $r > 0$, any better than it can to the function with $r = 0$. The posterior distribution $p(r)$ will then peak at $r = 0$, and it will indicate by its width the probable error in detecting the position of a very close planet. That is, if the star does have a planet of detectable brightness, it is extremely unlikely to be further from the star than the width of that distribution.

A major feature of this data analysis procedure is that, thanks to the elimination of nuisance parameters, we can combine the data from many different measurements into one grand final distribution. For example, suppose we have developed a computer program that eliminates atmospheric distortion as a nuisance parameter. Now we take 1000 successive data sets on the same star, in a time so short that the planet has not moved appreciably, perhaps a week. The computer will make allowance for the atmospheric distortion separately in each data set, and the total evidence concerning r will be given by the product of the separate $p(r)$ distributions; we merely add up all the $\log p(r)$ functions from the individual data sets.

The point of this is that, even though any one data set might not have a high enough signal/noise ratio to draw any definite conclusions, the totality of them will. But this evidence could not be extracted by a single analysis of the pooled data, because the atmospheric distortion will vary erratically from one data set to the next. From the pooled data one would be able to consider only a smeared-out average over those erratic variations, with resulting far poorer resolution. The difference between a long time exposure and what an observer can detect at a single instant of good seeing gives only a slight indication of how much this can help.

The computer is not only powerful, but flexible. In the course of a research project, one is almost sure to learn new things about the phenomenon being studied and the capabilities of the apparatus, which change one's views about how the data should be processed. If one is trying to do the processing by optical or mechanical engineering feats, this might involve rebuilding a telescope. If a computer is doing all the data processing, one needs only rewrite a few lines of the program code.

Intuitive Meaning of the Result

One problem we have is that the algorithm and final result (17) are so slick and efficient that, at first glance, it is far from obvious that this is really a sensible data analysis procedure; as it stands, (17) looks unpromising. Put differently, intuition alone would never have been powerful enough to tell us that this is the thing to do.

But intuition can be educated; so let's look at the result more closely to see some of the wonderful things that are hiding in (17); first by rewriting it in a form like the usual student t-distribution notation.

We are most interested in the region of the maximum. Let $\hat{a}, \hat{b}$ be a point where the sufficient statistic (15) reaches its absolute maximum (there are generally two such points, because the distribution is symmetric in a and b; so choose the one where $a < b$.) Let $Q(a, b)$ be the departure of the sufficient statistic from that maximum:

$$\sum h^2 = \sum h_{max}^2 - Q(a, b) \tag{21}$$

and define the quantity s^2 by

$$\sum d^2 - \sum h_{max}^2 = (N - m)s^2. \tag{22}$$

Now the joint posterior probability density (17) is, to within a normalization constant,

$$\left[\frac{1}{1 + \frac{Q(a,b)}{(N-m)s^2}} \right]^{(N-m)/2} \tag{23}$$

which is the form in which we are used to seeing the t-distribution. This is still exact everywhere, only written in different notation; but one sees that we have set it up for a Gaussian approximation. When N becomes large, (23) goes into

$$\exp\left[-\frac{Q(a, b)}{2s^2} \right]. \tag{24}$$

But in the neighborhood of the peak, $Q(a, b)$ can be expanded as a quadratic form:

$$Q(a, b) = Q_{11}(a - \hat{a})^2 + 2Q_{12}(a - \hat{a})(b - \hat{b}) + Q_{22}(b - \hat{b})^2 \tag{25}$$

and so, in spite of first appearances, the exact distribution (17) is very nearly, in the most important region, a bivariate Gaussian; quite accurately so if we have a lot of data.

The quantity

$$s^2 = \frac{\sum d^2 - \sum h^2_{max}}{N - m} \tag{26}$$

(where we are still writing m as the number of model functions being fitted to the data, to show the general formula) is just the estimate of the mean-square noise level that probability theory is making.

Let us explain this more fully. The computer finds a "best" model function, by which we mean the one that makes the best least-squares fit to the data, out of the class of possible model functions (*i.e.*, number and range of parameters) that we have specified in setting up the problem. The numerator of (26) is the sum of the squares of the residuals for that "best" model, a measure of how well the best model is able to fit the data.

Now as is clear from (4), anything in the data that the computer cannot fit to that best model, it is obliged to consider as "noise." The denominator of (26) indicates that this total mean-square noise is then ascribed equally to the remaining $(N - m)$ degrees of freedom that are not being fit to the model.

Generally, the integration (20) over R should be performed numerically by the computer. But in the Gaussian approximation (25) we can do it analytically, with the result that the posterior probability density $p(r)$ for r alone is proportional to

$$\exp\left[-\frac{1}{2s^2}\frac{Q_{11}Q_{22} - Q_{12}^2}{Q_{11} + 2Q_{12} - Q_{22}}(r - \hat{r})^2\right] \tag{27}$$

where $\hat{r} = \hat{b} - \hat{a}$ is the planetary distance at which $\sum h^2$ peaks. Thus the "best" estimate and probable error of that estimate would be given by the (mean) $\pm$ (standard deviation) of (27):

$$(r)_{est} = \hat{r} \pm s\left[\frac{Q_{11} + 2Q_{12} - Q_{22}}{Q_{11}Q_{22} - Q_{12}^2}\right]^{1/2}. \tag{28}$$

In this approximation, the accuracy of the distance estimate depends on the estimated RMS noise level s and the coefficients in the expansion (25). Roughly speaking, the larger those coefficients, the sharper the peak in the sufficient statistic, and therefore the more accurate the estimate, as common sense would lead us to expect.

Probability theory never makes gratuitous assumptions about other things that might be in the data. As noted, it simply dumps out everything that it cannot fit to the best model into a bin called "noise," without passing any judgement about whether it is "systematic" or "random."

This is the automatic built-in safety device that prevents probability theory from making over-optimistic claims about the accuracy of its estimates. As is shown in the "Bessel inequality" section of the Bretthorst thesis, anything in the data that cannot be fit to the best model, increases the estimate (26) and broadens the posterior distribution (24), increasing our error estimates for the quantities of interest.

But any further information that we have about other systematic effects that might be in the data, can be given to the computer in the form of more model functions, or more flexible model functions. Then it will be able to fit the data to the new model better, it will perceive by (26) that the noise is smaller than previously estimated, and so it will be justified in claiming smaller estimated errors for the quantities of interest.

Therefore, any information we have about systematic effects that might be in the data, whether or not they are of interest to us in the present problem, and whether we consider them to be part of the "model function" or part of the "noise," should be put into our model. In effect, this tells the computer to be on the lookout for such variations and make allowances for them, so that they do not deteriorate our estimates of the quantities of interest.

We stress the extreme importance of following this policy, which was revealed by the analysis of NMR data in the Bretthorst thesis. Here we are interested in determining the oscillation frequencies present; one would at first suppose that the way to do this is to take the Fourier transform of the data. But an additional systematic effect, exponential decay of the oscillations, is present. If the decay rate is put into the model, then integrated out as a nuisance parameter, one obtains orders of magnitude more accurate frequency determinations than are given by a Fourier transform. The implications of this for the planet problem must not be missed.

In particular, the computer can be taught to distinguish between atmospheric distortion and true noise. A human telescopic observer can to some extent make mental allowance for the atmospheric distortions he is seeing, and concentrate on what the shaky image is telling him about the real object. But a computer can do such a job far better than a human can, if it knows what specific kinds of distortions are to be expected, so that it can make allowance for them quantitatively.

This has been a fast tour through the solution for an oversimplified one-dimensional version of the problem, in order to give an intuitive feel for what is to be done, and what the results will mean.

Generalizations

In the real problem, the solution will need to be fixed up to allow for a dozen picky little details that we have left out above. We do not try to list them all here, because the realities are so complicated that one will not be able to anticipate all these details until a preliminary version of the program is running on real data. But we can indicate the more obvious ones.

The first is, of course, that all this must be restated in two-dimensional form. This is an entirely straightforward computer programming job; much more complicated programs than this have been written and run successfully with the Bretthorst algorithm. So we do not anticipate any problems here.

Perhaps next in importance is to deal with atmospheric turbulence effects. In the past, one has tried to capture instants of good seeing, or to stabilize the image position. But these are things that the computer can do much more easily.

A good example of this process, different in one detail but just the same in principle, is in the Bretthorst thesis where he considers how to eliminate trend distortion from the economic data, so as to detect periodicities if the data have evidence for them. Superposed on the periodic model function of interest, is a trend "nuisance function" $T(x)$ which the computer program estimates and removes.

In the present problem, the different detail is that the nuisance function would not be an additive term in the model function, of the form

$$f(x) + T(x),$$

but it would specify the likely distortions in the independent variable x that might occur:

$$f[x + q(x)].$$

Thus, just as Bretthorst expanded the trend function $T(t)$ in the orthogonal functions and eliminated the coefficients as nuisance parameters, we would now express the distortion function $q(x)$ in some suitable functions (probably a power series in the first tryout), and integrate out the coefficients.

We think that writing and testing a computer program to do this would be great fun because the details are new, and the result would be something never before seen. But some detailed information about the actual kinds of distortion $q(x)$ that occur with real telescopes would enable one to write a smarter program, that does it better or more efficiently.

Also, the computer can be told to deliver, for each data set analyzed, its best estimate of what the atmospheric distortion function $q(x)$ was for that data set. By comparing the planetary distance estimates and estimated distortions for different data sets on the same star, one could learn a great deal about the specific way in which atmospheric distortion affects information loss, that would help in optimizing the procedure to deal with the most distant possible stars.

One of the practical details ignored above is the size of the pixels for which we have data, and whether this is partially under our control. If it is, there is probably an optimal pixel size for these purposes, and some further theoretical analysis is needed to understand it. Needless to say, too coarse a pixel size will lose information by poor resolution. But too fine a pixel may, we suspect, also lose information by wasting some of the available light in the "fences" between the pixels in the detector or by leakage of charge from one pixel to another.

We stress that the only thing that is important is the *amount of information* contained in the data; its exact form and such things as signal/noise ratio for individual pixels do not matter because the computer can always extract the information if it is there.

Presumably, these considerations will not matter for the first tryout of the method; we think that if the near stars have planets – at least, on the Jovian scale – they will be found with the present detectors. But these considerations will become important in the final optimization to study more distant stars.

Also, in the final optimization one will need to take into account correlations in the noise of adjacent pixels. This will enable more sensitive detection and more accurate estimates of planetary distance (because for a given estimated mean-square error the noise vector will be confined by this information into a smaller volume of N-dimensional sample space, and some data vectors that lie just outside that volume will then carry significant information about the planet that they did not carry before we knew of the correlations).

If understanding of the noise mechanism is good enough to determine some respect in which it is known to depart from Gaussian, this will also enable better sensitivity and accuracy for a given mean-square noise level; but it is unlikely to help very much unless there is some known hard constraint on the possible magnitude of the noise.

The computer is readily programmed to take into account these different noise distributions. Fortunately, the change likely to help the most is also the easiest to carry out.

Nonwhite Noise

To take this into account, first get the computer program running which finds the optimal solution discussed above, given the model function matrix

$$G_{ij} = G_j(x_i), \qquad 1 \le i \le N, \quad 1 \le j \le m$$

and data vector $D = (d_1, d_2, \ldots)$ for white (*i.e.*, uncorrelated) noise values e_i. This is the program developed in the Bretthorst thesis.

But now suppose that the inverse correlation matrix of the noise is known to have instead the form M/σ^2 where σ is the RMS magnitude of the noise, known or unknown, and M is an $(N \times N)$ matrix indicating the correlation coefficients. No problem; just use the first program as a subroutine, and write a driver program that feeds it instead the massaged matrix and massaged data

$$G_0 = M^{1/2}G \quad \text{and} \quad D_0 = M^{1/2}D, \tag{29}$$

and it will generate the optimal solution for the nonwhite noise.

To prove this, we need only note that the basic sampling distribution for the nonwhite case, from which all else follows, is proportional to $\exp(-Q/2\sigma^2)$, where

$$\begin{aligned} Q &= \Sigma_{ir}[d_i - \Sigma_j G_{ij} A_j] \, M_{ir}[d_r - \Sigma_k G_{rk} A_k] \\ &= [D - GA]^T M[D - GA] \end{aligned} \tag{30}$$

and if M is the unit matrix, this reduces to the uncorrelated case, analyzed in the Bretthorst thesis. So if we had a computer programmed to do the bigger calculation defined by (30), but we gave it a problem in which M is the unit matrix, it would do just the Bretthorst calculation.

Now the crucial point is that M is guaranteed to be symmetric and positive definite, so it can be factored uniquely as

$$M = M^{1/2}M^{1/2} \tag{31}$$

and the two factors can be absorbed into the vectors on either side in (30). Doing this, and using the notation (29), we see that the quadratic form (30) is equal to

$$Q = [D_0 - G_0 A]^T[D_0 - G_0 A] \tag{32}$$

which is of the form used in the Bretthorst calculation. Therefore, if the "bigger" program is fed D, G, and the true matrix M; while the Bretthorst program is fed the massaged D_0, G_0, they will do just the same actual calculation. QED

Warning: Don't Apodize!

Apodizing was a pre-computer way of making the optical system do, crudely and inaccurately, a small part of this computation. But apodizing has serious limitations that a computer does not have.

Apodizing does suppress the wiggles in the point-spread function; but at a cost that is today not only unacceptable, but unnecessary. We note some of the difficulties with apodization, which could be overcome easily with computers.

Psychologically, apodizing always leaves us not knowing whether something better could have been done, because it is only an intuitive, *ad hoc* device not derived from any theoretical principles or optimality argument. For this same reason, it leaves us unable to judge the accuracy of our final results.

More serious, apodizing throws away valuable, relevant information, in two respects. The first is simply that any tampering with the pupil amplitude transmittance function throws away photons and sacrifices signal/noise ratio.

To see the second, more fundamental mode of information loss, note that if the pupil transmittance function of the original unapodized optical system is known, a computer can always calculate the apodized diffraction pattern for any apodizing scheme you please; there is no need to alter the optical system for this. But the transformation is not reversible; given an apodized diffraction ring, no computer can recover the original unapodized optical system.

Mathematically, the apodizing operator has no inverse; many different unapodized systems, which would give different information about the object being seen, all generate the same apodized image. Conversely, many different objects, which would be distinguishable by the original unapodized optical systems, all generate the same apodized image and become indistinguishable. That is what we mean by "information loss."

The purpose of a data analysis procedure is to extract as much information as it is possible to get out of the data, pertaining to the question of interest. A philosophy of data analysis that tells us to start by throwing away some of the information in the data, may be of some use as a "quick and dirty" expedient; but fundamentally it is just not a rational way of looking at the problem.

Some years ago, Ronald Bracewell pointed out to me a problem that is very similar both in topic and in theory. Let us recall it for the lesson it teaches us. In the 1950's, the Hanbury Brown-Twiss (HBT) interferometer for measuring stellar diameters was a wonderful, new, almost magical thing and few people understood correctly how it worked.

The Michelson interferometer had failed at rather short mirror separations because the difference in atmospheric turbulence at the two mirror positions washed out the interference fringes. But the HBT method rectified the signals before combining them, and instead of looking at optical interference fringes, looked at correlations in intensity fluctuations in light falling on the two mirrors. This still contained information about the stellar diameter, and in spite of an inherently lower signal/noise ratio for a given amount of light intercepted, the HBT interferometer was able to operate at greater separations, thus measuring smaller angular diameters.

Not surprisingly, some works appeared praising the virtues of the HBT way of looking at the problem, as a great advance in understanding. But Bracewell pointed out that the Michelson interferometer is in principle delivering far more information than is the HBT one; and it does not make sense to suppose that an instrument that yields less information about stellar diameters can be more informative about them.

The problem with the Michelson interferometer was that it was delivering information faster than the technology of the time could handle. The interference fringes were not absent; they were just moving rapidly. If one could record the details of the fringe position and visibility in real time, and analyze the resulting masses of data by computer, the Michelson interferometer would emerge as superior.

The HBT method was, like apodization, a pre-computer way of getting the optical/electronic system to do, crudely, a small part of the computation that an optimal data analysis method would perform. Both of these quick-and-dirty methods were good enough to be usable; but both deliver results far inferior to what a Bayesian computer analysis could give today.

The same lesson was learned again, ten years later, when the Maximum Entropy spectrum analysis method of John Parker Burg was announced. Previously, the Blackman-Tukey (BT) method had removed unwanted "side-lobes" in the Fourier transform of the data by a "lag window" which removed the sharp edges of the measured autocovariance at the end of the data record, shading it smoothly to zero. But the most popular "Hanning window," in removing most of the side-lobes, also threw away half the resolution; BT spectra have lines twice as wide as did the original Schuster periodogram.

In 1967, Burg pointed out that the BT method throws away crucially important information. His maximum entropy method, which conserves all the information in the data, can display spectrum lines orders of magnitude sharper than the BT ones, when the data contain evidence for them. The BT method rapidly became obsolete among those who tried the maximum entropy method, not for philosophical reasons, but because of the superior computer printouts.

But some holdouts remained. John Tukey, in a meeting in Princeton in December 1980, pointed out to the present writer that his windowing method for removing side-lobes is mathematically just a one-dimensional version of apodization – in the belief that this would persuade me of the merits of the BT method! Instead, this made me see clearly where the defects of apodization lie, and how to correct them.

When the application is important enough to justify tying a computer to the optical system, we don't have to put up with such limitations any more. The new rules of conduct are:

(1) Keep your optical system clear and open, gathering the maximum possible amount of light (*i.e.*, information).

(2) Don't worry abut wiggles in the point-spread function; the computer wil straighten them out far better than apodizing could ever have done, at a smal fraction of the cost.

(3) For the computer to do this job well, it needs only to know the actual point spread function $G(x)$, whatever it is. So get the best measurement of $G(x)$ tha you can, and let the computer worry about it from then on.

(4) What is important to the computer is not the spatial extent of the point-spread function, but its extent in Fourier transform space; over how large a "window" ir k-space does the PSF give signals above the noise level, thus delivering relevan information to the computer?

Apodizing contracts this window by denying us information in high spatia frequencies, associated with the sharp edge of the pupil function. But this i: just the information most crucial for resolving fine detail! In throwing awa} information, it is throwing away resolution. Apodizing does indeed "remove th‹ foot;" but it does it by shooting yourself in the foot.

STOCHASTIC COMPLEXITY AND THE MAXIMUM ENTROPY PRINCIPLE

Jorma Rissanen
IBM Almaden Research Ctr,
K52/802, San Jose, Ca. 95120-6099

ABSTRACT. This is an outline of a modeling principle based upon the search for the shortest code length of the data, defined to be the stochastic complexity. This principle is generally applicable to statistical problems, and when restricted to the special exponential family, arising in the maximum entropy formalism with a set of moment constraints, it provides a generalization which permits the set of the constraints or their number to be optimized as well.

1. INTRODUCTION

In this expository paper we outline the so-called *MDL* (Minimum Description Length) principle for statistical inference. This principle has its roots in the algorithmic theory of information, Solomonoff (1964), Kolmogorov (1965), and Chaitin (1975), (1987), where the complexity of a finite binary string is defined to be the length of a shortest program in a universal computer which generates the string. We have been particularly influenced by the version of Chaitin's, in which the programs are required to be self-delimiting forming a prefix-free set so that no program can be a prefix of another. The earliest work on code length minimization as applied to classification is due to Wallace and Boulton (1968), where a particular way of doing the required coding was discussed. In Rissanen (1978), (1983), (1984), (1985), (1986a,b), (1987a,b) the attachment to any coding technique was gradually removed, and the shortest code length - not in the mean sense as advocated in Wallace and Freeman (1987) but for the actually observed data - was defined to be the *Stochastic Complexity* of the string, relative to a probabilistic class of models.

The fact that the stochastic complexity is defined for the given set of observed data rather than for an imagined ensemble of them suggests a profound change in our view of statistical inquiry. No longer is there a need to postulate the existence of a "true" distribution, which we are trying to "estimate" from the data. Rather, the objective of statistical modeling becomes a search for better and better models and model classes, where any two classes, regardless of the number of parameters in them or regardless of their structure, may be compared by the stochastic complexity that they assign to the actually observed string. Neither is there any fundamental distinction between the "random" data and the "non-random" parameters; they all are just numbers that can be described

161

G. J. Erickson and C. R. Smith (eds.),
Maximum-Entropy and Bayesian Methods in Science and Engineering (Vol. 1), 161–171.
© 1988 by Kluwer Academic Publishers.

with finitely many binary digits. Because there is no unique model nor model class for the observations, the code will also have to include either explicitly or implicitly the description of any selected model, and we get a direct way to assess its complexity. By contrast, traditional statistics has no room for the complexity of the fitted models with the consequence that generally the best model will be the most complex one contemplated, unless prevented by subjective acts. Indeed, the several proposed model selection criteria do not admit a clear let alone a common data dependent interpretation, and therefore they cannot provide a rational basis for the comparison of different proposed models. Further, in our view prior knowledge consists in the selection of the model class which may well include distributions for the parameters, rather than in the choice of such "prior" distributions, as done traditionally. Hence, when we select such "priors", which frequently depend on nuisance parameters, our objective is to achieve a small stochastic complexity, and this often involves optimization of the nuisance parameters in the light of the observed data, which stands in a clear conflict with the idea of prior knowledge. Finally, a prefix code length for the parameters necessarily generates a "prior" distribution, and hence gives a justification for Bayesian approaches. Furthermore, the explicit formula we give for the stochastic complexity appears to strengthen a connection between our approach and the Bayesian philosophy, in which, one may perhaps say, the posterior distribution for the parameters plays a fundamental role. However, in many applications it appears to be completely hopeless to speculate prior distributions for complex objects such as decision trees with data at the nodes, while the construction of prefix codes for them is perfectly feasible. Accordingly, the interpretation of such code lengths as Bayesian priors is just empty formalism. Moreover, our fundamental approach to statistical problems can be stated simply as a search for a globally maximized likelihood, which is what the search for the smallest stochastic complexity amounts to, and in this the posterior either plays no role or only a partial one. There is, however, another idea, based upon prediction, due to Dawid (1984), which might be taken as Bayesian. As discussed in Rissanen (1986a) prediction is but one particular aspect of coding, and we have no conflict with such a "Bayesian" idea.

To illustrate a predictive form of the *MDL* principle we describe a numerical example, where we answer a question posed in Rao (1981), namely, how to justify certain optimal predictors for the weight growth of a number of mice, which were found in retrospect by looking at the predicted weights. These predictors are far better than the predictors found by traditional means.

The second objective in this paper is to study the *MDL* principle when the class of models is confined to the exponential family arising in the *Maximum Entropy (ME)* principle. When the number of the constraints in the entropy maximization problem is regarded as fixed, the *MDL* principle will be seen to degenerate to the *ME* principle. But since the former principle applies even when the set of the constraints is left open and subject to optimization, we get a potentially useful extension of the *ME* principle. This way, for example, in the spectrum estimation problem by Burg, as discussed in Jaynes (1982), it is perfectly feasible to let the number of the covariance constraints be free and to be determined by the observed data.

2. STOCHASTIC COMPLEXITY

Consider a class $M_k = \{f(x|\theta), \gamma(\theta)\}$ of parametric densities $f(x|\theta)$ defined on the data $x = x^n = x_1, x_2, \dots, x_n$, where $\theta = (\theta_1, \dots, \theta_k)$ denotes k parameters ranging over some set Ω^k with non-empty interior, reflecting the fact that the parameters are "free". The positive function $\gamma(\theta)$ may be taken as a "prior" distribution, but in general it need not be a distribution at all; it is mostly a technical device whose role is to make the integral

$$C(x) = \int f(x|\theta)\gamma(\theta)d\theta \tag{2.1}$$

finite. The function

$$\pi(\theta|x) = C^{-1}(x)f(x|\theta)\gamma(\theta) \tag{2.2}$$

defines a distribution for the parameters, and we may construct the marginal conditional density

$$f(x_t|x^{t-1}) = \int f(x_t|x^{t-1}, \theta)\pi(\theta|x^{t-1})d\theta, \tag{2.3}$$

where we use the same letter to denote different density functions, to be distinguished from each other by the nature of their arguments. This in turn defines the total density function

$$f(x) = \prod_{t=1}^{n} f(x_t|x^{t-1}), \tag{2.4}$$

except for the initial density $f(x_1) = f(x_1|x^0)$, which we select on prior grounds. We define the *stochastic complexity* of x, relative to the class M_k, to be

$$I(x|M_k) = -\log f(x) = -\log C(x) - \log[f(x_1)/C(x_1)]. \tag{2.5}$$

Again in contrast to Bayesian techniques, we see that here the important distribution is the marginal distribution on the data rather than the posterior. There exist compelling justifications for this to be a good definition for the shortest code length, some of them discussed in the cited references.

A closed form evaluation of the integral together with the requirement that $C(x)$ be finite and non-zero may force us to select $\gamma(\theta, \alpha)$ with nuisance parameters α, which changes the above derived quantities to $C(x, \alpha)$, $f(x, \alpha)$, and $I(x|M_k, \alpha)$, respectively. In principle, we could remove the nuisance parameters by integration just as we did with θ, for there exists a natural universal prior $q^*(y)$ for real numbers, Rissanen (1986a). In practice, however, we cannot evaluate the integral, and we proceed in one of three ways. The simplest is to fix the nuisance parameters to some values, which for long sequences of data is often justified in that the exact values do not make much of a difference to the result. But in general this is not true, and we seek to determine them so as to minimize the stochastic complexity. In order to regain the important normalization

$$\int e^{-I(x|M_k)}dx \leq 1,$$

which may be taken as a generalized Kraft-inequality, Rissanen (1984), to insure the existence of a prefix code for the data of the same length n, we define the *stochastic complexity* of x, relative to the class M_k, in this case to be

$$I(x|M_k) = \min_{\alpha} \{-\log f(x, \alpha)\} - \log q^*(\hat{\alpha}), \tag{2.6}$$

where $\hat{\alpha} = \hat{\alpha}(x)$ denotes the minimizing parameters. With just one nuisance parameter the second term is to a good approximation $\log \max \{1, |\hat{\alpha}(x)|\} + r$, where r denotes the number of fractional digits to which $\hat{\alpha}(x)$ is truncated. Provided that we do not intend to compare stochastic complexities relative to model classes with different number of nuisance parameters, we may drop the second term, which generally is also much smaller than the first term in (2.6). This gives the stochastic complexity approximately as

$$I(x|M_k) \simeq -\log C(x, \hat{\alpha}),\tag{2.7}$$

where we dropped even the terms $\log C(x_1, \hat{\alpha}) - \log f(x_1)$.

The third way to eliminate the nuisance parameter is to do it predictively. Let $\hat{\alpha}(x^t)$ denote the sequence of estimates defined by $x^t = x_1, \dots, x_t$, $t = 1, 2, \dots$, and put

$$I(x|M_k) = \sum_{t=0}^{n-1} I(x_{t+1}|x^t, \hat{\alpha}(x^t)),\tag{2.8}$$

where the incremental complexity is defined as

$$I(x_{t+1}|x^t, \hat{\alpha}(x^t)) = -\log[f(x^{t+1}, \hat{\alpha}(x^t))/f(x^t, \hat{\alpha}(x^{t-1}))].\tag{2.9}$$

Here again $f(x_1, \hat{\alpha}(x^0))$ is taken as the initial density $f(x_1)$.

The integral (2.1) can be evaluated in a closed form for the important class of gaussian distributions, where the parameters consist of the covariance matrix and of the coefficients of a linear combination defining the mean, such as in linear regression problems. It can also be evaluated for the class of multinomials and related distributions, which are central in modeling contingency tables, Rissanen (1987c), as well as in information theory. In general, however, we must resort to approximations, of which two are particularly important. For large amounts of data, the negative logarithm $-\log f(x|\theta)$ as a function of the parameters behaves generally as a very peaked quadratic function with the minimum at $\hat{\theta}$. Hence, a Taylor expansion up to second order terms is a good approximation, and we get

$$f(x|\theta) \simeq f(x|\hat{\theta})e^{-\frac{1}{2}(\theta - \hat{\theta})'\hat{J}(\theta - \hat{\theta})}.\tag{2.10}$$

With the conjugate priors for the parameters the integral (2.1) can be approximated, and the stochastic complexity admits the approximation

$$I(x|M_k) \simeq -\log f(x|\hat{\theta}) + \frac{1}{2}\log \det \hat{J}.\tag{2.11}$$

Moreover, the second term behaves as $\frac{k}{2}\log n$, where n denotes the number of observed data points, and this approximation gives the model selection criterion known either as the *MDL*, Rissanen (1978), or as the *BIC*, Schwarz (1978), the latter derived only for the exponential family. The right hand side with the added term nq, where each data point x_i is written to the precision e^{-q}, may also be interpreted as the length required to encode the data and the optimized parameters, each component truncated to an optimal precision. This represents a particular "two-pass" coding procedure, where the imagined encoder looks at all the data, finds the best parameters values, and then encodes the data using the best distribution in the class, while adding the code length required to encode the best parameter values as a preamble to the entire code. This second step is necessary, because the imagined decoder cannot determine the best parameter values which depend on the data he cannot decode without knowing these parameter values. It is as if the encoder had looked into the future, and he will have to tell the decoder what he saw for the decoding to be successful.

Such a preamble, then, is not needed if the encoder determines his parameter values always from the past by an algorithm which the decoder is thought to know, and this gives us an important second way to approximate the stochastic complexity. In fact, we already applied that technique above in eliminating the nuisance parameters. By letting $\hat{\theta}(t) = \hat{\theta}(x^t)$ denote the parameters that maximize $f(x_{t+1}|x^t, \theta)$, we define

$$L(x|M_k) = - \sum_{t=0}^{n-1} \log f(x_{t+1}|x^t, \hat{\theta}(t)), \qquad (2.12)$$

which except for the term nq represents the code length resulting from a predictive coding process. It is a rather remarkable fact that under suitable regularity conditions $L(x|M_k)$ too behaves asymptotically as (2.11), so that even though no preambles for the estimated parameters have been

added we still have not escaped the "penalty" $\frac{k}{2} \log n$, as it were. It is this implicit or explicit penalty on the complexity of the model that the stochastic complexity and all decodable codes for the data incorporate, and which makes the MDL principle sound. An important special case of (2.12) results if we pick the conditional densities $f(x_{t+1}|x^t, \theta)$ as normal, with variance τ and the mean a prediction of x_{t+1} made with some parametric predictor $\hat{x}_{t+1}(x^t, \hat{\theta}(x^t))$. In the further special case where the variance is fixed, $\tau = 1/2$, (2.12) becomes the sum of squared "honest" prediction errors, which unlike the usual quadratic deviations, can be meaningfully minimized with respect to the number of parameters. This represents then a natural extension of the classical least squares technique, Rissanen (1986b), and one can show that with the so-minimized number of parameters the mean prediction errors are minimized too in an asymptotic sense.

Unlike the last two approximations the stochastic complexity in the abstract form (2.5) does not involve any estimates of the parameters, although the information necessary to get such estimates is in it. Hence, whenever the problem calls for estimates we may use these approximations, above all the predictive one (2.12). One can, in fact, link the consistent estimability of the parameters to the behavior of $L(x|M_k)$ as an approximation of the stochastic complexity in that consistency of the parameter estimates imply that the difference $L(x^n|M_k) - I(x^n|M_k)$ remains uniformly bounded. This resolves for example the Neyman-Scott anomaly of the maximum likelihood estimates in the context where the number of nuisance parameters grows with the data and prevents the maximum likelihood estimates of the useful parameter from being consistent. Another important anomaly of the maximum likelihood estimates involves densities which can be made infinite. The insistence in truncated observations and parameters in the MDL principle automatically eliminates the source of such difficulties, because it is clear that nobody can encode a set of real truncated observations with a zero code length, which means that the search for the shortest code length, unlike the maximization of density functions, does not lead to meaningless results.

3. AN EXAMPLE

We illustrate the application of the MDL principle in its predictive version with an example from Rao (1981). Table 1 consists of the weights of 13 mice, obtained in seven consecutive time instances at 3-day intervals since the birth. The objective is to predict the weights in the last column from the observations in the first six columns using polynomial models. The prediction error is measured by the sum of the squares of the deviations for all the mice, while the polynomials may be fitted using any criterion.

mice \ days	3	6	9	12	15	18	21
1	0.109	0.388	0.621	0.823	1.078	1.132	1.191
2	0.218	0.393	0.568	0.729	0.839	0.852	1.004
3	0.211	0.394	0.549	0.700	0.783	0.870	0.925
4	0.209	0.419	0.645	0.850	1.001	1.026	1.069
5	0.193	0.362	0.520	0.530	0.641	0.640	0.751
6	0.201	0.361	0.502	0.530	0.657	0.762	0.888
7	0.202	0.370	0.498	0.650	0.795	0.858	0.910
8	0.190	0.350	0.510	0.666	0.819	0.879	0.929
9	0.219	0.399	0.578	0.699	0.709	0.822	0.953
10	0.255	0.400	0.545	0.690	0.796	0.825	0.836
11	0.224	0.381	0.577	0.756	0.869	0.929	0.999
12	0.187	0.329	0.441	0.525	0.589	0.621	0.796
13	0.278	0.471	0.606	0.770	0.888	1.001	1.105

Table 1. Weights of 13 mice at 7 time instances

Before describing the predictive *MDL* estimates we summarize the results obtained by traditional techniques as given in Rao (1981), which then may be contrasted with ours. The set of weights y_{it} of the mouse i at time t would be considered as a sample from a random process as follows:

$$y_{it} = \sum_{j=0}^{k} \theta_j t^j + \varepsilon_{it}, i = 1, \dots, 13, \ t \geq 1,$$

where ε_{it} denotes a family of zero-mean uncorrelated random variables with some variance $E\varepsilon_{it}^2 = \sigma^2$. The parameters to be estimated are k, σ, and $\theta_0, \dots, \theta_k$. The estimate of the degree k will be common for all the mice, but the coefficients may be estimated for each mouse individually.

The traditional criteria for the estimation of the degree will give different values, and none of them is based upon arguments that instill great confidence. However, a bit of common sense or prior knowledge suggests that the weight growth ought to taper out eventually, which behavior could be captured by a second degree polynomial that curves downwards. In fact, with the hindsight offered by a peek at the weights in the last column to be predicted, Rao in (1981) verified that such a polynomial, indeed, will do the best prediction. For the first row, for example, the best least squares polynomial fitted to the first six measurements is $f_1(t) = -0.219 + 0.341\,t - 0.0185\,t^2$, where the coefficients are so calculated that t denotes the tth weighing instance. Hence the prediction of the last weight is then $f_1(7) = 1.158$.

With the optimal degree $k = 2$ thus established, sophisticated techniques developed in the traditional statistics can now be put to work in order to find the best way to estimate the coefficients of the quadratic polynomial, which by no means needs to be the least squares fit. In Rao (1981) five methods and the resulting predictors for the weights in the last column of Table 1 were worked out, namely, BLUP (Best Linear Unbiased Predictor), JSRP (James-Stein Regression Predictor), James and Stein (1961), SRP (Shrunken Regression Predictor), EBP (Empirical Bayes Predictor),

Rao (1975), and RRP (Ridge Regression Predictor), Hoel and Kennart (1970), which gave the prediction errors as follows:

Method	BLUP	JSRP	SRP	EBP	RRP
Pred. Error	.1042	.1044	.0972	.0951	.1047

We see that the difference between the best and the worst is about 5%, which means that we are talking about marginal improvements.

Quite interestingly, Rao in (1981) also noticed that one could fit the polynomials only to the past q columns, where $q \in \{1, \dots, 6\}$ should be determined optimally. Hence, in our terminology, he wanted to examine different classes of models. Indeed, the former class is the special case of the latter corresponding to the fixed value $q = 6$. When this is done, then again with the knowledge of the weights to be predicted, the best predictions were found to result from the line that for each mouse; i.e., for each row in Table 1, passes through the 5'th and the 6'th weights. Hence, for example, for the first row the best prediction is 1.132 + (1.132-1.078) = 1.186. This time the average prediction error was much better, namely, 0.0567. (This differs from the number 0.0552 given in Rao (1981), which we could not duplicate; the difference is unimportant, however.)

Hence, we see that the initial traditional assumption of the parent population and distribution was not too good after all, not even when the increasingly sophisticated refinements developed in the field have been invoked. A much better predictor was found to lie in the second model class, although its superiority could be established only by hindsight. The important question arises whether hindsight is necessary if we apply our predictive *MDL* principle. After all, the hindsight provided by the weights in the last column of Table 1 is not needed to tentatively consider the second model class. In traditional thinking there is hardly any point to consider different classes, for there is no rational basis to select the best.

For each degree k and each number q of the past observations, which restricts the range from which the least squares estimates of the coefficients are computed, we can calculate the sum of the prediction errors for each mouse, (2.12), when the data in the first six columns of Table 1 are predicted,

$$L_i(y|k,q) = \sum_{t=1}^{6} \hat{e}_{it}^2, \ i = 1, \dots, 13.$$

The sum of these over all the mice, $L(y|k,q) = \sum_{i=1}^{13} L_i(y|k,q)$, is listed in the following Table 2.

q \ k	0	1	2	3	4
5	.498	.101	.095	.122	.222
4	.448	.0977	.0963	.148	-
3	.359	.0944	.108	-	-
2	.253	.0944	-	-	-

Table 2. Past prediction errors for various models

We see that fixing $q = 5$, which means that the least squares estimates of the coefficients for each t are determined from all the past data, the best degree polynomial is 2, which gives the best prediction error of the past data relative to the first class of polynomials as 0.095. But when q is set free, then the best degree of polynomial is 1, and it is obtained by either fitting the line to the past 2 or 3 weights. The corresponding optimal prediction errors for the past data are equal, namely, 0.0944. Except for the tie regarding the line, both of these optimal degrees of polynomials in the two considered model classes agree with those found in Rao (1981) by help of hindsight.

Regarding the preference of the two model classes, the latter class gives the smaller prediction error for the observed data. Therefore, these models are the ones with which we should do the prediction of the weights in the last column of Table 1. When this is done with the two equally optimal lines, the sum of the final prediction errors is 0.0567, obtained with the lines passing through the past two points, and 0.0793, obtained with the lines fitted to the past three points for each mouse. Both values are well below those obtained with the quadratic polynomial, fitted to all the past data, regardless of how the values of the coefficients are determined, of which the best was seen to be 0.0951, obtained with the empirical Bayes predictor. These findings also highlight another myth, generally hailed in traditional statistics. It has been claimed that one of the foremost achievements in modern statistics is the ability to provide a measure of confidence in the parameter estimates. Without actually having calculated the confidence intervals for the coefficients of the second degree polynomials in the five above listed traditional methods, chances are that they are quite tight, providing reasonable confidence in the findings. However, we know now that the confidence in the parameter estimates, provided by such intervals no matter how tight they were, is quite irrelevant to the key question, how well we can predict, because much better predictors exist in a different model class. This is generally true, and even though we may derive a lot of confidence in the estimates of our parameters, the traditional theory offers no way to calculate the confidence in what counts, the relevance of the parameters.

4. CONNECTION TO MAXIMUM ENTROPY

The investigation in this section was initiated by an observation by R. Arps, namely, that for the gaussian distribution the negative logarithm of the maximized likelihood coincides exactly with Shannon entropy for the same distribution, evaluated at the maximum likelihood parameter values. The same is generally known to be true for the Bernoulli distribution for binary strings. The question that arises naturally is to characterize the class of distributions for which the same fact holds true. A solution was promptly offered by T. Speed in the form of a special exponential family, and the link with the maximum entropy formalism became immediate.

Consider the maximum entropy problem

$$\max_{f} - \int f(x) \log f(x) dx, \tag{4.1}$$

where the density function is subject to the moment constraints $\int f(x) A_i(x) dx = d_i, i = 1, \dots, k$, written collectively as $E_f A = d$. The solution, which follows at once from the main inequality in information theory without variational arguments, van Campenhout and Cover (1981), is the exponential distribution

$$p(x|\lambda) = Z^{-1}(\lambda) e^{-\lambda' A(x)}, \tag{4.2}$$

where $\lambda' A(x) = \sum_{i=1}^{\kappa} \lambda_i A_i(x)$, and $\lambda = \lambda_d$ is the solution of the equation

$$- grad \log Z(\lambda) = d. \tag{4.3}$$

If we now pick the constant vector as $d = A(x)$, x denoting the given set of data, then the maximized entropy satisfies

$$\hat{H} = \log Z(\lambda_d) + \lambda'_d d = \min_{\lambda} - \log p(x|\lambda). \tag{4.4}$$

Hence, the constrained maximum entropy problem is equivalent with an unconstrained maximum likelihood problem for the exponential distributions (4.2) involving the function $A(x)$ appearing in the constraints.

We may think of the right-most expression in (4.4) as the shortest code length for the data, when the constraint vector d is taken as fixed and known to the decoder, so that he can compute the optimized parameters from the model class and these constraints, and the code for the optimized parameters is not needed. In this sense, then, we may regard the maximum entropy technique as a special degenerate instance of the *MDL* technique. It seems to us, however, that the constraints in the maximum entropy problem, at least in the statistical applications, cannot be taken as granted. In fact, in our view they just define a more or less arbitrary class of models, and hence they ought to be subjected to alterations. The maximum entropy principle in its present formulation certainly cannot be applied to determine the number of constraints k, but the *MDL* principle can. The result is an extended maximum entropy principle, where we would either compute the stochastic complexity of the observed data relative to the model class M_k, defined by the exponential family (4.2) with k functions in $A(x)$, or we would resort to the approximations discussed in Section 2. Then, the optimal number of constraints can be found by minimizing the stochastic complexity in the usual fashion.

References

van Campenhout, J.M. and Cover, T.M. (1981), 'Maximum Entropy and Conditional Probability', *IEEE Trans.Inf.Thy*, **IT-27**, Nr. 4, 483-489.

Chaitin, G.J. (1975), 'A Theory of Program Size Formally Identical to to Information Theory', *J.ACM*, **22** , 329-340.

Chaitin, G.J. (1987), *Algorithmic Information Theory*, Cambridge University Press, Cambridge, 175 pages.

Dawid, A.P. (1984), 'Present Position and Potential Developments: Some Personal Views, Statistical Theory, The Prequential Approach', *J. Royal Stat. Soc.* Series A, **147** , Part 2, 278-292 (with discussions).

Hoel, A.E. and Kennard, R.W. (1970), 'Ridge regression: Biased estimation for nonorthogonal problems', *Technometrics*, **12** , 55-68.

James, W. and Stein, C. (1961), 'Estimation with quadratic loss', *Proc. 4th Berkeley Symp.* **1**, 363-379.

Jaynes, E. T. (1982), *Papers on Probability, Statistics, and Statistical Physics*, a reprint collection, D. Reidel, Dordrecht-Holland.

Rao, C.R. (1975), 'Simultaneous Estimation of Parameters in Different Linear Models and Applications to Biometric Problems', *Biometrics*, **31**, 545-554.

Rao, C.R. (1981), 'Prediction of future observations in polynomial growth curve models', *Proc. Indian Stat. Inst. Golden Jubilee Int. Conf. on Statistics: Applications and New Directions.*, 512-520.

Kolmogorov, A.N. (1965), 'Three Approaches to the Quantitative Definition of Information', *Problems of Information Transmission* **1**, 4-7.

Rissanen, J. (1978), 'Modeling by shortest data description', *Automatica,* **14**, pp. 465-471.

Rissanen, J. (1983), 'A Universal Prior for Integers and Estimation by Minimum Description Length', *Annals of Statistics,* **11**, No. 2, 416-431.

Rissanen, J. (1984), 'Universal Coding, Information, Prediction, and Estimation', *IEEE Trans. Inf. Thy*, **IT-30**, Nr. 4, 629-636.

Rissanen, J. (1986a), 'Stochastic Complexity and Modeling', *Annals of Statistics,* **14**, No 3, 1080-1100.

Rissanen, J. (1986b), 'A Predictive Least Squares Principle', *IMA J. of Math. Control and Information,* **3**, Nos 2-3, 211-222.

Rissanen, J. (1987a), 'Stochastic Complexity', *Journal of the Royal Statistical Society, Series B,* **49**, No. 3, 223-265 (with discussions).

Rissanen, J. (1987b), 'Stochastic Complexity and the MDL Principle', *Econometric Reviews,* **6**, nr 1, 85-102.

Rissanen, J. (1987c), 'Complexity and Information in Contingency Tables', Proceedings of *The Second International Tampere Conference in Statistics*, June 1-4, Tampere, Finland.

Schwarz, G. (1978), 'Estimating the Dimension of a Model', *Annals of Statistics,* **6**, 461-464.

Solomonoff, R.J. (1964), 'A Formal Theory of Inductive Inference'. Part I, *Information and Control* **7**, 1-22; Part II, *Information and Control* **7**, 224-254.

Wallace, C.S. and Boulton, D.M. (1968), 'An Information Measure for Classification', *The Computer Journal*, **11**, No. 2, 185-194.

Wallace, C.S. and Freeman, P.R. (1987), 'Estimation and Inference by Compact Coding', *Journal of the Royal Statistical Society, Series B,* **49,**   No. 3, 239-265 (with discussions).

THE AXIOMS OF MAXIMUM ENTROPY

John Skilling
Department of Applied Mathematics
            and Theoretical Physics
Silver Street
Cambridge CB3 9EW, England

Abstract. Maximum entropy is presented as a universal method of finding a "best" positive distribution constrained by incomplete data. The generalised entropy $\Sigma(f - m - f \log(f/m))$ is the only form which selects acceptable distributions f in particular cases. It holds even if f is not normalised, so that maximum entropy applies directly to physical distributions other than probabilities. Furthermore, maximum entropy should also be used to select "best" parameters if the underlying model m has such freedom.

INTRODUCTION

Many quantities of interest are positive distributions. Typical examples are the pattern of light intensity arriving on a photographic plate, and the power spectrum of some radiative field. Following Jaynes (1984), we shall call such a distribution a "scene", and an estimate of it an "image". Usually a scene is a real function f of a continuous spatial or temporal argument x (which may itself have more than one dimension), requiring an infinite number of bits of information to specify it fully. Our knowledge of it, gleaned ultimately from observation, will only be finite.

Even though our knowledge is incomplete, we still wish to obtain a single image from the many which are consistent with our knowledge. In the first half of this paper, we discuss some guidelines (axioms) for finding a single "best" image, based purely on realistic selection criteria, and not relying on probability or information theory. These axioms lead to the maximum entropy (MaxEnt) method for selecting the best image consistent with our knowledge.

MaxEnt needs a given prior model of the scene. This is useful, because it allows prior insight into the nature of the scene to be incorporated into the formalism. However, the insight may be somewhat vague, and contain unknown parameters, such as the positions and intensities of point stars in an astronomical photograph, or lines in a spectrum. In the second half of this paper, we discuss a related set of guideline axioms for finding the best values of any such

G. J. Erickson and C. R. Smith (eds.),
Maximum-Entropy and Bayesian Methods in Science and Engineering (Vol. 1), 173–187.
© *1988 by Kluwer Academic Publishers.*

parameters. Again, these rules are based purely upon selection criteria.

Remarkably, the <u>same</u> entropy formula is derived. Thus MaxEnt should be used <u>both</u> to find the best single image <u>and</u> to find the best set of parameters underlying it.

SELECTING AN IMAGE
We aim to provide an image which is the "best" according to an agreed criterion. This involves setting up a ranking procedure which determines which of two images is "better". To avoid circularity, and to ensure that there is always some image which is not "bettered" by any other, we impose the transitivity requirement

(f better than g) and (g better than h)
$$\Rightarrow \text{(f better than h).}$$

Any transitive ranking can be described by real numbers, assigning a number $S(\underline{f})$ to each image f, such that

"f better than g" $\Longleftrightarrow$ $S(\underline{f}) > S(\underline{g})$ (1)

Choosing the "best" image is equivalent to regularising f by maximising $S(\underline{f})$. However, the form of $S(\underline{f})$ remains to be defined.

A fundamental requirement is universal applicability, that S should be independent of the type of data we are given. This assumption is useful because it allows a unified approach to data analysis.

The axioms
Remarkably, a few very simple examples of accept-able reconstructed images suffice to determine the form of S. These examples, considered as axioms, progressively restrict the form of S, until only one form remains (or equivalently a monotonic function of it). Anticipating the result, S is the entropy of f.

For the special case of a probability distribution, it seems that Jaynes (1957a,b) was the first to suppose that consistency arguments alone might suffice to determine the entropy formula in the context of inference. His conjecture was proved by Shore and Johnson (1980) who gave a formal axiomatic derivation. Independently, Tikochinsky, Tishby and Levine (1984) arrived at the same formula from a somewhat more physical viewpoint. Earlier derivations of entropy as an uncertainty or information measure (Shannon 1948, Shannon and Weaver 1949, Kullback 1959, Cox 1961) also treated it as a property of a probability distribution.

However, the theorems are more generally applicable, and indeed the proofs are simpler without the normalisation $\int f(x)dx = 1$ which is imposed on probability distributions.

In the following presentation, the formal statement of each axiom is followed by a justification, then by its consequence, a proof thereof, and a comment. Greek letters denote functions appearing in S except that λ and μ are reserved for Lagrange multipliers such as that appearing in the archetype variational equation

$$\delta(S - \lambda.\text{constraint}) = 0$$

The symbol $f[I,m]$ represents the image f reconstructed by maximising S with respect to constraint information I, over a Lebesgue measure m on x.

Axiom I.   Subset independence.
Let I_1 be information pertaining only to $f(x)$ for $x \in D_1$ and similarly let I_2 pertain only to $f(x)$ for $x \in D_2$. Then, if D_1 and D_2 are disjoint,

$$f[I_1,m] \cup f[I_2,m] = f[I_1 \cup I_2 , m] \tag{2}$$

where "U" is the union operator.

Justification:
Information about one domain should not affect the reconstruction in a different domain, provided there is no constraint directly linking the domains.

Consequence:
S must be of the form

$$S(\underline{f}) = \int dx \; m(x) \; \theta(f(x),m(x),x) \tag{3}$$

where θ is an arbitrary function.

Proof:
Consider first the discrete case. Let D_1 and D_2 be non-intersecting domains with union D. Let there be a linear constraint

$$\sum_{i \in D_1} a_{1i} f_i = b_1 \quad , \quad \sum_{i \in D_2} a_{2i} f_i = b_2 \tag{4}$$

on each domain. In D_1 and D_2 respectively, f is separately determined by the variational equations

$$\delta S/\delta f_i = \lambda_1 a_{1i} \quad , \quad \delta S/\delta f_i = \lambda_2 a_{2i} \tag{5}$$

Using both constraints together, f is determined by

$$\delta S/\delta f_i = \begin{cases} \mu_1 a_{1i} & , \quad i \epsilon D_1 \\ \mu_2 a_{2i} & , \quad i \epsilon D_2 \end{cases} \qquad (6)$$

Taking two cells j,k both in D_1, the reconstruction is to be independent of the constraints and values of f in the other domain D_2. Accordingly, $(\delta S/\delta f_j)/(\delta S/\delta f_k)$ is independent of all f_i for $i \epsilon D_2$. This must hold for arbitrary decomposition of D into D_1 and D_2. Hence

$$(\delta S/\delta f_j)/(\delta S/\delta f_k) = \alpha_{jk}(f_j,f_k) \qquad (7)$$

where α_{jk} is a function which might depend on coordinates j,k but which does not depend on any f_i other than f_j and f_k themselves.

A technical argument now leads from this to the result (3). Consideration of a third cell l yields

$$(\delta S/\delta f_k)/(\delta S/\delta f_l) = \alpha_{kl}(f_k,f_l) \qquad (8)$$

$$(\delta S/\delta f_l)/(\delta S/\delta f_j) = \alpha_{lj}(f_l,f_j) \qquad (9)$$

Multiplying the latter three equations,

$$1 = \alpha_{jk}(f_j,f_k) \; \alpha_{kl}(f_k,f_l) \; \alpha_{lj}(f_l,f_j) \qquad (10)$$

Hence

$$0 = \delta^2(\log \alpha_{jk})/\delta f_j \delta f_k \qquad (11)$$

so that, on using the antisymmetry (7) of $\log \alpha$ in j and k,

$$\alpha_{jk}(f_j,f_k) = \beta_{jk}(f_j)/\beta_{kj}(f_k) \qquad (12)$$

where β_{jk} is an as yet un-determined function. Substituting in (10) and differentiating with respect to f_j yields

$$(\log \beta_{jk}(f_j))' = (\log \beta_{jl}(f_j))' \qquad (13)$$

for arbitrary k and l, so that the differential $\beta_{jk}'(f_j)$ does not depend on k. The arbitrary constant which appears on integration can be absorbed in the definition (12), so that the second suffix on β may be dropped. Equation (7) can then be rewritten as

$$(\delta S/\delta f_j)/(\delta S/\delta f_k) = \beta_j(f_j)/\beta_k(f_k) \qquad (14)$$

Define

$$R(\underline{f}) = \sum_i \theta_i(f_i) \quad \text{where} \quad \theta_i'(x) = \beta_i(x) \qquad (15)$$

Then $\partial R/\partial f_i = \beta_i(f_i)$ for all i, and (14) shows that

$$(\partial S/\partial f_j)/(\partial S/\partial f_k) = (\partial R/\partial f_j)/(\partial R/\partial f_k) \qquad (16)$$

This means that the gradients $(\partial/\partial f_i)$ of R and S are parallel. Accordingly, $R(\underline{f})$ and $S(\underline{f})$ produce exactly the same reconstructions from given constraints, because any difference in the gradient magnitudes is absorbed in the Lagrange multiplier(s) of the constraint(s). Hence, without loss of generality, S can be restricted to the form

$$S(\underline{f}) = \sum_i \theta_i(f_i) \qquad (17)$$

Passage to the continuum limit requires a Lebesgue measure m to be introduced on the coordinate x, as $\theta_i(f_i)$ is replaced by its continuum equivalent $\theta(f(x),x)$. This completes the proof of (3), in which θ is assigned an explicit additional argument m(x), separate from x, for later convenience.

Comment:
It is not surprising that the axiom is only satisfied by a simple sum over the individual cells i of the scene. The effect of the axiom is precisely to exclude cross-terms between different points.

Axiom II. Coordinate invariance.
Let Γ be a coordinate transformation from x to Γx. Then f[I,m] transforms to

$$\Gamma(f[I,m]) = f[\Gamma I,\Gamma m] \qquad (18)$$

Justification:
We expect the same answer when we solve the same problem in two different coordinate systems, in that the reconstructed images in the two systems should be related by the coordinate transformation.

Consequence:
S must be of the form

$$S(\underline{f}) = \int dx\ m(x)\ \phi(f(x)/m(x)) \qquad (19)$$

Proof:
Write (3) in the form

$$S(\underline{f}) = \int dx\ m(x)\ \phi(f(x)/m(x),m(x),x) \qquad (20)$$

First, let I be the simple linear constraint $\int dx\ f(x) = 1$. The variational equation gives

$$\sigma(f(x)/m(x),m(x),x) = \lambda \qquad (21)$$

where σ is the derivative of ϕ with respect to its first argument, and λ is the Lagrange multiplier. Suppose there are two points x_1 and x_2 at which m takes the same value, and in the neighbourhood of which m is continuous. Let Γ be the transformation which exchanges equal-volume neighbourhoods D_1 of x_1 and D_2 of x_2. Then $\Gamma I = I$ and $\Gamma m = m$, hence $\Gamma f = f$. Also $\Gamma \lambda = \lambda$ because (by axiom I) operation of Γ leaves all points other than those in D_1 and D_2 unaffected. Substitution in (21) yields

$$\sigma(f/m,m,x_1) = \sigma(f/m,m,x_2) \tag{22}$$

showing that σ does not depend on its third argument. Integrating, ϕ itself is also independent of its third argument, except for an additive term which does not affect the maximisation over f and may be dropped.

Next, let I be the more general linear constraint

$$\int dx \; a(x)f(x) = 1 \tag{23}$$

for which the variational equation is

$$\sigma(f(x)/m(x),m(x)) = \lambda a(x) \tag{24}$$

Apply a coordinate transformation $x \rightarrow \Gamma x$ with Jacobian $\gamma(x) = \delta(\Gamma x/\partial x)$. This gives

$$dx \rightarrow \gamma \; dx \quad , \quad m \rightarrow \gamma^{-1}m \quad , \quad f \rightarrow \gamma^{-1}f \quad , \quad a \rightarrow a \tag{25}$$

and the variational equation becomes

$$\sigma(f(x)/m(x),m(x)/\gamma(x)) = \mu a(x) \tag{26}$$

where μ is a (possibly different) Lagrange multiplier. Dividing the two forms, we see that

$$\frac{\sigma(f(x)/m(x),m(x)/\gamma(x))}{\sigma(f(x)/m(x),m(x))} \quad \text{is constant in x .} \tag{27}$$

This holds for arbitrary $\gamma(x)$, so σ can not depend on its second argument. Integrating, neither does ϕ depend on its second argument, except for a term which does not affect the maximisation over f and can be omitted. This proves the required result (19).

Because S is now constructed purely from invariants m(x)dx and f(x)/m(x), reconstructions obtained from it must clearly satisfy axiom II as well as axiom I. The proof would have been shorter if S itself had been assumed to be invariant. Such an assumption would be plausible, but its truth is a consequence of the prime requirement that the reconstructed image should be invariant.

Comment:
It is via this axiom that the additive nature of f is
introduced. For example, in incoherent optics, it makes
sense to treat the radiative energy flux $\int$ intensity(x) dx
in a domain as an invariant quantity under coordinate
transformation, whereas it would not make sense to treat the
corresponding integral of wave amplitude $\int$ amplitude(x) dx
as an invariant. Accordingly, we would identify f(x) with
the additive flux density rather than some other function of
it such as its square root.

Axiom III. System Independence.
We now restrict the form of S($\underline{f}$) by requiring a specific
reconstruction for a particularly simple problem.

Let $m(x_1,x_2) = 1$ on the unit square $0 \leq x_1 \leq 1$, $0 \leq x_2 \leq 1$. Let the
constraints I be values of the marginals

$$\int dx_2\ f(x_1,x_2) = a_1(x_1) \quad , \quad \int dx_1\ f(x_1,x_2) = a_2(x_2) \qquad (28)$$

themselves obeying the consistent normalisation condition

$$\int dx_1\ a_1(x_1) \quad = \quad \int dx_2\ a_2(x_2) = 1 \qquad (29)$$

Then we require the reconstructed image to be the direct
product

$$f(x_1,x_2) = a_1(x_1)\ a_2(x_2) \qquad (30)$$

Gull, reported in Gull and Skilling (1984) and Livesey and
Skilling (1985), has presented a less abstract formulation
of this axiom.

Justification:
$f(x_1,x_2)$ represents a distribution of proportions, because
clearly it is constrained to satisfy $\iint dx_1 dx_2\ f(x_1,x_2) = 1$.
If all we know about f are its marginal distributions $a_1(x_1)$
and $a_2(x_2)$, then (in the absence m = 1 of any contrary bias)
we wish to recover the uncorrelated reconstruction $f = a_1 a_2$.
Any other choice of $f(x_1,x_2)$ would imply correlations for
which there is evidence neither in the data nor in the
measure. Good (1963) showed that this lack of correlation
in contingency tables would be a consequence of MaxEnt, but
here we reverse the argument and use the lack of correlation
in a derivation of MaxEnt.

Consequence:
S must be of the form

$$S(\underline{f}) = -\int dx\ f(x)\ (\log(f(x)/m(x)) + c) \qquad (31)$$

where c is a constant.

Proof:
With the given constraints, the variational equation

$$\delta(S - \int dx_1 \lambda_1(x_1) a_1(x_1) - \int dx_2 \lambda_2(x_2) a_2(x_2)) = 0 \qquad (32)$$

yields

$$\sigma(f(x_1,x_2)) = \lambda_1(x_1) a_1(x_1) + \lambda_2(x_2) a_2(x_2) \qquad (33)$$

in which σ is the derivative of ϕ as before, and where $f(x_1,x_2)$ is given in terms of a_1 and a_2 by the axiom (30). Applying $\delta^2/\delta x_1 \delta x_2$ yields

$$y \sigma''(y) + \sigma'(y) = 0 \qquad (34)$$

in which $y=f(x_1,x_2)$ can be chosen to take arbitrary values by suitable choice of constraint functions. Integrating twice,

$$\sigma(y) = A \log y + B \qquad (35)$$

where A and B are constants. Integrating again,

$$\phi(y) = A y \log y + (B-A) y \qquad (36)$$

plus another constant which does not affect the maximisation of S over f and may be dropped. A should be negative, to ensure that the extremum of S is a maximum ($\delta^2 S < 0$), but is otherwise merely an arbitrary scaling of S. Choosing $A = -1$, we have

$$\phi(y) = - y (\log y + c) \qquad (37)$$

(c = constant), which immediately gives the required form (31).

Comment:
This is the crucial axiom, which reduces S to the entropic form. The basic point is that when we seek an uncorrelated image from marginal data in two (or more) dimensions, we need to multiply the marginal distributions. On the other hand, the variational equation tells us to add constraints through their Lagrange multipliers. Hence the gradient $\delta S/\delta f$ must be the logarithm,

$$\delta S/\delta f = \log m - \log f \qquad (38)$$

which is the only function which converts a product into a sum. Integrating log f yields the "f log f" entropic form.

Axiom IV: Scaling.

$$f[\emptyset,m] = m \tag{39}$$

where $\emptyset$ represents the absence of any information.

Justification:
In the absence of any additional information, we wish to recover the initial measure.

Consequence:
The last ambiguity is resolved, and

$$S(\underline{f}) = \int dx (f(x) - f(x) \log(f(x)/m(x))) \tag{40}$$

Proof:
Unconstrained maximisation of (31) over f yields

$$f(x) = m(x) e^{-1-c} . \tag{41}$$

It would not actually be inconsistent to have a universal scaling factor e^{-1-c} between initial measure m and reconstruction f, but it would be arbitrary and often inconvenient. The value c = -1 avoids the difficulty.

Comment:
This choice may be viewed as a convention defining the units of f to be those of m.

The entropic regularisation function (40), properly written with two arguments as

$$S(\underline{f},\underline{m}) = \int dx (f(x) - f(x) \log(f(x)/m(x))) \tag{42}$$

is the form to be maximised when selecting an optimal image f. We should note that S does obey all four axioms, so that the axioms are mutually consistent. Before defining S in (42) to be the "entropy of f", we shall investigate the role of the measure m more closely.

SELECTING A MODEL
The global maximum of S over f, attained in the absence of further constraints, occurs at f=m, when the image equals the measure. This suggests a useful interpretation of m. As well as being an abstract Lebesgue measure, m can also be thought of as a prior model for the image. Imposition of further constraints will modify the selected image in such a way that it will always be as "close" (in the sense of maximising S) to the model as possible.

It often happens that a scene contains particular features,
such as point sources or spectral lines, which can be
described by a fairly small number of parameters. It would
be helpful to provide model parameters which were "best"
according to an agreed criterion. This involves setting up
a ranking of models which itself implies the existence of a
functional H($\underline{m}$,$\underline{f}$) , or H($\underline{m}$) for short, to be maximised
over m.

The Axioms
        Again, we will use simple properties of selected
models to restrict the form of H until only one form remains
(or equivalently a monotonic function of it). We assign the
symbol   J   to such freedom, by analogy with possible
constraint information I on f. The symbol m{J,f} represents
the optimal model allowed by J, on the basis of scene f.

Axiom I'. Subset independence.
Let J_1 be freedom allowed to m(x) in x$\in D_1$, and let J_2 be
freedom independently allowed to m(x) in x$\in D_2$. Then, if D_1
and D2 are disjoint,

$$m\{J_1,f\} \ U \ m\{J_2,f\} \ = \ m\{J_1 \ U \ J_2 \ , \ f\} \qquad (43)$$

Justification:
The model fitted to one domain should not affect the model
in a different domain, provided there are no parameters
directly linking the domains.

Consequence:
H must be of the form

$$H(\underline{m}) = \int dx \ f(x) \ \theta(m(x)/f(x),f(x),x) \qquad (44)$$

Proof:
This follows exactly as before (axiom I), with S replaced by
H, and f replaced by m, though the function θ may be
different.

Axiom II'. Coordinate invariance.
Let Γ be a coordinate transformation from x to Γx.   Then
m{J,f} transforms to

$$\Gamma(m\{J,f\}) = m\{\Gamma J,\Gamma f\} \qquad (45)$$

Justification:
We expect the same answer when we solve the same problem in
two different coordinate systems, in that the models in the
two   system   should   be   related   by   the   coordinate
transformation.

Consequence:
H must be of the form

$$H(\underline{m}) = \int dx \; f(x) \; \phi(m(x)/f(x)) \tag{46}$$

Proof:
This follows exactly as before (axiom II), with S replaced by H, and f replaced by m, though the function ϕ may be different.

Axiom III'. System independence.
We now restrict the form of $H(\underline{m})$ by requiring a specific model for a particularly simple problem. Let $\Xi(x_1,x_2)$, defined on the unit square $0 \leq x_1 \leq 1$, $0 \leq x_2 \leq 1$, be normalised $\iint dx_1 dx_2 \; f(x_1,x_2) = 1$. Let it be modelled by the factorised form

$$m(x_1,x_2) = n_1(x_1) \; n_2(x_2) \tag{47}$$

with normalisation

$$\int dx_1 \; n_1(x_1) = \int dx_2 \; n_2(x_2) = 1 \tag{48}$$

For this, we require

$$n_1(x_1) = \int dx_2 \; f(x_1,x_2) \quad , \quad n_2(x_2) = \int dx_1 \; f(x_1,x_2) \tag{49}$$

Justification:
$f(x_1,x_2)$ represents a distribution of proportions, because of its normalisation. In the model, $n_1(x_1)$ represents proportional structure in the x_1 dimension, as does $n_2(x_2)$ in the x_2 dimension, and we wish to recover the correct marginals. The model has no way of displaying correlations between x_1 and x_2, and we do not wish east-west (x_1) knowledge to influence our reconstruction of overall north-south structure (x_2), neither should x_2 structure influence x_1.

Consequence:
H must be of the form

$$H(\underline{m}) = \int dx \; [\; f(x) \; \log(m(x)/f(x)) - c \; m(x) \;] \tag{50}$$

Proof:
Perturbing the model in (46) yields, in this two-dimensional example

$$\delta H = \iint dx_1 dx_2 \; \sigma(m(x_1,x_2)/f(x_1,x_2)) \; \delta m(x_1,x_2) \tag{51}$$

where σ is the differential of ϕ. From the factorised form (47),

$$\delta m(x_1,x_2) = n_1(x_1)\ \delta n_2(x_2) + \delta n_1(x_1)\ n_2(x_2) \qquad (52)$$

where n_1 and n_2 obey normalisation (48) but are otherwise unrestricted. Substituting this into the variational equation

$$\delta(\ H - \lambda_1 \int dx_1\ n_1(x_1) - \lambda_2 \int dx_2\ n_2(x_2)\ ) = 0 \qquad (53)$$

for model parameter perturbations δn_1 and δn_2 gives

$$\int dx_2\ \sigma(\ n_1(x_1)n_2(x_2)/f(x_1,x_2)\ )\ n_2(x_2) = \lambda_1$$
$$\int dx_1\ \sigma(\ n_1(x_1)n_2(x_2)/f(x_1,x_2)\ )\ n_1(x_1) = \lambda_2 \qquad (54)$$

from δn_1 and δn_2 respectively. In (54), $n_1(x_1)$ and $n_2(x_2)$ are to equal the marginals (49), for arbitrary scene f. Perturb f in such a way that the marginals are unchanged. This gives

$$\int dx_2\ \tau(\ f(x_1,x_2)/n_1(x_1)n_2(x_2)\ )\ \delta f(x_1,x_2) = \delta\lambda_1\ n_1(x_1)$$
$$\int dx_1\ \tau(\ f(x_1,x_2)/n_1(x_1)n_2(x_2)\ )\ \delta f(x_1,x_2) = \delta\lambda_2\ n_2(x_2) \qquad (55)$$

where $\tau(y)=d\sigma(y^{-1})/dy$. Select the particular marginal-preserving perturbation proportional to

$$\delta f(x_1,x_2) = \delta(x_1-a_1)\delta(x_2-a_2) - \delta(x_1-b_1)\delta(x_2-a_2)$$
$$- \delta(x_1-a_1)\delta(x_2-b_2) + \delta(x_1-b_1)\delta(x_2-b_2) \qquad (56)$$

where δ on the right-hand side is the Dirac delta function and a_1, a_2, b_1, b_2 are coordinates between 0 and 1, so that the rectangle of points (x_1,x_2) selected by the four delta functions lies within the given unit square. Values of x_1 other than a_1 and b_1 show that $\delta\lambda_1 = 0$. Putting $x_1 = a_1$ gives

$$\tau(f(a_1,a_2)/n_1(a_1)n_2(a_2))$$
$$- \tau(f(a_1,b_2)/n_1(a_1)n_2(b_2)) = 0 \qquad (57)$$

This holds for arbitrary a_1, a_2, b_2 so that the two arguments of τ can each take arbitrarily different values. The only way of satisfying this is to set

$$\tau(y) = A \quad , \quad A = constant \qquad (58)$$

from which

$$\sigma(y) = A\ y^{-1} + B \qquad (59)$$

Integrating again,

$$\phi(y) = A \log y + B\ y \qquad (60)$$

plus a constant which does not affect the maximisation over
m and may be omitted. A should be positive, to ensure that
the extremum of H is a maximum ($\delta^2 H < 0$), but is otherwise
merely a scaling factor. Choosing A = 1 and setting B = -c
gives the quoted form (50).

Axiom IV': Scaling.

$$m\{\emptyset, f\} = f \qquad (61)$$

where $\emptyset$ represents the absence of any restriction on the
model.

Justification:
In the absence of any restriction, we seek to recover the
starting scene.

Consequence:
The last ambiguity is resolved, and

$$H(\underline{m},\underline{f}) = \int dx \ [\ f(x) \ \log(m(x)/f(x)) - m(x) \ ] \qquad (62)$$

Proof:
Unconstrained maximisation of (50) yields

$$m(x) = f(x)/c \qquad (63)$$

whereas m=f is required. Thus c=1 and (62) is obtained.

Finally, we should note that H($\underline{m}$) defined in (62) does obey
all four axioms, so that the axioms are mutually consistent.

SYNTHESIS
    S($\underline{f},\underline{m}$) from (42) and H($\underline{m},\underline{f}$) from (62) are the same
function (apart from additive terms in each which do not
affect maximisation of the other). They can be combined
into the joint form (denoted by S, because there is now no
need for separate symbols):

$$S(\underline{f},\underline{m}) = \int dx \ [\ f(x) - m(x) - f(x) \ \log(f(x)/m(x)) \ ] \qquad (64)$$

in the continuous case, and in the discrete case

$$S(\underline{f},\underline{m}) = \sum_i [\ f_i - m_i - f_i \ \log(f_i/m_i) \ ] \qquad (65)$$

This can be used to rank and thence to select image-model
($\underline{f},\underline{m}$) pairs, and is the only function which obeys all the
axioms above. We shall call S "entropy" because of its
close connection with the classic "-Σ p log p" form. The
decrease of S from its global maximum of zero quantifies the
deviation of f from its model m.

The "maximum entropy method" in data analysis, then, consists of maximising the entropy S, either over an image f subject to given constraints, or over the model m within given degrees of freedom, or both.

CONCLUSIONS
        Any universally applicable method of selecting a single positive image ought to give acceptable, sensible results in particular cases. Four such cases, codified as axioms, lead to MaxEnt as the only consistent selection procedure. The MaxEnt method is valid for any type of data, regardless of the normalisation of the image. Of fundamental importance is that the entropy gradient (38) is logarithmic. This neatly ensures that any reconstructed image is both positive and finite, and as close as possible to some prior model. The more complicated formula (64)

$$S(\underline{f},\underline{m}) = \int dx \ [\ f(x) - m(x) - f(x) \ \log(f(x)/m(x)) \ ]$$

for the entropy itself is the integral of this gradient.

Even though MaxEnt has already had considerable practical success in reconstructing various types of positive distribution, theory indicates that the method should be yet more powerful, because the same entropy formula should be used to select optimal parameters in the underlying prior model. Thus MaxEnt should also be used to estimate parameters pertaining to positive distributions, which opens a particularly promising avenue to future research.

In this paper, no attempt has been made to quantify the probabilistic reliability of the MaxEnt estimates. Maximum entropy can stand in its own right as a selection procedure.

References

Cox, R.T. (1961). The algebra of probable inference.
        Johns Hopkins Press, Baltimore, MD.
Good, I.J. (1963). Maximum entropy for hypothesis
        formulation, especially for multi-dimensional
        contingency tables. Annals.Math.Stat.,34, 911-934.
Gull, S.F. & Skilling, J. (1984). The maximum entropy
        method. In Indirect imaging, ed. J.A. Roberts.
        Cambridge: Cambridge University Press.
Jaynes, E.T. (1957a). Information theory and statistical
        mechanics I. Phys. Rev.,106, 620-630.
Jaynes, E.T. (1957b). Information theory and statistical
        mechanics II. Phys. Rev.,108, 171-190.
Jaynes, E.T. (1984). Monkeys, Kangaroos and N. Presented
        at fourth maximum entropy workshop, Calgary,
        ed. J.H. Justice, Dordrecht: Reidel.
Kullback, S. (1959). Information theory and statistics.
        New York: Wiley.
Livesey, A.K. & Skilling, J. (1985). Maximum entropy theory
        Acta Cryst.,A41, 113-122.
Shannon, C.F. (1948). A mathematical theory of
        communication. Bell System Tech. J.,27, 379-423
        and 623-656.
Shannon, C.E. & Weaver, W. (1949). The mathematical theory
        of communication. Urbana, Illinois: University
        Illinois Press.
Shore, J.E. & Johnson, R.W. (1980). Axiomatic derivation of
        the principle of maximum entropy and the principle
        of minimum cros-entropy. IEEE Trans.Info.Theory,
        IT-26, 26-37 and IT-29, 942-943.
Tikochinsky, Y., Tishby, N.Z. & Levine, R.D. (1984).
        Consistent inference of probabilities for
        reproducible experiments. Phys.Rev.Lett.,
        52, 1357-1360.

UNDERSTANDING IGNORANCE

C.C. Rodriguez*
State University of New York at Albany
Department of Mathematics and Statistics
Albany, New York 12222

*This research was supported in part by PHS
grant number 1-R01-CA41171-01A1 awarded by
the National Cancer Institute, DHHS.

1 INTRODUCTION

Among the several schools of thought of the
theory of inference two main groups are now clearly
recognizable: Those that define the probability of an
event as the limit of its relative frequency when
independent identical trials of a random experiment are
performed, and those that define probability in some
other way. We shall agree to call the former group
Frequentist and the latter Bayesian, only for
identification purposes. The need for an alternative
theory of probability arose in practice to handle
situations when there was relevant prior information
concerning the problem under scrutiny or when repeated
trials were practically or theoretically impossible to
perform.

The passage from the Frequentist to the
Bayesian point of view is remarkably analogous to the
passage from classical to modern physics. In
statistics, as in physics, (or in any other science) it
is conceptually appealing to have new theories that are
extensions of old ones. That is, the new theory
coincides with the old one on the phenomena that the
old theory was predicting correctly. This kind of
theoretical advance saves the trouble of starting from
scratch and gives to the theoretician the feeling
(possible illusory) of convergence to an ultimate
truth. Relativistic and quantum mechanics are dramatic
examples of theoretical advance in this inclusive way.
Formally we have:

$$\lim_{c \to \infty} (\text{Relativistic Mechanics}) = (\text{Classical non-Relav.}$$
$$\text{Mechanics})$$

and similarly

$$\lim_{h \to 0} (\text{Quantum Mech.}) = (\text{Classical Mech.}) \qquad (1)$$

189

G. J. Erickson and C. R. Smith (eds.),
Maximum-Entropy and Bayesian Methods in Science and Engineering (Vol. 1), 189–204.
© 1988 by Kluwer Academic Publishers.

In words: If there were no speed limit in the universe
(i.e., if the speed of light c, were +∞) then the
relativistic and non-relativistic equations would be
the same. Similarly, classical mechanics is formally
obtained from quantum mechanics as the limit when
Plank's constant h tends to zero (i.e., if there were
no quantization and all values of energy were
possible).

       The Frequentist and Bayesian theories of
inference can also be related by an equation analogous
to (1). Formally we have:

$$\lim_{I \to I_0} \text{(Bayesian)} = \text{(Frequentist)} , \qquad (2)$$

where "I" represents prior information and "I_0" the
state of knowledge of "Total Ignorance" i.e., absence
of prior information. Equation (2) holds only when the
inference problem under consideration is meaningful
from the Frequentist point of view. We shall see later
that there are problems that are meaningless for the
Frequentist eye but nevertheless are well defined in
the Bayesian setting. In such cases the right hand
side of equation (2) should be replaced by "(Bayesian
with no prior information)". It seems that the
Bayesian theory constitutes a radical departure from
the Frequentist theory as it was modern physics from
its classical counterpart.

       As an illustration of equation (2) consider
the simple but important case of inference about the
mean μ of a normal $N(\mu, \sigma^2)$ when σ^2 is assumed to be
known and the data $x_1, x_2, \ldots, x_n$ from a $N(\mu, \sigma^2)$
population are available. The $100(1-\alpha)\%$ frequentist
confidence interval for the unknown, but fix, parameter
μ is a random interval (A, B) satisfying

$$P\left[(A,B) \ni \mu \right] = 1 - \alpha .$$

Under the assumption that $x_1, x_2, \ldots, x_n$ are i.i.d.
$N(\mu, \sigma^2)$ we obtain

$$(A,B) = \left\{ \overline{x} - z_{\alpha/2} \frac{\sigma}{\sqrt{n}} , \overline{x} + z_{\alpha/2} \frac{\sigma}{\sqrt{n}} \right\}, \qquad (3)$$

where $\overline{x}$ is the sample mean and $P[Z \geq z_{\alpha/2}] = \alpha/2$ with
$Z \sim N(0,1)$. On the other hand, the concept of
probability used by the Bayesian school makes

meaningful the assignment of probabilities to the unknown parameter μ. It should be recalled here that we are grouping under the same label of "Bayesians" schools of thought with different concepts of probability; but whether the interpretation given to probability is subjective, logically objective, or something in between is irrelevant for our discussion at this point. However, what is important is that all these Bayesians will agree in the possibility of assignments of prior probabilities to parameters that are regarded as deterministic but unknown by the frequentists.

The natural way to specify the uncertainty about μ is by assuming a conjugate prior distribution for it (i.e., the posterior distribution remains in the same parametric family). In the problem under consideration the conjugate prior is:

$$\mu \sim N(\mu_0, \sigma_0^2) \quad . \tag{4}$$

Conjugate priors are appealing for treating prior information and data points symmetrically. The Bayesian with a conjugate prior acts "as if" extra virtual data points would have been observed. After the specification of a prior distribution for μ the Bayesian regards the data points $x = (x_1, x_2, \ldots, x_n)$ conditional on μ as i.i.d. $N(\mu, \sigma^2)$; and updates the uncertainty about μ conditioning on the data x. Hence, after application of Bayes' theorem, the posterior distribution of μ is given by

$$[\mu|x] \sim N(\mu_1, \sigma_1^2) \tag{5}$$

where the posterior parameters are updated with the formulas

$$\mu_1 = \frac{a\overline{x} + b\mu_0}{a + b} \quad \text{and} \quad \sigma_1^{-2} = a+b \tag{6}$$

with $a = n/\sigma^2$ and $b = 1/\sigma_0^2$. The $100(1-\alpha)\%$ Bayesian estimation interval is a deterministic interval (a,b) satisfying

$$P[\mu \in (a,b)|x] = 1-\alpha$$

where the above probability is computed using the
posterior distribution of μ given by (5). In this case
the Bayesian confidence interval is given by

$$(a,b) = (\mu_1 - z_{\alpha/2}\sigma_1 \;,\; \mu_1 + z_{\alpha/2}\sigma_1) \; . \tag{7}$$

It is clear from (3), (6) and (7) that

$$\lim_{\sigma_0 \to \infty} (a,b) = (A,B) \; . \tag{8}$$

Therefore, in the limit of vague prior knowledge about
μ (i.e., $\sigma_0 \to \infty$) the Bayesian and the Frequentist both
express the same uncertainty about μ on the basis of
the same data x, regardless of what were their
justifications for doing what they did. The parameter
"I" used in (2) could be identified with the prior
density for μ in (8), $\Pi_{\sigma_0}(\mu)$. If we specify only the
dependence on μ with the proportionality sign α, we can
write from (4)

$$\Pi_{\sigma_0}(\mu) \;\; \alpha \;\; \exp\left\{\frac{-(\mu-\mu_0)^2}{2\sigma_0}\right\}$$

from where we can identify I_0 in (2) with Π_∞ i.e.,

$$\Pi_\infty(\mu) = \lim_{\sigma_0 \to \infty} \Pi_{\sigma_0}(\mu) \;\; \alpha \;\; 1 \tag{9}$$

Equation (9) shows that the state of knowledge of
"Total Ignorance" I_0 about the location parameter μ can
be identified with the Laplace flat prior (9) for μ.
The above equation (8), has been known for a long time
but it was not until Jeffreys (1939) showed that there
were noninformative priors different from the uniform
in (9) (see below) that the subject was given any
serious theoretical attention.

We consider now the inference about the
unknown parameters θ and Φ of a $N(\theta,\Phi)$ ($\theta \in R$, $\Phi \in R_+$).
The conjugate priors for θ and Φ are given by

$$[\theta|\Phi] \sim N(\mu_0, \Phi/\lambda_0) \text{ and } (\upsilon_0 u_0/\Phi) \sim \chi^2_{\upsilon_0} \;,\; \tag{10}$$

where $\mu_0 \in R$ and $\upsilon_0, u_0, \lambda_0 > 0$ are the prior parameters.
Applying Bayes theorem under the assumptions that given

θ and Φ, the data $x=(x_1, x_2,..., x_n)$ are i.i.d. $N(\theta,\Phi)$ we obtain the posterior distributions

$$[\theta|\Phi,x] \sim N(\mu_1,\Phi/\lambda_1) \text{ and } [\upsilon_0 u_0/\Phi|x] \doteq \upsilon_1 u_1/\Phi \sim \chi^2_{\upsilon_1}$$

$$(11)$$

where $\doteq$ means equality in distribution and the posterior parameters are obtained from the formulas,

$$\lambda_1 = \lambda_0+n, \quad \upsilon_1 = \upsilon_0+n-1, \quad \mu_1 = (\lambda_0\mu_0+n\bar{x})/\lambda_1 \quad (12a)$$

and

$$\upsilon_1 u_1 = \upsilon_0 u_0 + \sum_{i=1}^{n} (x_1-\bar{x})^2 + \frac{n\lambda_0}{\lambda_1} (\bar{x}-\mu_0)^2 \quad . \quad (12b)$$

Moreover, it can be shown (after a change of variables and integrating over Φ) from (11) that the posterior distribution of t, with

$$t = \sqrt{(\lambda_1/u_1)} \; (\theta-\mu_1) \quad ,$$

is student with υ_1 degrees of freedom i.e.,

$$\Pi(t|x) \; \alpha \; (1+t^2/\upsilon_1)^{-(\upsilon_1+1)/2} \quad . \quad (13)$$

Hence, the $100(1-\alpha)\%$ Bayesian C.I. for θ, (a_θ,b_θ) and Φ, (a_Φ,b_Φ) are given by

$$(a_\theta,b_\theta) = (\mu_1 - \sqrt{(\lambda_1/u_1)} \; t^{\alpha/2}, \; \mu_1 + \sqrt{(\lambda_1/u_1)} \; t^{\alpha/2})$$

and

$$(a_\Phi,b_\Phi) = (\upsilon_1 u_1/\chi^2_{\upsilon_1}(\alpha/2), \; \upsilon_1 u_1/\chi^2_{\upsilon_1}(1-\alpha/2)) \quad ,$$

where, $P[T \geq t^{\alpha/2}] = \alpha/2$ with $T \sim t_{\upsilon_1}$ (Student-t with

υ_1 d.f. see (13)) and $P[X^2 > \chi^2_{\upsilon_1}(\beta)] = \beta$ with $X^2 \sim \chi^2_{\upsilon_1}$ [Chi-squared with υ_1 d.f.'s i.e., $\Gamma(\upsilon_1/2, 1/2)$]. As in (8), in the limit of diffuse prior knowledge about both θ and Φ (i.e., when $\lambda_0+\upsilon_0\rightarrow0$; absence of virtual observations) we obtain:

$$\lim_{\lambda_0, \upsilon_0 \to 0} (a_\theta, b_\theta) = (\overline{x} - \frac{s}{\sqrt{n}} t_{n-1}^{\alpha/2} \ , \ \overline{x} + \frac{s}{\sqrt{n}} t_{n-1}^{\alpha/2}) \ ,$$

where s denotes the sample standard deviation computed from x. Hence, the classical frequentist $100(1-\alpha)\%$ confidence intervals for the mean and variance of a $N(\mu, \sigma^2)$ are recovered. In this case the a priori information "I" in (2) is associated to the joint pdf of θ and Φ. The marginal pdf's $\Pi_{\lambda_0, \upsilon_0}(\theta)$ and $\Pi_{\lambda_0, \upsilon_0}(\Phi)$ are given, from (10), by

$$\Pi_{\lambda_0, \upsilon_0}(\theta) \ \alpha \ [\upsilon_0 u_0 + \lambda_0(\theta - \mu_0)^2]^{-\upsilon_0/2 - 1} \qquad (14a)$$

and,

$$\Pi_{\lambda_0, \upsilon_0}(\Phi) \ \alpha \ \Phi^{-\upsilon_0/2 - 1} \exp\{-\upsilon_0 u_0/(2\Phi)\} \qquad (14b)$$

Therefore, I_0 in (2) should be identified with $\Pi_{0,0}$ i.e.,

$$\Pi_{0,0}(\theta) = \lim_{\lambda_0 + \upsilon_0 \to 0} \Pi_{\lambda_0, \upsilon_0}(\theta) \ \alpha \ 1 \qquad (15)$$

and

$$\Pi_{0,0}(\Phi) = \lim_{\lambda_0 + \upsilon_0 \to 0} \Pi_{\lambda_0, \upsilon_0}(\Phi) \ \alpha \ 1/\Phi \ . \qquad (15b)$$

Equations (15a) and (15b) show that the state of knowledge of "Total Ignorance", I_0, about both θ and Φ is associated with the joint pdf $\Pi_{0,0}(\theta, \Phi)$ given by:

$$\Pi_{0,0}(\theta, \Phi) \ \alpha \ 1/\Phi \ . \qquad (16)$$

Equation (16) is known as the Jeffreys' rule. Numerous applications of this rule can be found in Box and Tiao (1973). Jeffreys' rule shows that in order to describe mathematically the state of knowledge of "Total Ignorance" distributions other than the Laplace-flat prior (see (9)) are sometimes necessary.

2 A SIMPLE PARADOX

I shall make use of a very simple paradox to introduce new noninformative priors and to point out the connection between the state of knowledge of "Total-Ignorance" and self-similarity. The origins of the paradox are at the present time unclear to me. All that I know is that it was informally presented to I.J. Good as an amusing problem at a recent conference on the Foundations of Probability. The following analysis in terms of noninformative priors is, as far as I know, new. I state the problem in what I call "The Choice", an intuitive analysis is what I call "The Reasoning" and its paradoxical consequences are labelled by "The Paradox".

The Choice

Suppose that you and another person are presented with the following dilemma. There are two identical envelopes in front of you and you are given the following information: "one envelope contains a check for an amount (a real number) twice as much as the other". Each of you pick up one envelope. You open yours and you find $a. (Note: we use the dollar sign only to motivate the problem. It should always be kept in mind that the checks are not in US currency but that real numbers are written on them. This ideal situation can be approximated in practice by truncating or rounding to two decimal places numbers that are initially written down to ten decimal places.

In this manner the last digit in the decimal expression for $\underline{a}$ will give only negligible evidence about which card contains the "double".)

We want to consider the following two questions: i) should you exchange envelopes with the other person? and ii) what is the probability of doubling the amount by exchanging envelopes?

The Reasoning

You reason in the following way: Given that I have $a in my envelope and that I know that one envelope has twice as much as the other then, if I exchange, I could get either $2a (if my envelope contains the smallest amount) or $a/2 (if my envelope contains the largest amount). My total ignorance about these two events ($2a, $a/2 in the other envelope) makes me assign equal probabilities to them. Hence, my expected return (if I exchange) is:

$$\frac{1}{2}(2a) + \frac{1}{2}(\frac{a}{2}) = \frac{5}{4}a > a$$

Therefore, you conclude, "I should exchange!". Notice that the conclusion "you should exchange" is independent of "a" and therefore the same conclusion is reached if you had not looked in your envelope.

The Paradox

By symmetry, exactly the same conclusion is reached by the other person. This cannot be the case. If you both win, then money is being created and we have a money pump. On the other hand, if you choose not to look in your envelope you still conclude that you must exchange. But after exchanging you are at the same state of knowledge as before and therefore, you must exchange again! Hence, you enter a nonsense of never ending exchanges.

Let us introduce some notation to handle this problem more formally. Define the r.v.'s:

Z = "amount in your envelope"

Y = "amount in the other envelope"

X = "amount in the envelope with less money"

and the events:

E_1 = "You get the envelope with \$X"

E_2 = "You get the envelope with \$2X" .

Notice that the two schools of inference, Frequentist and Bayesian, could give different interpretations to the variables X,Y,Z. For the Bayesian the quantities X,Y and Z are always random variables but for a Frequentist, X might not be a r.v. it might be a parameter whose value ($x_0 \in R_+$) happens to be unknown at the moment. Hence,

$$X = x_0, \quad Z = \begin{cases} x_0 \text{ with prob. } 1/2 \\ \\ 2x_0 \text{ with prob. } 1/2 \end{cases}$$

and

$$Y = \begin{cases} 2x_0 \text{ if } E_2 \text{ occurs} \\ \\ x_0 \text{ if } E_2 \text{ occurs} \end{cases}$$

Therefore, if we are given that $P(E_1) = P(E_2)$ we obtain

$$EZ = EY = (1.5)x_0 \tag{1}$$

Hence, if you do not look in your envelope then the frequentist analysis would correctly conclude in this case that one should be indifferent to exchange. Equation (1) shows that what we expect to observe in our envelope i.e., EZ, is equal to what we expect to obtain as the result of exchanging i.e., EY, and therefore you should be indifferent to exchange. But it is arbitrary to choose X as the fixed unknown parameter x_0 and let Y and Z to be random variables whose possible values depend on x_0. The Frequentist could have chosen Z to be the parameter and in this case,

$$Z = z_0, \quad X = \begin{cases} z_0 \text{ if } E_1 \\ \\ z_0/2 \text{ if } E_2 \end{cases} \quad \text{and} \quad Y = \begin{cases} 2z_0 \text{ if } E_1 \\ \\ z_0 \text{ if } E_2 \end{cases}$$

from where $EY = (1.5)z_0 > z_0 = EZ$ and therefore, we should exchange. If the frequentist chooses Y to be the parameter, then.

$$
Y = y_0 \quad \text{and} \quad Z = \begin{cases} y_0/2 & \text{if } E_1 \\ \\ \\ 2y_0 & \text{if } E_2 \end{cases}
$$

from where $EZ = (5/4)y_0 > y_0 = EY$ and therefore, we should not exchange. Hence, by using the Frequentist analysis we can obtain any of the possible answers to question (i) by just changing the way in which we choose to solve the problem! This seems to me totally inadmissible as a serious method. So, we had better go back to the Bayesian setting and consider all 3 variables X,Y and Z to be random variables. We have,

$$
P[Y=2a|Z=a] = P[Y=2a|Z=a,E_1] \ P[E_1|Z=a]
$$

$$
+ P[Y=2a|Z=a,E_2] \ P[E_2|Z=a]
$$

Clearly $P[Y=2a|Z=a,E_1] = 1$ and $P[Y=2a|Z=a,E_2] = 0$; but the numerical values of the other quantities i.e., $P[E_1|Z=a]=p$ and $P[E_2|Z=a] = 1-P[E_1|Z=a]=q$ depend on the distribution of the r.v. X, defined above. To see this clearly, consider the extreme degenerate case when the pdf of X is given by

$$
\Pi(x) = \delta(x-a) \ ,
$$

where δ denotes the Dirac delta function. In such a case, p=1. Hence, p cannot be obtained directly from the information given, unless we use "meta-probabilistic" arguments. In other words the problem: (ii) What is the probability of doubling the money by exchanging envelopes?, is ill-posed from a Frequentist point of view until the distribution of X, $\Pi(x)$, is specified. However, as we shall see below, the symmetries of the information given in the problem can be exploited to logically constraint the possible a priori pdf's $\Pi(x)$.

The assignment of p=q=1/2 in "The Reasoning" is tied up with the following, plausible, meta-probabilistic argument:

The complete ignorance about the unspecified distribution of X makes us assume that the knowledge of what we have in our envelope i.e., the occurrence of the event [Z=a], does not affect the probability of E_1, because if it does, it _means_ that we have some idea about the distribution of X and we are claiming information that we do _not_ have.

I shall now prove that the above meta-probabilistic argument is related to a specific form of $\Pi(x)$. To see this, notice that

$$Z = \begin{cases} X \text{ with probability } 1/2 \\ \\ 2X \text{ with probability } 1/2 \end{cases}$$

hence,

$$P[Y=2a\,|\,Z=a] = \int_0^\infty P[Y=2a\,|\,Z=a,X=x]\ \Pi_{X|Z}(x\,|\,a)\ dx\ ,\quad (3)$$

where

$$\Pi_{X|Z}(x\,|\,a) = f_{X,Z}(x,a)/f_Z(a) \tag{4}$$

where $f_{X,Z}$ and f_Z denote the joint pdf of X and Z and the marginal pdf of Z respectively. The following equations are also easily obtained:

$$\Pi_{2X}(y) = (1/2)\ \Pi_X(y/2) \tag{5}$$

and

$$f_{X,Z}(x,z) = (1/2)\Pi(x)\delta(x-z) + (1/4)\Pi(x)\delta(x-z/2) \tag{6}$$

and from (6) we can write

$$f_Z(z) = \int_0^\infty f_{X,Z}(x,z)\ dx = (1/2)\Pi(z) + (1/4)\Pi(z/2) \tag{7}$$

Applying Bayes' theorem we have,

$$P[Y=2a\,|\,Z=a] = P[E_1\,|\,Z=a] = \{\Pi_{Z|E}(a\,|\,E_1)\ P[E_1]\}/f_Z(a)$$

and using (7) we obtain,

$$P[E_1|Z=a] = (1/2 \; \Pi(a))/\{(1/2) \; \Pi(a) + (1/4) \; \Pi(a/2)\} \quad (8)$$

Notice that the same result is obtained from (3) by using (4), (6) and (7). Hence, from (8), $P[E_1|Z=a] = p$ for all values of a if and only if Π satisfies the self-similarity relation,

$$\Pi(a) = \frac{p}{2(1-p)} \; \Pi\left[\frac{a}{2}\right] \qquad \forall a > 0 \qquad (9)$$

which is a particular case of the general equation

$$\Pi(x) = \theta\Pi(x/\mu) \qquad \forall x > 0 \qquad (10)$$

for θ and μ given positive constants.

We shall now show that equation (10) together with the assumptions that $\Pi(x) > 0$ and the existence of $\Pi''(x) \; \forall x > 0$ drastically reduce the possible forms of Π. To see this, define

$$f(t) = \Pi(e^t) \qquad \forall t \in R$$

we have from (10) that,

$$f(t) = \theta f(t - \log\mu) = \theta f(t-c)$$

and doing $c = \log \mu$ we obtain,

$$h(t) = \log f(t) = \log \theta + h(t-c) \quad . \qquad (11)$$

Let us take the derivative with respect to t in equation (11). We can write

$$g(t) = h'(t) = g(t-c) \qquad \forall t \in R \quad . \qquad (12)$$

Therefore, g must be a simple periodic function on the real line with period c. Hence, we can write the Fourier series expansion of g as,

$$g(t) = \sum_{n=-\infty}^{+\infty} a_n \exp\{2\pi itn/c\} = a_0 + R(\exp\{2\pi it/c\})$$

$$(13)$$

where R is a convergent trigonometric series of period c and no constant term i.e., $R(0) = 0$. Notice that

$$g(t) = \frac{d}{dt}\{\log \circ \Pi \circ \exp(t)\}$$

where $\circ$ denotes the composition of functions.
Therefore, the existence of $\Pi''(x)$, $\forall x>0$ assures the
uniform convergence of the Fourier series (13). Hence,

$$h(t) = \int_0^t g(y)\,dy + K'$$

$$= a_0 t + \int_0^t R(\exp\{2\pi iy/c\})\,dy + K'$$

with K' an arbitrary constant. The constant a_0 is
obtained by using (11). From (12) we have,

$$h(t-c) = \int_0^{t-c} g(y)\,dy + K' = \int_c^t g(y)\,dy + K' \quad .$$

Hence,

$$h(t) = \int_0^c g(y)\,dy + \int_c^t g(y)\,dy + K'$$

$$= \int_0^c g(y)\,dy + h(t-c) \quad .$$

Therefore, (11) and the mean value property applied to
the function R imply that,

$$\int_0^c g(y)\,dy = a_0 c = \log \theta \quad ,$$

from where we obtain

$$a_0 = \log \theta \,/\, \log \mu = \eta \tag{14}$$

and the most general representation of $\Pi(x)$ is given
by,

$$\Pi(x) = Kx^\eta \, \exp\left[\int_0^{\log x} R(\exp[2\pi iy/c])\,dy\right] = Kx^\eta S(x) \tag{15}$$

where K is an arbitrary positive constant. We have

then shown the following

Theorem: If $\forall x > 0$, $\Pi''(x)$ exists and $\Pi(x) > 0$ then
equation (10) is equivalent to

$$\Pi(x) = K\, x^{\eta}\, S(x) \qquad x > 0$$

where $K > 0$ and $S(x)$ are defined as in (15).

Notice that S satisfies $S(x/\mu) = S(x)$ $x > 0$.
Moreover, $S(x\mu^{n}) = S(x)$ $\forall n = 0, \pm 1, \pm 2, \ldots$ and $\forall x > 0$.   This
shows that S takes the same value $S(x)$ on each point in
the set $\{x\mu^{n}: n = 0, \pm 1, \pm 2, \ldots\}$.   Therefore, $\Pi(x)$ has
infinitely many "bumps" unless $S(x)$ is constant and
equal to one (since S is constant iff R is zero).
Hence, the only function able to represent ignorance in
the class of functions Π of the form given in (15) is

$$\Pi(x) \propto x^{\eta} \ . \qquad\qquad (16)$$

This function could be characterized as the only one in
the class defined by (15) that has a convex tail; all
the others contain an infinite amount of "information"
(bumps) in the tail and should not be used to represent
lack of information.   Hence, we have been able to prove
that an assignment of $P[E_1 | Z=a] = p$ $\forall a > 0$ basically
commits you to a unique noninformative prior given by
(16).   In the special case of our paradox we have
$p = 1/2$.   Using (9) and (14) we obtain $\eta = -1$ and from
(16) we have,

$$\Pi(x) \propto x^{-1}$$

which is the Jeffreys' prior obtained in the
introduction.   We already know that the selection of
$p = 1/2$ was a bad one (it produced the paradox).   But
we can eliminate the inconsistencies by replacing the
Jeffreys' noninformative prior by the only one
associated to the p^{*} such that

$$E[Y | Z=a] = p^{*}(2a) + (1-p^{*})(a/2) = a = E[Z | Z=a]$$

which implies $p^{*} = 1/3$ and from (9) and (14) $\eta^{*} = -2$.
Hence, the symmetries of the information stated in the
problem have been sufficient to reduce the class of
possible noninformative priors of X to only one:

$$\Pi(x) \propto x^{-2} \qquad (17)$$

which in turn is the only noninformative prior
distribution for X that is able to eliminate the
paradox.

3 SELF-SIMILARITY AND FRACTAL REPRESENTATION

The self-similarity relation (2.10) satisfied
by the noninformative prior distributions (2.16) can be
studied from a geometric point of view. I shall point
out informally some interesting connections between
fractals and our paradox.

Consider a noninformative prior Π that
satisfies (2.10). We have that the area under the
curve $\Pi(x)$ is a measure of length (since Π is
interpreted as a pdf) $L(x)$. Hence, using (2.10)

$$L(x) = \int \Pi(x)\,dx = \theta \int \Pi(x/\mu)\,dx \qquad (1)$$

and doing $y=x/\mu$ in (1) we obtain,

$$L(x) = \theta\mu L(x/\mu) \qquad (2)$$

and the fractal dimension, D, associated to the
self-similarity relation (2) is such that (see
B. Mandelbrot, 1977)

$$L(x) = x^{1-D}$$

and replacing in (2) we obtain that D is,

$$D = -\log \theta \; / \; \log \mu = -\eta \qquad (3)$$

therefore the fractal dimension associated is equal to
the absolute value of the exponent of the
noninformative prior in (2.16). This could be
exploited to obtain concrete pictorial representations
of the symmetries of the actual information given in a
probabilistic problem. Moreover, notice that the
fractal dimension D(p) as a function of p (see (2.9))
is given by

$$D(p) = 1 + \log((1-p)/p) \; / \; \log 2 \qquad (4)$$

when p varies from 0 to 1 D(p) varies from $+\infty$ to $-\infty$.
The interesting dimensions for a curve are in between 1

and 2. In this case we have,

$$1 \leq D(p) \leq 2 \quad iff \quad (1/2) \geq p \geq (1/3) \qquad (5)$$

Equation (5) shows why $p^* = 1/3$ was a good choice for p (see (2.17)). This $p^* = 1/3$ has associated the least informative fractal i.e., a curve plane-filling with D=2!

It is also interesting to look at this connection between fractals and our paradox as a way to associate information in pictures of fractals with information about what is in the envelopes. It is easy to show that a fractal curve of dimension D is associated to the information about X in our paradox when the following statement is added to the problem:

"The odds in favor of E_2 are $(1/2)^{D-1}$"

More research along these lines could produce useful results for the AI problem of Knowledge Representation.

Acknowledgments: I would like to thank I.J. Good for his comments and corrections and M. Range for helpful discussion about the proof of the theorem.

REFERENCES
Box, G.E.P., & Tiao,G.C. 1973. Bayesian Inference in Statistical Analysis. Addison-Wesley, Reading.
Jeffreys, H. 1961. Theory of Probability. Oxford University Press, London.
Mandelbrot, B.B. (1977). Fractals: Form, chance, and dimension. San Francisco, Freeman.

MAXIMUM ENTROPY CALCULATIONS ON A DISCRETE PROBABILITY SPACE

P. F. Fougere
AFGL/LIS
Hanscom AFB, Bedford, MA

To Ed Jaynes, who started it 30 years ago and whose
clarity of exposition is an inspiration to us all.

I. The Maximum Entropy Principle

In a remarkable series of papers beginning in 1957, E. T.
Jaynes (1957) began a revolution in inductive thinking with his
principle of maximum entropy. He defined probability as a degree of
plausibility, a much more general and useful definition than the
frequentist definition as the limit of the ratio of two frequencies in
some imaginary experiment. He then used Shannon's definition of
entropy and stated that in any situation in which we have incomplete
information, the probability assignment which expresses all known
information and is maximally non-committal with respect to all unknown
information is that unique probability distribution with maximum
entropy (ME). It is also a combinatorial theorem that the unique ME
probability distribution is the one which can be realized in the
greatest number of ways. The ME principle also provides the fairest
description of our state of knowledge. When further information is
obtained, if that information is pertinent then a new ME calculation
can be performed with a consequent reduction in entropy and an
increase in our total information. It must be emphasized that the ME
solution is not necessarily the "correct" solution; it is simply the
best that can be done with whatever data are available. There is no
one "correct solution", but an infinity of possible solutions. These
ideas will now be made quite concrete and expressed mathematically.

(a) Discrete Probability Space.

We have n propositions or statements, S_1, S_2 . . . S_n, each
of which can be assigned a probability p_i, i = 1,n. The number p_i
runs from zero when our information tells us that S_i is not true to
one when we assume that S_i is true. In the case of a die, S_i might be
the proposition that on the next throw of the die face i will be up.
If the die has not yet been cast then our belief that face i will come
up next is described by assigning a number to p_i. If the die were
perfectly symmetric and thrown in a fair way, making no attempt to
favor any face, then every face would be equally likely to occur and
then since one of them must occur, the probability of the statement
"some i will occur" is 1. Thus the probabilities would each be set
to 1/n; in the case of a die (p_i = 1/6, i=1,6). This is a simple

205

expression of Laplace's "principle of insufficient reason " which has
been attacked by many but has never been replaced. It is essentially
a symmetry principle. If the mechanism of selecting a number at
random from the possible set of n is symmetric with respect to all
members of the set then the probability of each is 1/n. There are
many practical realizations of this mechanism of selection. All of
the resulting problems are isomorphic and all can be solved in
precisely the same way.

1.   There are n distinguishable but otherwise identical objects
numbered 1, 2, ...n in an opaque container. An experiment consists
of selecting an object, noting its number and replacing the object in
the container.

2.   A roulette wheel containing 36 numbered slots is spun and a small
ball is set in motion in the opposite direction. When both wheel and
ball slow down sufficiently the ball drops into one of the slots. The
number is recorded.

3.   An ordinary 6 sided die is thrown. The number of spots facing up
is recorded.

4.   A deck of 52 playing cards is shuffled face down. A card is
selected and its value noted.

Note that there may be <u>bias</u> introduced either accidently or
deliberately (to cheat) in any of these games. But also note that if
the bias (a favoring of any outcome over the others) becomes large
enough, the players of the game will almost certainly notice, with
retribution to the perpetrator soon to follow. Cheats at poker, craps
(dice) and roulette have often met an untimely end!

We will soon see that the ME method is admirably suited to detecting
such biases, even very tiny ones. Every time a correctly calculated
ME probability distribution fails to reproduce an observed frequency
distribution accurately enough, the conclusion can be drawn that a
bias which has not yet been taken into account is operating. In just
this way was quantum mechanics discovered!

The principle of insufficient reason will be derived as the maximum
entropy <u>assignment</u>: given only an enumeration of the possibilities
and normalization:

<u>and nothing else.</u>
$$\Sigma\ p_i = 1, \qquad\qquad (1)$$

Throughout this article, sums on i will always run from 1 to n, and
for simplicity of notation the limits will not be typed. The ME
probability distribution given only the above information is (p_i =
1/n, i=1, 2 ...n). This statement will be proved in Section b.[i] This

expresses exactly the known information and nothing more. Any
subsequent information which is provided, for example: "the die is
not symmetric", will lower the entropy and change the probabilities
accordingly.

    (b)   Entropy.
    In his wonderful little book on information theory Shannon
(1948) first set forth the axioms or elementary desiderata of
consistency as follows: if S is the measure of information or
uncertainty and p_i = probability of the i'th outcome:

1. $S = S(p_1, p_2, \ldots p_n)$

The information depends upon the entire set of probabilities.

2. If all p_i are equal then S is a monotone increasing function of n.
With more possibilities to choose from the information in a choice is
greater.

3. S is additive for compound independent events. If events A and B
are independent, $S(AB) = S(A) + S(B)$. The information contained in
the statement "it is raining and today is Tuesday" is exactly equal
to the information contained in the statement "it is raining" plus the
information contained in the statment "today is Tuesday".

4. S does not depend upon how the problem is setup. See Figure 1.

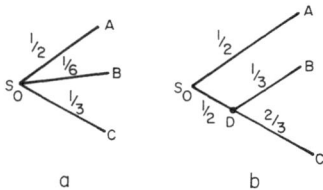

a   b

Figure 1. Two sets of probability assignments. In 1a there are three
events A, B, C with probabilities 1/2, 1/6, 1/3 respectively. In 1b
the final state A, B, C is reached via an intermediate state D with
probability 1/2. The information in both diagrams at stage A, B, C
must be the same.

The information in the probability assignment A = 1/2, B = 1/6, C =
1/3 in Figure 1a must be the same as that in Figure 1b where we have
used the intermediate point D.

Shannon then proved [see also Tribus (1961, 1969)] that this measure of information has the form:

$$H = - K \sum p_i \log p_i \qquad (2)$$

and furthermore that this functional form is unique: it is the only form capable of satisfying the four axioms. The constant K is merely a scale factor and the base of the logarithm is arbitrary; for convenience the constant K is set to 1 and the base of the logarithm is taken to be natural. Thus we have:

$$H = - \sum p_i \ln p_i \qquad (3)$$

Since the p_i are all in [0,1], $H \geq 0$, if we agree that $0 \ln 0 = 0$, (a proposition which has zero probability conveys no information). As an elementary exercise let us prove that the probability assignment with maximum entropy is one with $p_i = 1/n$.

We have $\sum p_i = 1,$ $H = - \sum p_i \ln p_i \qquad (4)$

Form the expression $Q = - \sum p_i \ln p_i + \lambda (\sum p_i - 1)$

Where λ is a Lagrange multiplier used to enforce normalization.

Now differentiate with respect to p_1:

$$\frac{\partial Q}{\partial p_j} = - (\ln p_j + 1) + \lambda = 0$$

thus $\ln p_j = \lambda - 1$

then $p_j = \exp (\lambda - 1) \qquad (5)$

But this is independent of j. Thus all p_j are equal and by normalization they sum to 1; therefore $p_j = 1/n$, j=1,n. Thus with only an enumeration of the possibilities which are exhaustive (one must occur) and exclusive (only one can occur) and normalization, the probability assignment which maximizes the entropy brings us back to Laplace's principle of insufficient reason. Any further information would change the probabilities and lower the entropy. We do not need Laplace's principle of insufficient reason; entropy maximization subject only to normalization produces Laplace's principle as a theorem or result.

(c)   Maximum Entropy Formalism.
          Since we will be maximizing entropy under a variety of
constraints, it is helpful to have "cookbook recipe" or a "crank to
turn".

In addition to normalization (Eq. 1) we may have M constraints in the
form of expectation values or averages in the form:

$$\Sigma \; p_i \; f_m \; (x_i) \; = \; <f_m> = \; F_m, \quad m = 1, 2 \dots M \qquad (6)$$

We use the calculus of variations now and take variations of our
important equations 4 and 6 to get:

$$\delta H = - \Sigma \, (1 + \ln p_i) \, \delta p_i = 0$$

$$(\lambda_0 - 1) \, \Sigma \, \delta p_i = 0$$

$$\Sigma_m \lambda_m \, \Sigma_i \, f_m \, (x_i) \, \delta p_i = 0 \qquad (7)$$

$\lambda_0, \lambda_1 \dots \lambda_M$ are, of course, Lagrange multipliers. Now add the
three equations and factor δp_i:

$$\Sigma_i \left[1 + \ln p_i + \lambda_0 - 1 + \Sigma_m \lambda_m \, f_m \, (x_i) \right] \delta p_i = 0 \qquad (8)$$

For any arbitrary variation, δp_i, the expression in brackets must
vanish for every value of i. Solving for $\ln p_i$ we get

Thus
$$\ln p_i = - \lambda_0 - \Sigma_m \lambda_m \, f_m \, (x_i)$$

$$p_i = \exp \left[- \lambda_0 - \Sigma_m \lambda_m \, f_m \, (x_i) \right] \qquad (9)$$

Now for normalization we have that

$$\Sigma \, p_i = 1 = \Sigma_i \, \exp \left[- \lambda_0 - \Sigma_m \lambda_m \, f_m \, (x_i) \right] \qquad (10)$$

Solving for $\exp (\lambda_o)$, which we call the partition function Z:

$$Z = \exp (\lambda_0) = \Sigma_i \, \exp \left[- \Sigma_m \lambda_m \, f_m \, (x_i) \right] \qquad (11)$$

Taking logs of both sides

$$\lambda_0 = \ln \Sigma_i \, \exp \left[- \Sigma_m \lambda_m \, f_m \, (x_i) \right] \qquad (12)$$

Thus λ_0 is the log of the partition function Z; for reasons which will become clear immediately we call λ_0 the potential function.

Now differentiate λ_0 with respect to r

$$\frac{\partial \lambda_0}{\partial \lambda_r} = \frac{- \sum_i f_r(x_i) \exp\left[- \sum_m \lambda_m f_m(x_i)\right]}{\sum_i \exp\left[- \sum_m \lambda_m f_m(x_i)\right]} \quad (13)$$

Multiply numerator and denominator by $\exp(-\lambda_0)$

Then

$$\frac{\partial \lambda_0}{\partial \lambda_r} = \frac{- \sum_i f_r(x_i) \exp\left[- \lambda_0 - \sum_m \lambda_m f_m(x_i)\right]}{\sum_i \exp\left[- \lambda_0 - \sum_m \lambda_m f_m(x_i)\right]} \quad (14)$$

Now notice from Eq. 9 that the exponential of the bracketed term in numerator and denominator is just the probability p_i. Thus

$$\frac{\partial \lambda_0}{\partial \lambda_r} = \frac{- \sum_i f_r(x_i) p_i}{\sum_i p_i} = -<f_r> \quad (15)$$

We now see that λ_0 is called the potential function because the constraints are given as derivatives of λ_0 with respect to all the other λ's.

For convenience we now summarize the important formulas:

$$Z = \sum_i \exp\left[- \sum_m \lambda_m f_m(x_i)\right]$$

$$\frac{\partial \ln Z}{\partial \lambda_m} = -<f_m> \quad (16)$$

$$p_i = \exp\left[- \sum \lambda_m f_m(x_i)\right] / Z$$

We have exactly one Lagrange Multiplier λi for each constraint and we determine the set of λ's by solving the MXM set of equations

$$\frac{\partial}{\partial \lambda_m} \ln Z (\lambda_1, \lambda_2, \ldots, \lambda_M) = -F_m \qquad (17)$$

Finally the probabilities are given by:

$$p_i = 1/Z \exp \left[-\lambda_1 f_1(x_i) - \lambda_2 f_2(x_i) \ldots - \lambda_M f_M(x_i) \right] \qquad (18)$$

We can seee immediately that $\sum p_i = Z/Z=1$ and thus the formalism automatically produces a normalized set of p_i.

II. Wolf's Dice Data

To make the foregoing ideas as concrete as possible we will now examine in detail a remarkable series of experiments performed about 100 years ago by the Swiss scientist Rudolf Wolf who is known well for his work on sunspots. One of the experiments, reported by Czuber(1908), consisted of throwing a pair of dice, one red, the "ROTER WÜRFEL" and the other white, the "WEISSER WÜRFEL", a total of 20,000 times. The dice were thrown carefully in such a way as to avoid as much as possible introducing any bias, any artificial favoring of any of the 6 sides. Evidently (as we shall see) the dice were made using ordinary care but not extraordinary care - they were in fact quite noticeably biased.

Ed Jaynes has written extensively on dice in general and on Wolf's dice data in particular in no less than four publications (1963a, 1978, 1979, 1982). I would urge the reader to look up and read this exciting scientific saga. I freely acknowledge my deep indebtedness to Ed Jaynes for my inspiration in writing this paper but of course any mistakes which I may have made in interpretation, emphasis, algebra or arithmetic are mine alone.

Table I lists the totals obtained by Wolf for the 36 distinct possibilities - that is: white 1 red 1; white 1 red 2; . . . up to white 6 red 6.

Table I Wolf's Dice Data:

		Weisser Würfel						RM	RF
	NR.	1	2	3	4	5	6		
	1	547	587	500	462	621	690	3407	0.17035
	2	609	655	497	535	651	684	3631	0.18155
Roter Würfel	3	514	540	468	438	587	629	3176	0.15880
	4	462	507	414	413	509	611	2916	0.14580
	5	551	562	499	506	658	672	3448	0.17240
	6	563	598	519	487	609	646	3422	0.17110
WM		3246	3449	2897	2841	3635	3932	20,000	
WF		.16230	.17245	.14485	.14205	.18175	.19660		1.0

RM and WM are the red and white marginals, respectively.
RF and WF are the red and white relative frequencies, respectively.

Since there is no evidence for and no reason to expect that the two
dice were correlated, the results for the white die are independent of
those for the red die, and Table I also lists the white marginals, the
total number of times that the white die came up a given number of
spots independent of which red spot was showing. Similary the red
marginals are listed. It can be seen at once that the dice were
indeed biased; for example W6 appeared 3932 times, almost 600 times
more than expected if the die were fair; W4 appears only 2841 times,
492 times less than expected. The relative frequencies given in Table
I are just the marginals divided by 20,000.

a. The White Die
Let us now, following Ed Jaynes, try to account for some
of the discrepancies or biases using ME. At this point, it is
important to know what a conventional playing "die" is. It is a solid
cubical object, made of a machineable substance such as ivory.
Hemispherical depressions or excavations (spots) are made
symmetrically in each face, with the number of spots on opposite faces
totaling 7. The spots are painted in a contrasting color. Thus 1 is
opposite 6, 2 opposite 5 and 3 opposite 4. If face 6 is "up" and face
2 is visible, then face 4 is to the right of face 2. The reader's
intuition will be aided by actually examining a real die.

1. One constraint. The most obvious physical asymmetry is now
apparent. Whereas six spots are removed from face 6 only one is
removed from face 1 and thus the center of gravity of the die is
shifted very slightly toward the 1 face. Similarly the 2 and 3 faces
are slightly heavier than their opposites 5 and 4 respectively.
Quantitatively, the center of gravity will be shifted toward the "3"
face by small distance ϵ corresponding to a one-spot discrepancy.
Similarly the center of gravity will be shifted toward the "2" face by
3 ϵ and towards the "1" face by 5 ϵ. Thus the spot frequencies should

be shifted proportionally (frequency shift = a times center of gravity shift = $\alpha\epsilon$). Then the spot frequencies should vary linearly with i:

$$g_i = 1/6 + \alpha\epsilon\, f_1\,(i) \tag{19}$$

Where $f_1\,(i) = i-3.5$.

Thus the expected number of spots would be shifted to (all of the sums on i will now run from 1 to 6.)

$$<i>= \Sigma\, i\, g_i = 3.5 + 17.5\,\alpha\epsilon \tag{20}$$

or the function $f_1(i)$ has a non-zero expectation:

$$<f_1> = 17.5\,\alpha\epsilon \tag{21}$$

We note by calculating from Table I that the average number of spots showing on the white die was 3.5983. This was larger than 3.5 as expected on the physical grounds just discussed and not equal 3.5 as would have been expected from a fair die. Let us use this one piece of information as a constraint and find the six p_i's which yield maximum entropy. The complete statement of the problem at this stage is: We are given 1: an enumeration of the possibilities, namely i = 1,2,3,4,5,6 and 2 : $<i>$ = A and nothing else. It is thus simpler to use h (x_i) = i as constraint function, rather than $f_1(x_i)$ = i -3.5, because we are given the average value of h = A. The ME equations become:

$$Z = \Sigma\, \exp\, \lambda h(x_i); \quad h(x_i) = i;$$
$$\Sigma\, p_i\, h(x_i) = \Sigma\, i\, p_i = A \tag{22}$$

Let   y = exp (λ)

$$Z = \Sigma\, (\exp\,(\lambda))^i$$
$$= \Sigma\, y^i = y + y^2 + y^3 + y^4 + y^5 + y^6 \tag{23}$$

$$\frac{\partial\, \ln Z}{\partial\, \lambda} = y/Z\left[1 + 2y + 3y^2 + 4y^3 + 5y^4 + 6y^5\right] = A \tag{24}$$

Expanding and simplifying we get:

$$(1 - A) + (2 - A)y + (3 - A)y^2 + (4 - A)y^3$$
$$+ (5 - A)y^4 + (6 - A)y^5 = 0 \tag{25}$$

This 5'th degree equation has one real root; Table II gives the value of the real root y versus the average A. Here we have used the IMSL subroutine "ZPOLY".

Table II. Root of Eq. 20 (y) versus average value (A).

A	y	A	y	A	y	A	y	A	y
1.0	.000000	2.0	.532820	3.0	.839769	4.0	1.190804	5.0	1.876805
1.1	.090912	2.1	.565943	3.1	.870434	4.1	1.235307	5.1	2.006740
1.2	.166756	2.2	.597991	3.2	.901644	4.2	1.282800	5.2	2.164185
1.3	.231313	2.3	.629215	3.3	.933540	4.3	1.333821	5.3	2.360807
1.4	.287438	2.4	.659827	3.4	.966271	4.4	1.389030	5.4	2.616096
1.5	.337239	2.5	.690010	3.5	1.000000	4.5	1.449254	5.5	2.965257
1.6	.382249	2.6	.719927	3.6	1.034906	4.6	1.515549	5.6	3.479017
1.7	.423584	2.7	.749726	3.7	1.071191	4.7	1.589282	5.7	4.323151
1.8	.462068	2.8	.779545	3.8	1.109085	4.8	1.672267	5.8	5.996777
1.9	.498321	2.9	.809516	3.9	1.148853	4.9	1.766964	5.9	10.999661

For Wolf's white die, we had $A = 3.5983$ giving $y = 1.034302$, $Z = 6.76292$. The ME probabilities are $p_i = y^i/Z$ and are given in Table III.

Table III. Wolf's dice data with one constraint (white die)

i	g_i	p_i	$\Delta_i = g_i - p_i$	C_i
1	0.16230	0.15294	0.0094	11.46
2	0.17245	0.15818	0.0143	25.75
3	0.14485	0.16361	− 0.0188	43.02
4	0.14205	0.16922	− 0.0272	87.25
5	0.18175	0.17502	0.0067	5.18
6	0.19660	0.18103	0.0156	26.78
				199.43

g_i are the relative frequencies (WF) from Table I.

p_i are the ME probabilities based on the constraint: $A = \langle i \rangle = 3.5983$.

$C_i = 20,000_2(g_i - p_i)^2/p_i$ = Partial contribution to Chi^2. The critical value: $Chi^2_c (0.05) = 9.49$ on 4 degrees of freedom. The concept of degrees of freedom will be discussed later.

Examining Table III carefully we see that the deviations, $\Delta_i = g_i - p_i$ between observed relative frequencies, g_i, and ME probabilities, p_i, are negative for faces 3 and 4 and positive for faces 1, 2, 5, 6 and the C_i tell us that these deviations are highly significant. This does not mean that ME has failed but that there is a further physical constraint. At this point in Jaynes' paper he again demonstrates his genius as a practical working physicist, who as Enrico Fermi did, now delights in going into the machine shop to make things work. Jaynes explains to us just how to turn a lump of ivory into as perfect a cube as possible. A milling machine used by an expert would have no trouble in cutting 5 sides of the die all accurately plane with all angles accurately $90°$ and the top face accurately square. But then the die would have to be removed from the machine and turned upside down to finish to final face. It would be extremely difficult to ajust the work table height so that the final dimension is exactly equal to the other two: The result of the difficulty would be a die which is either: (i) slightly "oblate" with one dimension shorter than the other two or (ii) slightly prolate with one dimension slightly greater than the other two. Of course either type of imperfection would constitute a "constraint" and would change the relative frequencies.

2. Two Constraints. We can now see, quite clearly, that the white die must have been <u>prolate</u> with the 3 - 4 dimension being slightly greater than the 1 - 6 and 2 - 5 dimensions! See Figure 2 for an exaggerated sketch of a prolate die. Such a die is more likely to fall "flat" with a 1, 2, 5 or 6 showing and thus frequencies cf 3 and 4 spots would be lower than the frequencies of 1, 2, 5 or 6 spots.

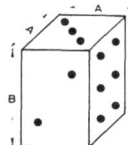

Figure 2. A prolate die with the 3-4 (top - bottom) dimension B slightly larger than the other two equal dimensions A (1-6 and 2-5).

Suppose that the 3 - 4 dimension were greater than the other two by an amount δ. This would increase the frequencies g_1, g_2, g_5, g_6 by a proportional amount: $\beta\delta$ and decrease the frequencies g_3 and g_4 by an amount $2\beta\delta$ (this preserves normalization).

Thus we now define a new constraint function:

$$f_2 (i) = 1, 1, -2, -2, 1, 1, \tag{26}$$

and we find
$$\langle f_2 \rangle = \Sigma\, g_i\, f_2 (i) = g_1 + g_2 - 2(g_3 + g_4)$$
$$+ g_5 + g_6 = 0.1393 \tag{27}$$

from Wolf's data on the white die given in Table I. We will have two Lagrange multipliers and the partition function Z will now be:

$$Z (\lambda_1,\, \lambda_2) = \Sigma\, \exp\left[-\, \lambda_1 f_1 (i) - \lambda_2 f_2 (i) \right] \tag{28}$$

where $f_1(i) = i - 3.5$ from Eq. 19 and $f_2(i)$ is given in Eq. 26.

letting $x = \exp(-\lambda_1)$; $y = \exp(-\lambda_2)$

Then $Z(\lambda_1, \lambda_2) = x^{-5/2} y (1 + x + x^2 y^{-3} + x^3 y^{-3} + x^4 + x^5)$

We now have two constraint equations:

$$Z F_1 - x \frac{\partial Z}{\partial x} = 0, \quad Z F_2 - y \frac{\partial Z}{\partial y} = 0 \qquad (29)$$

These yield two coupled equations in x and y:

$$(2F_1+5) + (2F_1+3) x + (2F_1+1) x^2 y^{-3} +$$
$$(2F_1-1) x^3 y^{-3} + (2F_1-3) x^4 + (2F_1-5) x^5 = 0 \qquad (30)$$

and $\quad (F_2-1) (1+x+x^4+x^5) + (F_2+2) (x^2+x^3) y^{-3} = 0$

The IMSL library now comes to our aid with a very nice subroutine ZSPOW, which solves n simultaneous non-linear equations in n unknowns. For x and y we get 1.03223 and 1.07442 and the resulting ME probabilities are given in Table IV. $Z = 6.08530 \, x^{-2.5} \, y$.

Table IV Wolf's dice data with two constraints (white die)

i	g_i	p_i	$\Delta_i = g_i - p_i$	c_i
1	0.16230	0.16433	-0.00203	0.50
2	0.17245	0.16963	0.00282	0.94
3	0.14485	0.14117	0.00368	1.91
4	0.14205	0.14573	-0.00368	1.85
5	0.18175	0.18656	-0.00481	2.48
6	0.19660	0.19258	0.00402	1.68
				9.37

See the footnote for Table III. Chi^2_c (0.05) on 3 degrees of freedom is 7.81.

Table IV agrees with Ed Jaynes' results except that he used 5 degrees of freedom and the critical value of Chi^2 at the 5% level is 11.07. He thus concluded that "there is now no statistically significant evidence for any further imperfection. . ..". In a later paper Jaynes (1979) discusses the number of degrees of freedom he should have been using and concludes unequivocally that the correct formulation is:

$$df = n - 1 - m$$

where df = number of degrees of freedom in Chi^2, n = number of
possibilities (= 6 for a die) and m = number of constraints. We
subtract one more for normalization. Simply put, the number cf
degrees of freedom is the number of independent values of the
probability which can be assigned. In the case of two constraints
plus normalization (essentially three constraints) we could assign
only three probabilities lying on the range 0 to 1 and then the other
three would be uniquely determined.

Thus we see that for the white die there is still a statistically
significant (at the 95% level) imperfection not explained by misplaced
center of mass or oblateness. Jaynes 1979 says now that: "To assume
a further very tiny imperfection [(the 2-3-6) corner chipped off] we
could make even this discrepancy disappear; but in view of the (great)
number of trials one will probably not consider the result as
sufficiently strong evidence for this." The word "great" probably was
intended to be "small".

Let us disagree midly with Jaynes at this point and actually look for
this tiny third imperfection.

3.   Three Constraints.   Figure 3 gives a sketch of a die with the
imperfection suggested by Jaynes.

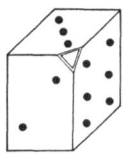

Figure 3.   A die with a small chip broken off the 2, 3, 6 corner.
Such an imperfection would tend to increase the probability of the die
landing with the 2, 3, or 6 face showing "up".

By shifting its center of gravity, such a die would slightly favor the
2, 3 and 6 faces. Let us express this constraint as:

$$f_3 (i) = -1, 1, 1, -1, -1, 1 \tag{31}$$

Table V summarizes all three constraints we are now considering and
attempts to simplify the algebra.

Let $w = \exp (- \lambda_3)$

Table V Summary of the three constraints

Factor out

i	$f_1(i)$	$f_2(i)$	$f_3(i)$	Contribution to Z	$x^{-2.5}y\,w^{-1}$
1	-2.5	1	-1	$x^{-2.5}y\,w^{-1}$	1
2	-1.5	1	1	$x^{-1.5}y\,w$	$x\,w^2$
3	-0.5	-2	1	$x^{-0.5}y^{-2}\,w$	$x^2\,y^{-3}\,w^2$
4	0.5	-2	-1	$x^{0.5}\,y^{-2}\,w^{-1}$	$x^3\,y^{-3}$
5	1.5	1	-1	$x^{1.5}\,y\,w^{-1}$	x^4
6	2.5	1	1	$x^{2.5}\,y\,w$	$x^5\,w^2$

Since the algebra gets a little tedious and mistakes are likely, the
use of such a table is recommended in general. As a footnote,
programs capable of simple algebra and differential calculus exist
now. Use of such programs would be really beneficial. The three
non-linear coupled equations for the constraints are now:

$$(2F_1+5) + (2F_1+3)\,x^2 w^2 + (2F_1+1)\,x^2 y^{-3} w^2 +$$
$$(2F_1-1)x^3 y^{-3} + (2F_1-3)x^4 + (2F_1-5)x^5\,w^2 = 0. \tag{32}$$
$$(F_2-1)\,(1+xw^2 + x^4 + x^5 w^2) + (F_2+2)\,x^2\,y^{-3}\,(w^2+x) = 0.$$
$$(F_3+1)\,(1+x^3 y^{-3}+x^4) + (F_3-1)\,w^2\,x\,(1+xy^{-3} + x^4) = 0.$$

With values: $F_1 = 0.0983$; $F_2 = 0.1393$; $F_3 = 0.0278$ the three coupled
equations can be solved to give $x = 1.03072$; $y = 1.07425$; $w = 1.02159$
and $Z = 6.196106\,x^{-2.5}\,y\,w^{-1}$. Thus we get Table VI summarizing the
resulting maximum entropy probabilities.

Table VI. Wolf's dice data with three constraints (white die)

i	g_i	P_i	$\Delta_i = g_i-P_i$	C_i
1	.16230	.16139	.00091	0.10
2	.17245	.17361	$-.00116$	0.16
3	.14485	.14434	.00051	0.04
4	.14205	.14256	$-.00051$	0.04
5	.18175	.18215	$-.00040$	0.02
6	.19660	.19594	.00066	0.04
				$\overline{0.39}$

See footnote to Table II. Chi_c^2 on 2 degrees of freedom is 5.99.

The agreement between the observed <u>frequencies</u> g_i and the maximum entropy <u>probabilities</u> p_i is now essentially perfect. In fact it is too good! The agreement is much better than would be expected if Wolf's experiments had been repeated many times. The observed frequencies in many sets of experiments, each 20,000 tosses long, would differ from each other by much more than the g_i-p_i from Table VI. Jaynes (1978) calculates that the fluctuations in the observed frequencies ought to be of order $(g_i/N)^{1/2}$. For $g_i = 1/6$, $\Delta g_i \sim$ 0.003. All of the deviations g_i-p_i in Table VI are smaller than this and all but g_2-p_2 are about an order of magnitude smaller. Nevertheless, looking at Table IV again, with only two constraints, four of the deviations are larger than 0.003. In summary the observed frequencies for the white die can be completely explained by three physical constraints:

> The largest is No 2, the oblateness,
> The next largest is No 1, the center of gravity shift
> by spot removal and:
> The smallest is a tiny chip off the 2 - 3 - 6 corner.

The first two are required - the evidence for them is overwhelming. The evidence for the third is much weaker. From Table IV again, for two constraints, $Chi^2 = 9.37$ which is just significant at the 95% level but not significant at the 97.5% level.

<u>Further thoughts on the white die</u>. The computer program which solves the three constraint problem has been generalized (quite simply) to solve all of the imbedded problems:
> No constraints
> any one of the three acting by itself
> any two acting together
> all three.

The first case is trivial and reduces to $p_i = 1/6$. The last case has just been described. We summarize the results of all cases in Table VII.

Table VII. Chi squared for the white die. 1 = constraint on; 0 = off.

	Constraints			
No. 1	No. 2	No. 3	df	Chi Square
1	1	1	2	0.39
1	1	0	3	9.37
0	1	1	3	56.28
0	1	0	4	72.01
1	0	1	3	189.77
1	0	0	4	199.42
0	0	1	4	253.85
0	0	0	5	270.96

In summary the most important single constraint is No. 2 (oblateness) the next important single is No. 1 center of gravity shift and the least important single is No. 3, corner chip. The best 2 constraints are 1 and 2 acting together followed by 2 and 3 and then 1 and 3. As a final footnote it is not sufficient to set one of the F's and its λ equal to zero and then solve the three equations. The equation for the inactive constraint must be dropped altogether and the corresponding λ set to zero. This has been done in the program.

b. The Red Die

To the best of my knowledge no one has ever attempted a complete analysis of the red die but with a simple program in place it becomes a trivial task to see if the same kind of thinking works just as well in this case. It had better! But we must be quite careful because although we expect similar kinds of asymmetries they need not be identical.

1. One Constraint. The first constraint as in the case of the white die, simply requires the average spot number. For the red die this value is: $\langle i \rangle = 3.49165$ which is less than 3.5. Even though this is less than 3.5 and not greater than 3.5 as expected we run the ME calculation with the one constraint:

$$\langle i-3.5 \rangle = -0.01835.$$

We get $x = 0.993728$ and $Z = 5.86966$. The ME probabilities are given in Table VIII.

Table VIII. Wolf's dice data with one constraint (red die)

i	g_i	P_i	$\Delta_i = g_i - p_i$	C_i
1	.17035	.16930	.00105	.13
2	.18155	.16824	.01331	21.07
3	.15880	.16718	−.00838	8.41
4	.14580	.16613	−.02033	49.77
5	.17240	.16509	.00731	6.47
6	.17110	.16406	.00704	6.05
				91.90

See footnotes to Table III.

Looking at $\Delta_i = g_i - p_i$ from Table VIII we see at once that Δ_3 and Δ_4 are negative while the others are all positive. This is precisely the same situation we found in Table III for the white die. The red die is also prolate in exactly the same way as the white die! This situation is not really as bizarre as might first be thought. Given that the die maker was prone to err on the prolate side, the only real coincidence is in the numbering of the faces. If he started his numbering (carving of spots) at the one spot he would be twice as likely to start with one of the four faces which are a short distance apart as on either of the two "long" faces. Having done so, the two spots would be on short faces just as often as on a long face. Don't forget that once a one spot has been carved, the six must be on the opposite face. Thus the appearance of identical asymmetries on the two dice is not very surprising at all.

2. Two Constraints. We may now use the same program again to incorporate the first two constraints with values $F_1 = \langle f_1 \rangle = -0.01835$; $F_2 = \langle f_2 \rangle = 0.0862$. We get $x = 0.993965$; $y = 1.04508$; $Z = 5.66614 x^{-2.52} y$ and Table IX gives the resulting probabilities.

Table IX Wolf's die data with two constraints (red die)

i	g_i	p_i	$\Delta_i = g_i - p_i$	C_i
1	.17035	.17649	−.00614	4.27
2	.18155	.17542	.00613	4.28
3	.15880	.15276	.00604	4.77
4	.14580	.15184	−.00604	4.80
5	.17240	.17227	.00013	0.00
6	.17110	.17123	−.00013	0.00
				18.13

See footnotes to Table III.

3. Three Constraints. We see here a tremendous improvement with an added bonus. Now that we have removed, by ME, the effects of the first two constraints, a third, smaller, but significant, constraint is now very obvious. Sides 5 and 6 have been fit very well indeed and the other four discrepancies are all of the same magnitude but with two plus signs and two minus signs. A possible physical explanation will be discussed later but the constraint to use now instead of the third constraint we used for the white die is:

$$f_3 (i) = -1,\ 1,\ 1,\ -1,\ 0,\ 0 \qquad \textbf{(33)}$$

we now modify the master program slightly to accomodate this new constraint. Once again we can solve all of the imbedded problems. Table X shows the results.

Table X. Chi squared for the red die. 1 = constraint on: 0 = off.

| Constraint | | | Degrees of | |
1	2	3	freedom	Chi Square
1	1	1	2	0.08
0	1	1	3	2.40
1	1	0	3	18.13
0	1	0	4	20.44
1	0	1	3	74.86
0	0	1	4	77.16
1	0	0	4	91.90
0	0	0	5	94.19

Summarizing our results for the red die we have seen, that:

The red die was no more fair than the white die.

The excavation of spots and the subsequent shift of the center of gravity was not an important constraint for this die as it was for the white die. Other (unknown) compensatory constraints must have been at work.

The red die was oblate in essentially the same way that the white die was. For both dice this was the most important constraint.

There was no evidence of a corner chip here as there was for the white die but a constraint of the mathematical form $-1, 1, 1, -1, 0, 0$ was operating. No simple physical explanation seems in order but perhaps two simple constraints were acting in concert. A small wear spot on the $2 - 3$ edge and a small excess of material on the $1 - 4$ edge would make 2 and 3 more likely and 1 and 4 less likely.

After removing the most important constraint (oblateness) the misfit as expressed by $Chi^2 = 20.44$ is quite significant. Critical value Chi^2 on 4 df is 9.5 at 5% level.

When constraints number 2 and 3 are used together Chi^2 drops way down to 2.40 and the agreement between the observed frequencies g_i and the ME probabilities p_i is too good! Repetitions of the 20,000 toss experiment would very likely produce departures larger than the Δ_i obtained from these two constraints.

The final conclusion from our exhaustive analysis of the two dice is that the maximum entropy principle allows us to discover physical imperfections in a pair of dice from data over 100 years old. At least as far as real dice are concerned, the principle of ME works and works brilliantly!

III. Published Criticisms
There have been many published papers which criticize the maximum entropy principle in general and Jayne's treatment of dice experiments in particular. Most of these attacks have been answered in the literature, some of them many times.

a. Older Criticism
For some of the earlier criticism see for example the paper by Rowlinson (1970) and Jaynes's (1978) answer. For a particularly virulent set of attacks see Friedman and Shimony (1971) and for defenses see Jaynes (1978) p 53, Tribus and Motroni (1972) Gage and Hestenes (1973) and Hobson (1972). See also Friedman (1973) and Shimony (1973) for their replies.

b. Frieden's Paper
The latest adventure in "anti-maximum-entropism" comes from B. Roy Frieden (1985) who professes to be "quite happy with (his) empirical results" using the maximum entropy formalism. The careful reader of Frieden's "Dice, Entropy and Likelihood" hereinafter referred to as DEL, might take pause at some of the statements to be quoted now.

Statement 1:

> "For example, this author originally believed ME to provide a maximum probable answer. However, at least for photon images, this is usually wrong. ...... Or, if it were required to estimate the most probable roll occurrences for an unknown die, the die would have to be known A priori to be fair, a rather restrictive assumption."

Wolf's dice were not fair. A priori, there is no requirement for fairness.

Statement 2:

> "Usually an engineer wants to know how probable his answer is, not how degenerate it is. The two concepts differ in general, and only coincide when every outcome has the same probability (i.e. when the die is fair."

The maximum-entropy die is fair only if there are no constraints acting besides normalization.

Statement 3:

> "The aim of this paper is to show that the die experiment just spoken of has solutions by classical, Bayesian estimation; that the probability of these solutions may be computed, as with any Bayesian problem; that therefore, there is no need to introduce a new

concept such as maximum entropy in this most basic of
problems; and that maximum entropy is not coincident with
these solutions. In fact maximum entropy not only gives
the wrong answer, it gives an answer that is very far
from right."

Note the glee in the last sentence. Note also that the entire purpose
of ME is to determine a prior probability assignment. This prior can
then be used in any subsequent Bayesian analysis.

Statement 4:

"We shall solve this problem in a purely classical way,
without the need for recourse to any exotic estimator,
such as ME."

Note the pejorative word "exotic".

Statement 5:

"As we shall see, the most valid objection to the use of
[Frieden's Eq.] (7) is that, although it describes
'maximum ignorance,' it does not describe the user's
state for a die in particular. The wrong experiment is
being performed to model maximum ignorance".

Frieden changes Jaynes' die problem brutally and then complains that
his new problem is not the right problem.

Statement 6:

"What this means is that we are not in a state of maximum
ignorance when given an unknown die. We know what to
expect a priori of its biases. For the particular case
of a die, a real one, it would be wrong to assume maximum
ignorance present. Hence, rolling a die is the wrong
experiment to use when attempting to model 'maximum
ignorance' situations. No wonder the result [Frieden's
Eq.] (17) goes against intuition."

Once again, Frieden, having changed the problem, complains that this
new problem is the wrong problem.

Statement 7:

"We suggest that in the past readers have been seduced
into a belief in ME principally because of this confusion
between what constitutes maximum ignorance on one hand,
and what constitutes the state of ignorance in a real die
experiment on the other. If you want maximum ignorance
do not consider a die experiment!"

Did you catch the truly pejorative word "seduced"?

Note in Statement 3, the use of the word "new" in connection with ME, and in Statement 4 the even more revealing word "exotic" which also appears again later. Note also the word "seduced" in Statement 7. A psychologist examining this paper might conclude that something other than pure scientific discourse is going on here. There is a pervasive feeling here that the author thinks he has found a fundamental flaw in the use of the ME principle and he is downright gleeful about it! Just reread Statement 3.

At this point we will examine the substance of the Frieden paper DEL.

Recall that in Jaynes' formulation of the problem, we are given:

> An enumeration of the possibilities,
>
> The average value of some linear constraint (e.g. the average spot values) measured in some previous experiment
>
> Normalization
>
> And nothing more.

In DEL, Frieden now changes the problem from that of a six sided real die to that of a three-sided imaginary die formed by combining rolls of one and six to yield one; two and five to yield two and three and four to yield three. He then calls the unknowns "biases" and labels them x_1, x_2, x_3. Then the real heart of the paper is introduced with Statement 8.

Statement 8:

> "By 'nothing' the user usually means that a priori every possible set of numbers x_1, x_2, x_3 (obeying normalization equation (1)) may be present with equal probability or frequency. Such a flat or uniform law is widely used in estimation problems. for example: when x_1, x_2, x_3 are the spatial coordinates of a material object whose location in a finite box is completely unknown a priori. Or, when a uniformly glowing planar image emits photons from unknown positions $(x,y) = x_1, x_2$. Or, when a distant aircraft of unknown coordinates (x,y) is being tracked; etc. This is also MacQueen and Marschak's (1975) definition of maximum ignorance, and we shall use it as well."

Here we go off the deep end! Frieden has changed an essentially discrete problem into an essentially continuous problem!

Recall the discussion in section Ia to the effect that Jaynes' die problem is _isomorphic_ to any number of essentially _discrete_ games, eg roulette, drawing a ball from a bag, drawing a card from a pack, etc. The essential features of these games are two in number: they are _discrete_ and there is a _symmetry_ principle operating. While small biases may be present in any of these games, large biases would be self defeating; they would be too easily detected. What the "user usually means" is, not only mathematically so vague as to be useless but also is completely irrelevent! Frieden can set up and attempt to solve any problem be choses. What he must not do is call his problem "Jaynes' problem"!

This Statement 8 changes Jaynes' problem by adding an enormous amount of information nowhere present in Jaynes' statement of the problem quoted above. Let us ask the question "how many bits would be required to encode the possible answers to Jaynes' problem"? Clearly for the three sided die, not even two bits would be necessary to encode the possible outcomes "1", "2" or "3". But if we are to take Statement 8 seriously we need another layer of information to discover which one of the infinite number of possible dice we are, in fact, shooting. Frieden, later in the paper, tries to simulate his continuous problem on a computer as follows:

Statement 9:

> "In other words, the prediction is that only roll outcomes 2 occurred! Actually this result can be explained in hindsight. Suppose we try to simulate the situation by repeatedly selecting sets of biases for a die, rolling the die, and only counting those biases which give rise to the required n. In this way, $p(x_1, x_2, x_3)$ is built up as a histogram, event by event. Let the biases be selected on a fine grid so that "every" triplet x_1, x_2, x_3 is sampled only once. This accomplishes the flat prior probability law [Frieden's Eq.] (7). Which such triplet will most often give rise to a value n = 2? It is obvious that the triplet (0,1,0) can only give rise to value n = 2."

Clearly B. Roy Frieden changed the problem – and drastically so. Frieden's problem now becomes: given an entire urn full of dice, all different, made very carefully by some imaginary machinist, so that each one will exhibit a different set of probabilities for the three faces. For a very crude set, with 11 possible probabilities for each face our patient die maker would manufacture 66 dice. Sixty-six is the number of normalizable triplets with a granularity of 0.1.

One real die for Jaynes, 66 imaginary dice for Frieden! And if
Frieden wanted 101 possible probabilities for each face, our die maker
would need to produce 5151 precisely carved dice! No wonder Frieden
further changed the problem so that our old fashioned real six sided
die lost half of its faces! Three - sided die indeed!

Now with our new three - sided die we are told that the average toss
in a previous experiment was 2.0. Frieden now goes through some
calculations to show that out of our urn containing a large number of
dice, we have indeed selected the rare die with probabilities 0, 1, 0!
Of course this screwball die would give an average toss of 2 - it had
no choice. It had zero entropy - it always showed a 2 because it had
to. Tossing this die yielded no new information, it couldn't. It was
always pointless to toss it at all. What an enormous constraint to
lower our entropy from a maximum to zero! Where in the original
statement of the problem by Jaynes did it ever say that any face was
impossible?

Frieden insists that his new problem represents a state of true
ignorance and that the one single real Jaynes' die does not. We do
not achieve a state of ignorance by making thousands of unnecessary
assumptions! What we do is put in an enormous amount of prior
information. Is it any wonder at all that Frieden's answer is wildly
different from Jaynes?

Returning to the question asked about how many bits would be required
for encoding the Frieden die, we see that we would first of all
require $\log_2$ (5151) or about 7 bits to encode the information "one die
out of 5151 dice has been selected".

Let us examine Frieden's Monte Carlo calculation in a little more
detail. If we use a granularity of 0.1 we will get 11 possible
"biases" or probabilities for each face for a total of $(11)^3 = 1331$
dice. Of the 1331 dice only 66 can be normalized and of the 66
permissable dice only 6 will yield an expectation value of 2.0. These
six have probabilities of (0,1,0), (.1,.8,.1) (.2,.6,.2), (.3,.4,.3),
(.4,.2,.4), (.5,0,.5). The middle member of this set (.3,.4,.3) is
the closest we can come to a "fair die" with probabilities
(1/3,1/3,1/3).

For a granularity of 0.01, there will be 101 possible biases for each
face (0., 0.01 ... 1.00). Thus there will be $(101)^3$ or 1,030,301
possible triplets, of which only 5151 can be normalized. From this
set, any single choice will occur with probability 1/5151.

Of these 5151 dice only 51 would yield an expectation value of 2.0.
These 51 would be (0.00,1.00,0.00), (0.01, 0.98, 0.01) ....(0.50,
0.00, 0.50). The closest to "fair" of any of these dice would be
(.33, .34, .33).

Not only does Frieden change Jaynes' discrete problem into a
continuous one to apply Bayes' Theorem, but he changes back to the
discrete case when he "explains" Jaynes' ME approach. He says:

Statement 10:

> "Jaynes' ME approach [Frieden's refs] to the die problem is
> as follows. Assume that N is <u>large enough</u> [Frieden's
> Emphasis] that the law of large numbers [refs] holds, so
> that the die biases can be well approximated by values g_i
> $= n_i/N$.".

Did Frieden ever read Jaynes' paper? Where does Jaynes ever talk
about N being large enough?

The only effect that N has is to determine the <u>variance</u> of the ME
probabilities, not the probabilities themselves $(p_i$, i = 1,n). In
fact in the same paper referenced by Frieden, Jaynes (1982) discusses
an experiment with only N = 50 throws of a die in which we were given
the average number of spots as 4.5 instead of 3.5 as expected from a
fair die. Rowlinson (1970) advocated a binominal distribution instead
of the ME distribution. We now quote Jaynes exactly: "Even if we
come down to N = 50, we find the following. The sample numbers which
agree most closely with (10, 16) while summing to N_k = 50 are $\{N_k\}$ =
$\{3,4,6,8,12,17\}$ and $\{N'_k\}$ = $\{0,1,7,16,18,8\}$ respectively. With such
small numbers, we no longer need asymptotic formulas. For every way
in which Rowlinson's binominal distribution can be realized, there are
exactly W/W' = (7!16!18!)/(3!4!6!12!17!) = 38,220 ways in which the
maximum–entropy distribution $\{N_k\}$ can be realized". In the above
statement, equations (10 and 16) are Jaynes' ME <u>probabilities</u> and
Rowlinson's binominal <u>probabilities</u> respectively.

c. <u>Musicus' Paper</u>
    The paper DEL by Frieden elicited a comment by Bruce
Musicus (1986). Musicus accepted the Frieden transmogrification of
Jaynes' discrete problem into the continuous problem we have already
discussed. But Musicus made the excellent point that is nowhere
mentioned in DEL that Frieden is discussing not <u>probabilities</u> but
<u>probability</u> densities. Musicus proceeded to integrate Frieden's
densities to generate marginal densities. With these marginal
densities Musicus makes the point that no single point estimate would
be at all useful or meaningful without a confidence region. Musicus
then finds several "unreasonable" point estimates which he calls:

Statement 1:

$$MAP - A: \quad x_1, x_2, x_3 = (0,1,0)$$

$$MAP - B: \quad x_1, x_2, x_3 = \begin{array}{l} (0,0.5,0), \quad \text{for N even} \\ (0,0,0) \text{ (sic) for N odd} \end{array}$$

We certainly agree with Musicus that these estimates are unreasonable.

Musicus adds:

Statement 2:

>"The fact that these point estimators all give radically
>different estimates is hardly surprising, given that the
>probability density in Frieden's problem is not unimodal,
>and is not strongly clustered around the center."

Musicus then proceeds to discuss Maximum Entropy as follows:

Statement 3:

>"Note that Maximum Entropy is thus justified for a problem
>involving known a priori biases x_1, x_2, x_3 and incomplete
>observation data (we only know the mean $\bar{n}$ of the throws
>of the dice, n_1, n_2, n_3) with asymptotically infinite
>numbers of throws N. Frieden's paper reverses the
>problem, asking for estimates of x_1, x_2, x_3 given the
>observation mean $\bar{n}$; it is not surprising that he gets a
>very different answer."

Fact: Using ME we are not given "a priori biases". It is the duty of
the ME caluclation to convert information – the given mean $\bar{n}$ – into a
probability distribution. No asymptotically infinite numbers of
throws are necessary. Frieden's paper doesn't reverse the problem at
all! Frieden changes an essentially discrete problem into an
essentially continuous problem. We agree with Musicus' last statement
"it is not surprising that Frieden gets a different answer".

    d.   Makhoul's Paper.
       The Frieden paper we have been discussing was first
pointed out to me at the Third ASSP Workshop on Spectrum Estimation
and Modelling in a paper entitled "Maximum Confusion Spectral
Analysis" by John Makhoul (1986). The content of this paper, which is
available in the proceedings, was not quite as whimsical as its title
suggested; at least two scientists in the audience seem to have been
convinced by its attacks on the ME method, one of which was a simply a
recounting of Frieden's paper. It was this presentation that
stimulated me to study the subject of Jaynes' die in depth and
ultimately to write this present paper. I am really indebted to John
Makhoul for the stimulation. The Makhoul paper was limited in length
to four pages of which only the first two are devoted to an
"explanation" of ME and to the dice problem. The concentration of
error per page in this paper is truly astounding!

Statement 1:

> "We assume that a random experiment has r possible events
> at each trial and that each event i, $1 \le i \le r$, has an a
> priori known probability x_i."

Fact: The prior probabilities are not known but <u>unknown</u>. The whole
point of ME is to determine a set of prior probabilities consistent
with all known information and maximally non-committal with respect to
everything else!

Statement 2:

> "Perhaps the greatest contributing factor to the confusion
> surrounding ME is the claim or allusion by some that ME
> provides a posterior estimate of the a priori
> probabilities x."

Fact: ME is used to determine the prior probabilities. No competent
ME practitioner, and certainly not Ed Jaynes, ever claims that ME
produces posterior probabilities. As in the die experiment a <u>sequence</u>
of ME calculations can produce sets of <u>probabilities</u> which agree
better and better with observed <u>frequencies</u>, but each set of
probabilities is essentially a prior probability assignment. If
another experiment were then performed, Bayes equation would then use
the ME probabilities and the experimental information to produce a set
of posterior probabilities which might be better than the ME
probabilities if the new information were neither redundant nor
contradictory but cogent.

Statement 3:

> "Furthermore, it is claimed that this estimate is the most
> probable or most likely solution, ie, it is a maximum a
> posteriori (MAP) estimate. Also, it is claimed to be the
> solution that is 'maximally noncommittal' and makes the
> fewest assumptions in regard to the unknown data."

Fact: The first statement is untrue. The second is precisely
correct, and the claim is also precisely correct.

Statement 4:

> "Far from being maximally noncommittal, the ME solution is
> based on a very specific and hightly committal assumption
> of an equiprobable prior."

Fact: No equiprobable prior is ever claimed by competent ME
practitioners. We have demonstrated in section Ib that under the
assumption of discreteness (we have an enumeration of the

possibilities) and normalization and <u>nothing</u> more, equal probabilities
for all possibilities is a consequence of ME, not an assumption. As
soon as more information, perhaps in the form of expectation values,
is provided, the ME probabilities become unequal in order to fit the
observed constraints.

Statement 5:

> "The ME principle is then invoked to obtain the most
> likely vector of frequencies f that obey the constraint
> [Makhoul's Eq.] (10). Using our intepretation of the ME
> principle, we in effect assume that the die is a priori
> fair (unbiased) and then we compute the most likely
> frequencies for which (10) is true. If u = 4.5, which is
> very different from the expected value of 3.5 for a fair
> die, then the ME solution is given by [Makhoul's Eq.]
> (1)."

Fact: The primary goal of ME is to obtain a set of <u>probabilities</u> not
<u>frequencies</u>. Ed Jaynes and other competent ME practitioners are
always careful to distinguish between probabilities which can be
assigned or calculated by ME or other valid procedures, and
frequencies which can be measured in a laboratory. Under certain
conditions which are elaborated in Jaynes (1968, 1978), there is a
very strong correspondence between ME probabilities and measured
frequencies but they are still quite distinct ideas conceptually.
Once again the die is <u>never</u> assumed to be fair! Where does this
gratuitous nonsense come from?

Statement 6:

> "While it is true that if N is large, having u = 4.5 is a
> good indicator that the die is most likely loaded because
> the probability of having u = 4.5 for a fair die is
> extremely small, the ME principle cannot be used
> productively to estimate the biases of the die. <u>The ME
> die is simply not loaded</u>. To name the problem the
> 'loaded die' problem has been a major source of confusion
> because it implies that the die is loaded and that the
> estimated frequencies are somehow related to the biases
> of the die. In ME, the die is known to be fair, but in
> an actual experiment the value of u comes out to be 4.5
> for example instead of 3.5, which is a unlikely but
> possible event. We then use ME to compute the
> frequencies that most likely occurred from this most
> unlikely event."

Fact: N large (small, medium, known or unknown) is completely
irrelevent for the solution of the ME problem! If N trials had been
used to <u>estimate</u> frequencies then N would have a very large effect on
the <u>variance</u> of the ME probabilities but none whatever on the
probabilities themselves.

Fact: The straight jacket which says that ME die is not loaded is a complete fiction! It exists only in the mind of the author and has nothing to do with the theory and practice of ME methods. The reader is asked to refer again to the exhaustive analysis of the Wolf dice data. If this doesn't convince the reader that ME works beautifully to discover physical biases which were present in dice thrown repeatedly over 100 years ago, then nothing will.

The essential difficulty in Makhoul's paper in addition to his complete and total misunderstanding of ME, is his transformation, in agreement with Frieden and Musicus of our basically discrete dice problem into a strange unrecognizable continuous problem with objects which no one should ever call "dice".

ACKNOWLEDGEMENTS

The author is very pleased to acknowledge his gratitude to Elizabeth Galligan for her expert typing of so many drafts of this manuscript and to Theresa Walker of the AFGL Art Department for her very patient and expert setting of the many equations.

References

Cox. R., (1974). Probability, Frequency and Reasonable Expectation, Am. J. Physics, 17, 1.

Cox, R., (1961). The Algebra of Probable Inference, Johns Hopkins University Press, Baltimore, MD.

Czuber, E., (1908) Wahrscheinlichkeitsrechnung.

Hobson, A. (1972). The Interpretation of Inductive Probabilities, J. Stat. Phys 6, 189.

Frieden, B. Roy (1985). Dice Entropy and Likelihood, Proc. IEEE 73, 1764.

Friedman K., (1973), Replies to Tribus and Motroni and to Gage and Hestenes, J. Stat. Phys 9, 265.

Friedman K. and A. Shimony, (1971). Jaynes Maximum Entropy Prescription and Probability Theory, J. Stat. Phys. 3, 193.

Gage, D. W. and D. Hestenes, (1973). Comments on the paper "Jaynes Maximum Entropy Prescription and Probability Theory", J. Stat. Phys 7, 89.

Jaynes, E. T., (1957). Information Theory and Statistical Mechanics, Part I, Phys. Rev., 106, 620; Part II; ibid, 108,171.

Jaynes, E. T. (1963a), "Brandeis Lectures" in E. T. Jaynes Papers on Probability, Statistics and Statistical Physics, R. D. Rosenkrantz, Ed. D. Reidel Publishing Co., Boston, Mass.

Jaynes, E. T., (1968). "Prior Probabilities", IEEE Trans Syst. Sci Cybern., SSC4, 227.

Jaynes, E. T., (1978). Where do we stand on Maximum Entropy, in the Maximum Entropy Formalism, R. D. Levine and M. Tribus,Editors, MIT Press, Cambridge, Mass.

Jaynes, E. T., (1979). "Concentration of Distributions at Entropy Maxima" in E. T. Jaynes: Papers on Probability, Statistics and Statistical Physics, R. D. Rosenkrntz, Ed., D. Reidel Publishing Co. Boston, Mass.

Jaynes E. T. (1982). "On The Rationale of Maximum − Entropy Methods", Proc. IEEE, 70 939.

Keynes, J.M., (1952). A treatise on Probability. MacMillam & Co,
       London.

deLaplace, Pierre Simon, (1951). A Philosphical Essay on
       Probabilities,
       Dover, New York.

Makhoul, J. (1986), "Maximum Confusion Spectral Analysis", Proc.
       Third ASSP Workshop on Spectrum Estimation and Modelling;
       Boston, Mass.

MacQueen J. and J. Marschak, (1975) "Partial Knowledge, Entropy and
       Estimation", Proc. Nat. Acad. Sci., Vol 72, pp. 3819-3824.

Rowlinson, J. S., (1970). Probability, Information and Entropy,
       Nature 225, 1196.

Shannon, C. E. and W. Weaver, (1949). The Mathematical Theory of
       Communication, The University of Illinois Press: Urbana.

Shimony, A., (1973). Comment on the interpretation of inductive
       probabilities, J. Stat. Phys 9, 187.

Teubners, B. G., Sammlung Von Lehr Buchern Auf Dem Gebiete
       Der Mathematischen Wissenschaften, Band IX p. 149,
       Berlin.

Tribus, Myron (1961), Thermostatics and Thermodynamics, D Van
       Nostrand Co., Princeton, N. J.

Tribus, Myron (1969), Rational Descriptions, Decisions and Designs.
       Pergamon Press, Oxford.

Tribus, Myron and H. Motroni, (1977) Comments on the Paper,
       Jaynes Maximum Entropy Prescription and Probability
       Theory", J. Stat. Phys 4, 227.

Quantum Density Matrix and Entropic Uncertainty[*]

R. BLANKENBECLER

Stanford Linear Accelerator Center
Stanford University, Stanford, California, 94305

and

M. H. PARTOVI

Physics Department, California State
University, Sacramento, California, 95819

ABSTRACT

A discussion of the determination of the quantum density matrix from realistic measurements using the maximum entropy principle is presented.

1. Introduction

The application of MAXENT to quantum theory is not a well-developed subject. Indeed, there is only one paper to my knowledge that gives a satisfactory treatment (in the mathematical sense) of this approach.[1] While this treatment lends itself to a discussion of certain questions in Quantum Measurement Theory, a normally esoteric subject of little physical consequence, it was developed with a

A Talk given at the Fifth Workshop on Maximum Entropy and Bayesian Methods in Applied Statistics, Laramie, Wyoming, August, 1985

[*] Work supported by the Department of Energy, contract DE–AC03–76SF00515.

G. J. Erickson and C. R. Smith (eds.),
Maximum-Entropy and Bayesian Methods in Science and Engineering (Vol. 1), 235–244.
© *1988 by Kluwer Academic Publishers.*

very practical problem in mind. Namely, how does one choose a quantum density
matrix based on realistic and necessarily incomplete measurements. Before answering this, one must answer a simpler question. How does one properly describe
the results of a measurement of a variable with a continuous spectrum (such as
position or momentum) when the experimental device has finite resolution?

The answer to this question was recently given by Partovi,[2] who based his
treatment on a criticism and an idea by Deutsch.[3]

Entropic Uncertainty:

The basic 'trick' of the entropic formulation of uncertainty is to introduce the
relevant characteristics of the measuring device into the measure of uncertainty.
This inclusion will lead to unexpected physical consequencies. To that end let us
introduce the notation of a measuring devive D^A which is used to measure the
observable A. Thus a measurement will consist of a partitioning of the spectrum
(either continuous or discrete) of A into a collection of subsets α_i. Therefore the
state of the system, $|\psi\rangle$, is to be described by a corresponding set of probabilities

$$P_i^A = P_i^A(\psi \mid D^A) . \tag{1.1}$$

The number P_i^A is the probability that the measurement will yield a value in
the subset α_i. Since the most common type of subset will be an interval, they
will be called 'bins'. The spectrum is a property of A, but the manner in which
it is partioned into bins is a property of the measuring device.

The process of binning is represented by the expression for the probability
defined above

$$P_i^a = \langle\psi| \pi_i^a |\psi\rangle / \langle\psi | \psi\rangle . \tag{1.2}$$

The operator π_i^a is the projection onto the subspace relevant to the subset α_i.
The completeness relation is transformed into the operator statement

$$1 = \sum_i \pi_i^a . \tag{1.3}$$

The entropy associated with the measurement of the observable A using the device D including the uncertainty due to the effects of binning is

$$S(\psi \mid D^A) = -\sum_i P_i^A \ln P_i^A , \tag{1.4}$$

where P_i^A is defined in eqn (1.2) .

In the special case that the subset α_i includes only one eigenvalue of the operator A, then the projection operator simplifies to

$$\pi_i^a = |a_i\rangle \langle a_i| , \tag{1.5}$$

and the measurement entropy reduces to

$$S(\psi \mid D^A) = -\sum_i |\langle \psi \mid a_i \rangle|^2 \ln |\langle \psi \mid a_i \rangle|^2 \tag{1.6}$$

which is the form given in ref.3 . However, in most interesting cases, at least some of the observables of interest will be either continuous or will be a discrete spectra with a limit point.

Duetsch [3] and Partovi [2] showed that a proper definition of the uncertainty in the measurement of the two observables A and B in the state ψ is simply the total entropy

$$U(D^A, D^B; \psi) = S(\psi \mid D^A) + S(\psi \mid D^B), \tag{1.7}$$

This expression has the property that it possesses a lower bound that depends on the measuring devices D^A and D^B but not on the state ψ.

Since the probabilities $P_i^{A\,or\,B}$ are normalized, the 'uncertainty' can be rewritten in terms of the joint probability $P_i^A\,P_j^B$:

$$U(D^A, D^B; \psi) = -\sum_i \sum_j P_i^A\,P_j^B \ln P_i^A\,P_j^B \ . \qquad (1.8)$$

This form also provides the physical justification for adding the individual entropic uncertainties and considering that as the proper measure of uncertainty. Note that this has nothing to do with 'simultaneous measurements' but is related to the serial measurement of the observables A and B on an identically prepared beam.

This uncertainty should be interpreted in terms of a <u>higher sample space</u> . It is a generalization of the standard case in that the subsequent samples are not measurements of the same observable. It corresponds to independent measurements of different observables performed on an identically prepared beam. Therefore the product of the two probabilities is expected. The above form also generalizes in a natural way to the case of more than two observables (or in other words to more than two independent samples).

A comparison with the classic Heisenberg's uncertainty principle is facilitated by noting first that by some simple manipulations, we have

$$P_i^A\,P_j^B \leq\| \pi_i^A + \pi_j^B \|^2\ /4\ . \qquad (1.9)$$

The double brace means the norm of the operator sum. Therefore the uncertainty can easily be shown to satisfy the inequality

$$U(D^A, D^B; \psi) \geq 2 \ln[\,2\,/\,sup_{i,j} \| \pi_i^A + \pi_j^B \|\,] \ . \qquad (1.10)$$

Since the π_i^a are projection operators, it immediately follows that the left hand side of eqn. (1.9) is between the limits of $1/4$ to 1. It achieves unity only if there is a common eigenfunction of the operators A and B. It is only in this circumstance that the uncertainty as given in eqn. (1.8) can be zero; in all other circumstances it is positive. Any definition of the uncertainty should satisfy this condition if it is to be at all reasonable and even consistent.

REMARK: It is interesting to note that the most popular extension of the uncertainty principle (as judged by the literature), namely

$$[\langle A^2 \rangle - \langle A \rangle^2][\langle B^2 \rangle - \langle B \rangle^2] \geq |\langle [A,B] \rangle|^2 / 4 ,\qquad (1.11)$$

where the expectation values are taken in the state ψ, does not satisfy this condition (for example, let A and B have opposite symmetry. Then the right-hand side vanishes if the state has a definite symmetry).

The Density Matrix:

The object of this section will be to argue that the measurement entropy defined above is more general than the standard form yet will reduce to the familiar form for the ensemble entropy in the correct physical limit. The measurement entropy will, however, allow the proper inference of the density matrix ρ from any set of necessarily incomplete measurements.

We shall always consider that a measurement is performed only once on a particular copy of the system. We imagine an 'oven' that produces a beam of identical copies of the basic system that is overall stationary in its properties, and thereby reproducible. To describe such a system, we introduce the standard density matrix by the replacement

$$|\psi\rangle \langle\psi| \rightarrow \rho ,\qquad (1.12)$$

and

$$\langle\psi| \pi_i^a |\psi\rangle \rightarrow tr(\rho\pi_i^a) .\qquad (1.13)$$

There is a considerable amount of physics behind this innocent looking replacement involving an average over the output of the oven and possible correlations.

The measurement entropy of the ensemble corresponding to the operator A is then taken to be

$$S(\psi \mid D^A) = -\sum_i tr(\rho \pi_i^A) \ln tr(\rho \pi_i^A) \ , \qquad (1.14)$$

which is a joint property of the system and the measuring device D.

The question that must now be answered is **what operator should we choose for A so that S corresponds to the entropy of the ensemble?** It should come as no surprise (a similar result was proven by von Neumann- his argument can be carried over to the present case with only a minimum of thought) that the unique choice for A is ρ itself. Since the density matrix describes all that can be known about the system and contains nothing superfluous, it is the operator that has the least uncertainty/entropy about the system. The motivation behind this argument and this choice for the operator we term the **maximum uncertainty principle** .

Now note that $tr\rho = 1$ and $tr\rho^2 \leq 1$. Therefore the density matrix has a purely discrete spectrum. Following the discussion of eqn. (1.5) , if the measuring device is sufficiently fine-binned, then the projection is

$$\pi_i^\rho = |i\rangle \langle i| \ , \qquad (1.15)$$

where the state $|i\rangle$ is an eigenfunction of the density matrix:

$$\rho |i\rangle = \rho_i |i\rangle \ . \qquad (1.16)$$

The entropy then becomes

$$S = -\sum_i \rho_i \ln \rho_i \,, \tag{1.17}$$

which can be immediately written in the familiar abstract form

$$S = -tr[\rho \ln \rho] \,. \tag{1.18}$$

Thus we see that when the measuring device is made sufficiently accurate, the ensemble entropy defined above reduces to the usual form. However, for 'cruder' (and more physical) devices, the finite resolution implicit in the projection operators plays a crucial and new role. Thus we have given a formalism that provides a unified physical description of both the microscopic and macroscopic situations.

The procedure now follows in analogy to the previous sections. The ensemble entropy is maximized subject to the measurement constraints ($a = 1,..A$)

$$P_i^a = tr(\rho \pi_i^a) \,. \tag{1.19}$$

The result is

$$\rho = Z^{-1} \exp[-\sum_{a,i} \lambda_i^a \pi_i^a] \,. \tag{1.20}$$

The multipliers λ_i^a are determined from

$$Z = tr(\exp[-\sum_{a,i} \lambda_i^a \pi_i^a]) \,, \tag{1.21}$$

and

$$P_i^A = Z^{-1} tr(\pi_i^A \exp[-\sum_{a,i} \lambda_i^a \pi_i^a]) \,. \tag{1.22}$$

It should be stressed that while the form, eqn. (1.20) , of the density matrix resembles the classical one, it is an operator. The projection operators in the exponent can be quite complicated in form, since they do not in general commute with each other. The density matrix must be evaluated with some care.

Some Examples:

In this paragraph we shall give two examples that illustrate the utility and simplicity of the entropic uncertainty approach. The examples are chosen for reasons of simplicity of presentation and relevance.

The first is the well-discussed problem of the correct form of the uncertainty relation for compact variables. We choose for an example the polar angle and its conjugate angular momentum. Thus the relevant operators are $A = \phi$ and $B = L_z$. The apparatus will be assumed to be able to measure the angle only to the extent that the result can be assigned to N bins, where $N = 2\pi/\delta\phi$ with $\delta\phi$ being the width of the angular bins, while the angular momentum can be measured and resolved down to a single value (which is integer). From eqn. (1.9) it is seen that we need to evaluate the maximum eigenvalue of the operator $\pi_i^A + \pi_j^B$. This eigenvalue equation takes the form

$$(\pi_i^\phi + \pi_m^{L_z}) |v\rangle = \lambda |v\rangle \ , \tag{1.23}$$

where (the bin boundaries are located at ϕ_i)

$$\pi_i^\phi = \Theta(\phi - \phi_i)\Theta(\phi_{i+1} - \phi) \tag{1.24}$$

and the angular momentum projector is the integral operator

$$\pi_m^{L_z} = \frac{1}{2\pi} \int\limits_0^{2\pi} d\phi' \, exp[\, im(\phi - \phi')] \ . \tag{1.25}$$

The solution corresponding to the maximum eigenvalue is found to be

$$|v_{max}\rangle = [1 + (\delta\phi/2\pi)^{1/2}\pi_i^\phi]\,|m\rangle\,, \tag{1.26}$$

where the state $|m\rangle$ is an eigenfunction of the angular momentum operator with eigenvalue m, with

$$\lambda_{max} = 1 + (\delta\phi/2\pi)^{1/2}. \tag{1.27}$$

The final result for the minimum uncertainty is

$$U(D^A, D^B; \psi) \geq 2\ln[2\,/\,[1 + (\delta\phi/2\pi)^{1/2}]\,. \tag{1.28}$$

Note that $\delta\phi$ is the resulution of the measuring device, not the variance. It is helpful in interpreting this result to note the following: if the angle ϕ is found to be in a particular bin while the angular momentum is measured to have a definite value m, then the joint probability has an upper bound given by

$$P_i^\phi\,P_m^{L_z} \leq \frac{1}{4}[1 + (\delta\phi/2\pi)^{1/2}]^2\,. \tag{1.29}$$

This is a much more reasonable expression of the physics in the uncertainty relation than is given by the standard Heisenberg-type inequality involving variances.

Finally, I will just quote the corresponding result for continuous variables and for comparison purposes will chose the classic pair x and p. For details, see ref. 1 . The position bins are of width δx and the momentum bins are of width δp. The general case is not worth working out in detail, but in the limit that $(\delta x\,\delta p) \lesssim 1$, the joint probability satisfies

$$P_i^x\,P_j^p \leq \frac{1}{4}(1 + [(\delta x\,\delta p)/2\pi]^{1/2})^2\,. \tag{1.30}$$

It is interesting to note that the wave function that saturates the above bound is localized in the appropriate bins but falls off only as a power-law, hence the variances used in the Heisenberg form are infinite for this wave function that is optimum for the 'bin' case.

REFERENCES

1. R. Blankenbecler and M. H. Partovi, Phys. Rev. Lett. *54*, 373 (1985).

2. M. H. Partovi, Phys. Rev. Lett. *50*, 1883 (1983).

3. D. Deutsch, Phys. Rev. Lett. *50*, 631 (1983).

INFORMATION-THEORETICAL GENERALIZATION OF THE UNCERTAINTY PRINCIPLE

A.J.M. Garrett
The Sir Frank Packer Theoretical Department, School of
Physics, The University of Sydney, Sydney N.S.W. 2006
Australia

Abstract. Recent suggestions on how to extend the uncertainty principle, using the concept of information, are
reviewed. The Heisenberg variance uncertainty principle is
shown to be a special case for canonically conjugate continuous variables. The possibility of further generalization is considered.

Deutsch (1983) has proposed that the uncertainty principle
be viewed as the bounding of the information sum associated with the
eigenstates of two non-commuting observables. This is by far the
deepest view yet of the underlying idea: it is not subject to the
usual restriction to canonically conjugate observables, while (negative)
information is easily the most compelling mathematical formulation of
uncertainty in this and other contexts (Deutsch 1983; Everett 1973;
Wootters & Zurek 1979; Jaynes 1983).

Consider two Hermitian operators A, B with discrete spectra and nor-
malised eigenkets $|a_i>$, $|b_i>$, and let the decomposition coefficients of
an arbitrary state $|\Psi>$ with respect to these bases be α_i, β_i:

$$|\Psi> \quad = \quad \sum_i \alpha_i |a_i> \quad = \quad \sum_i \beta_i |b_i>. \tag{1}$$

Denote the unitary transformation matrix between the bases by $\underline{\underline{T}}$, so
that

$$\underline{\alpha} \quad = \quad \underline{\underline{T}} \cdot \underline{\beta}. \tag{2}$$

The information I_A of the state $|\Psi>$ with respect to the base $|a_i>$ is

$$I_A \quad = \quad \sum_i p_i^A \log p_i^A; \quad p_i^A \quad = \quad |<\Psi|a_i>|^2 \quad = \quad |\alpha_i|^2 \tag{3}$$

so that the negative information sum associated with the two observables
is

$$- I_A - I_B \quad = \quad - \sum_i |\alpha_i|^2 \log |\alpha_i|^2 - \sum_i |\beta_i|^2 \log |\beta_i|^2. \tag{4}$$

245

G. J. Erickson and C. R. Smith (eds.),
Maximum-Entropy and Bayesian Methods in Science and Engineering (Vol. 1), 245–248.
© 1988 by Kluwer Academic Publishers.

Our problem is to seek the smallest value of (4) obtained by varying
the αs and βs and their complex conjugates, subject to the constraints
of (2) and normalisation. Deutsch (1983) has already proved the
existence of a bound, but failed to obtain an expression for it.
Clearly it depends only on the components of the transformation matrix
$\underline{T}$. On applying the variational calculus, we readily find the simultaneous equations to be solved for the coefficients; substitution of
their solution into (4) gives the bound. Unfortunately the logarithmic
nonlinearity of the equations rules out analytical solution, explaining
Deutsch's failure. We must learn to be content with viewing the uncertainty principle as a *prescription* for finding the extremal function
and the bound.

If some eigenfunctions of A coincide with some of B, the bound attains
its smallest possible value, zero, when the wavefunction equals any one
coincident eigenfunction. If a single eigenfunction of each operator
is nearly coincident, so that one of the inner products $||a_i - b_j||$ is
much smaller than the rest, a perturbation expansion can be set up.

Our confidence in this formulation is increased by the fact that, for
the special case of canonically conjugate continuous variable pairs, we
can find the bound and show that it implies the Heisenberg variance
inequality. This is done in the Appendix. It is first necessary to
generalize (4) to the continuous case by calculating the probability
density of measurement of the eigenvalue in the continuum. For the
discrete, bound case, all eigenstates are clearly equally likely; no
information is at hand indicating preference for any particular one.
(We take the view that probability is a consistent and unique theory for
conducting inference from given information. Of course, individuals
possessing different information will make different predictions about
the same system.) The probability density function for a continuum is
more difficult to determine (Jaynes 1968, 1979). Deutsch (1983) has
asserted that it can be deduced by regarding the continuum as the limit
of a discrete spectrum. However, the limiting process is not unique,
and worse, no such discretum is known to exist for the position variable
and many other observables. Partovi (1983), Bialynicki-Birula (1985)
and Bialynicki-Birula & Madajczyk (1985) interpret the probability in
terms of the resolution of the observing apparatus. While the
probability of measuring the eigenvalue in a narrow bin is undoubtedly
proportional to the width of that bin, one might expect there to exist
a deeper probability distribution which is independent of the apparatus.

Ultimately, our mathematical goal is to learn to discern the existence
and form of a "master inequality" implying by itself any other from the
same original equations. In the present condition of functional
analysis, this is no easy task. The fundamental nature of information
as an uncertainty measure strongly suggests that the continuum generalization of Deutsch's proposal be the successful candidate.

As it stands, this work relates solely to the mathematics of non-
commuting Hermitian operators. Its physical importance is that the

stronger an inequality, the easier it is to test in the laboratory. Such experiments provide a check on the correspondence between the quantum physics and the mathematics.

APPENDIX
Consider canonically conjugate continuous variables in $(-\infty, \infty)$, say x,k; a standard inequality is

$$\left\{ \sqrt{\frac{\mu}{2\pi}} \int_{-\infty}^{\infty} dx \; P_x(x)^{\frac{1}{2}\mu} \right\}^{1/\mu} \gtreqless \left\{ \sqrt{\frac{\nu}{2\pi}} \int_{-\infty}^{\infty} dk \; P_k(k)^{\frac{1}{2}\nu} \right\}^{1/\nu} \quad \text{as} \quad \nu \gtreqless \mu \tag{5}$$

(Beckner 1975) where the positive numbers μ, ν are such that $(\mu-1)(\nu-1) = 1$ (or equivalently, $\mu^{-1} + \nu^{-1} = 1$) and $P_x = |\psi|^2$, $P_k = |\tilde{\psi}|^2$ are the conjugate probability distributions; a tilde denotes Fourier transform. Equality in (5) for $\mu = \nu = 2$ follows from Parseval's theorem, while reversal of the inequality as $\nu \lessgtr \mu$ is a consequence of symmetry. We show that (5) is stronger than the Heisenberg variance inequality: by expanding about $\mu = \nu = 2$, we find

$$- \int_{-\infty}^{\infty} dx \; P_x(x) \log P_x(x) - \int_{-\infty}^{\infty} dk \; P_k(k) \log P_k(k) \gtreqless \log(\pi e) \tag{6}$$

(Beckner 1975). Since the priors for x and k are equal by symmetry, they must be flat (Sibisi 1983); (6) is therefore the continuum generalization of (4) for this case, with bound $\log(\pi e)$. (6) implies the variance inequality; on combining it with the lemma

$$- \int_{-\infty}^{\infty} d\varepsilon \; P(\varepsilon) \log P(\varepsilon) \leq \frac{1}{2} \log \left\{ 2\pi e \int_{-\infty}^{\infty} d\varepsilon (\varepsilon - \langle\varepsilon\rangle)^2 \; P(\varepsilon) \right\} \tag{7}$$

we recover immediately the result $\sigma_x^2 \sigma_k^2 \geq \frac{1}{4}$ (Beckner 1975; Everett 1973, p.52). Neither stage of this derivation is reversible. Lemma (7) is established locally by variational means, and is extended globally by a method of Jaynes' (1963). Gaussians are the unique extremal functions of (5), (7) and hence of the variance inequality, and also presumably - but not trivially - of (6). Other conjugate variable inequalities have been studied by Uffink & Hilgevoord (1984,1985) and Levy-Leblond (1985). These do not attain equality for Gaussian wave functions; furthermore the margin by which equality is exceeded changes with the width of the Gaussian. Given the convincing demonstration of Wootters & Zurek (1979) that information is precisely what is measured in dual slit experiments (and presumably others), we expect these inequalities to be contained in the information approach.

Although (5) is stronger than the information uncertainty principle, our conjecture that the latter is in general the strongest possible is not impaired; it may be that the information principle is contained in another inequality only for the special case of conjugate variables.

For conjugate variables with different ranges, such as the discrete angular momentum component L_z and the angle $\phi \in [0, 2\pi]$, the results are different (Beckner 1975; Breitenberger 1985). The bound of $(-I_A - I_B)$ in the continuous case must be invariant under canonical transformations of a conjugate pair A and B.

REFERENCES

Beckner, W. (1975). Ann. Math. 102, 159–182.
Bialynicki-Birula, I. (1985). Entropic Uncertainty Relations in
        Quantum Mechanics. In Quantum Probability and Applications
        II, eds. L. Accardi and W. von Waldenfels, Lecture Notes in
        Mathematics Vol. 1136, pp. 90–103. Berlin: Springer-Verlag.
Bialynicki-Birula, I. & Madajczyk, J.L. (1985). Phys. Lett. 108A,
        384–386.
Breitenberger, E. (1985). Found Phys. 15, 353–364.
Deutsch, D. (1983). Phys. Rev. Lett. 50, 631–633.
Everett, H. (1973). The Theory of the Universal Wave Function. In
        The Many-Worlds Interpretation of Quantum Mechanics, eds.
        B. S. de Witt and N. Graham. Princeton, N.J.: Princeton
        University Press.
Jaynes, E.T. (1963). Information Theory and Statistical Mechanics. In
        Statistical Physics (1962 Brandeis Lectures), ed. K. W. Ford,
        pp. 181–218. New York: Benjamin. Reprinted as Chapter 4
        of Jaynes (1983), with the relevant analysis on p. 46.
Jaynes, E.T. (Sept. 1968). Prior Probabilities. In IEEE Trans. on
        Systems Science and Cybernetics, SSC-4, pp. 227–241.
        Reprinted as Chapter 7 of Jaynes (1983).
Jaynes, E.T. (1979). Marginalization and Prior Probabilities. In
        Bayesian Analysis in Econometrics and Statistics: Essays
        in Honor of Sir Harold Jeffreys, ed. A. Zellner, pp. 43–87.
        Amsterdam: North-Holland. Reprinted as Chapter 12 of
        Jaynes (1983).
Jaynes, E.T. (1983). Papers on Probability, Statistics and Statistical
        Physics, ed. R. D. Rosencrantz. Dordrecht: Reidel.
Levy-Leblond, J.-M. (1985). Phys. Lett. 111A, 353–355.
Partovi, M.H. (1983). Phys. Rev. Lett. 50, 1883–1885.
Sibisi, S. (1983). Nature 301, 134–136.
Uffink, J.B.M. & Hilgevoord, J. (1984). Phys. Lett. 105A, 176–178.
Uffink, J.B.M. & Hilgevoord, J. (1985). Found. Phys. 15, 925–944 .
Wootters, W.K. & Zurek, W.H. (1979). Phys. Rev. D19, 473–484.

TIME, ENERGY, AND THE LIMITS OF MEASUREMENT

M. H. Partovi
Department of Physics, California State University,
Sacramento, California 95819

R. Blankenbecler
Stanford Linear Accelerator Center, Stanford, California 94305

ABSTRACT
We present a discussion of our microstatistical formalism for multitime quantum measurements. We show that this formalism is capable of dealing with time in quantum mechanics in a rigorous way, and enables one to precisely state and derive time-energy uncertainty relations. Another application to the problem of the quantum limit of accuracy of position measurements in the context of gravitational wave detection is briefly discussed.

1 INTRODUCTION

The general problem of treating incomplete information occurs at the fundamental level of quantum measurements in an unavoidable manner. The measurement of the state of a microscopic system in general requires a determination of the $N^2 - 1$ elements of its density matrix, N being the dimensionality of the Hilbert space of the states of the system. Now in general N is infinite, implying that an exhaustive measurement is in principle impossible. Stated simply, measurements performed on the most irreducible systems in nature are necessarily incomplete; see our previous contribution to these proceedings for further discussion and for quantitative examples.

The maximum entropy principle (MEP) provides a natural solution to the above problem. Indeed, given a properly formulated measure of entropy, the solution is formally identical to that of the standard problem, well known in statistical mechanics (Jaynes, 1957), of maximizing entropy subject to a set of constraints. Precisely such an entropy for quantum measurements was proposed by Deutsch (1983) and developed by one of us (Partovi, 1983). Proposing a *maximum uncertainty principle* (MUP) as the quantum version of MEP, we developed the statistical mechanics of microscopic systems (Blankenbecler and Partovi, 1985, and these proceedings). It will be convenient for the following development to summarize this formalism here.

The state of a quantum system is in general specified by a density matrix, $\hat{\rho}$, which is a self-adjoint operator whose eigenvalues are a discrete set of probabilities; i.e., they lie between zero and one and add up to unity. A measurement of

G. J. Erickson and C. R. Smith (eds.),
Maximum-Entropy and Bayesian Methods in Science and Engineering (Vol. 1), 249–255.
© *1988 by Kluwer Academic Publishers.*

the state of the system in general entails the measurement of a number of observables of that system (e.g., energy, position, spin), say $\widehat{A}^\nu$, by producing a large number of copies of the system under identical conditions, and subjecting a sufficiently large fraction of these to interaction with the measuring devices D^ν. In general each D^ν breaks up the range of possible values of A^ν (i.e., the spectrum of A^ν) into a number of bins, α_i^ν, and measures the frequencies $\mathcal{P}_i^\nu$ with which the values of the observable A^ν are found to lie in the bin α_i^ν. For each bin α_i^ν one can introduce a projection operator $\widehat{\pi}_i^\nu$, so that in symbols one has

$$\mathcal{P}_i^\nu = \text{tr } \widehat{\rho}\,\widehat{\pi}_i^\nu \ . \tag{1.1}$$

We are now in a position to state the generic microstatistical problem and its solution: Given that measurements have yielded a set of $\mathcal{P}_i^\nu$ and that no other information is known about the system, how is $\widehat{\rho}$ to be determined? The answer is: maximize the ensemble entropy $-tr\,\widehat{\rho}\ell n\,\widehat{\rho}$ subject to the constraints expressed in (1.1). The solution is

$$\widehat{\rho} = Z^{-1} \exp\left[-\sum_{\nu,i} \lambda_i^\nu \widehat{\pi}_i^\nu\right] \ , \tag{1.2}$$

where

$$tr\,\widehat{\rho} = 1\,, \quad \mathcal{P}_i^\nu = -(\partial/\partial\lambda_i^\nu)\,\ell n\,Z \ . \tag{1.3}$$

Notice the similarity to the analogous expressions of equilibrium statistical mechanics as well as the very important difference that no constraint on the energy of the system such as would appear in, e.g., the canonical ensemble occurs here. Notice also the fact that in all of the above we have assumed the various measurements to be simultaneous in the sense that each copy of the quantum system is submitted to the measuring device at precisely the same relative time subsequent to its preparation

In summary, then, Eq. (1.2) and (1.3) specify the density matrix of a quantum system subsequent to a single-time measurement according to the maximum entropy/uncertainty principle. Next, we shall tern to a discussion of the meaning of time and subsequently to the treatment of multitime measurements.

2   TIME

Time has always played a rather elusive role in quantum mechanics. The reason is simply that time is a parameter, and not a dynamical variable, for any system that obeys Hamiltonian dynamics. In quantum mechanics, for example, the change in the mean value of any dynamical observable $\widehat{A}$ in the state $\widehat{\rho}$ of a system with Hamiltonian $\widehat{H}$ is proportional to $i\,tr\,\widehat{\rho}\,[\widehat{H},\widehat{A}]$, for sufficiently small changes dA, with the constant of proportionality independent of $\widehat{A}$. Hence, for a pair of observables $\widehat{A}$ and $\widehat{B}$, one has $dA/dB = (tr\,\widehat{\rho}\,[\widehat{H},\widehat{A}])/(tr\,\widehat{\rho}\,[\widehat{H},\widehat{B}])$ for the ratio of their respective rates of change. Clearly, it is natural as well as convenient to introduce a standard for this sort of comparison by parametrizing the evolution of the system in the usual way: $dA = dt \cdot i\,tr\,\widehat{\rho}\,[\widehat{H},\widehat{A}]$. The choice of this parameter is in principle an arbitrary matter, although any choice other than our present one (or a linear function of it) would appear odd and unnatural to us (e.g., a healthy person's heart beat would slow down indefinitely if we choose $t' = \exp[t/t_0]$ as the new parameter).

Now it would be extremely convenient if we had an observable $\widehat{C}$ for which dC/dt had a constant value *independent* of the state of the system. Indeed if there were such a $\widehat{C}$, we would use it (or a linear function of it) as the standard chronometric variable and could thereby deal with time as simply another dynamical variable of the system. Unfortunately, such a state-independent universal $\widehat{C}$ does not exist, so that time has to be dealt with as a parameter characterizing the evolution of a dynamical system as described above.

Having clarified the meaning of t, we will now proceed to find a precise measure of the accuracy with which it can be measured. Suppose we wish to measure the time of an event using a system in a state $\widehat{\rho}$ as the clock and the value of a dynamical observable $\widehat{A}$ as the chronometric variable. Now the expectation value $A = tr\,\widehat{\rho}\widehat{A}$ is a function of time, giving us the required mapping $A(t)$, or (assuming invertibility) its inverse $t(A)$, for determining the value of t when a measured value of A is obtained. In general, the measured values of A will have a distribution, $\mathcal{P}(A)\,dA$, so that one will have a corresponding distribution in the corresponding values of t given by $\mathcal{P}[A(t)](dA/dt)\,dt$. In particular, the variance in the measured values of t will be given by

$$(\delta t)^2 = \int dt\,\mathcal{P}[A(t)](dA/dt)[t - \bar{t}]^2\,, \tag{2.1}$$

where $\bar{t}$ is the mean value of t with respect to the above distribution. Other moments can be similarly calculated.

To proceed, we must relate $(\delta t)^2$ back to the state of the system $\widehat{\rho}$ and the operator $\widehat{A}$. To do so, we rewrite Eq. (2.1) as

$$(\delta t)^2 = \int dA\,\mathcal{P}(A)[t(A) - \bar{t}]^2\,, \tag{2.2}$$

and note that $\mathcal{P}(A)\,dA$ is simply the probability of finding the measured value of $\widehat{A}$ in the interval dA. But then using Eq. (1.1), we have $\mathcal{P}(A)\,dA = tr\,\widehat{\rho}\widehat{\pi}(dA)$, where $\widehat{\pi}(dA)$ is the projection operator corresponding to the spectral interval dA

centered around A. Thus we can rewrite Eq. (2.2) in the form

$$(\delta t)^2 = \int [t(A) - \bar{t}]^2 \, \mathrm{tr} \, \widehat{\rho}\,\widehat{\pi}(dA) \, . \tag{2.3}$$

Since $\widehat{A}\widehat{\pi}(dA) = A\widehat{\pi}(dA)$, we can use the completeness property $\int \widehat{\pi}(dA) = 1$ to recast (2.3) into the final form

$$(\delta t)^2 = \mathrm{tr}\,\widehat{\rho}\,[t(\widehat{A}) - \bar{t}]^2 \, . \tag{2.4}$$

Note that $t(\widehat{A})$, now a function of an operator, is itself an operator. Equation (2.4) is a remarkably simple formula expressing the dispersion in the measured values of time in terms of the clock state $\widehat{\rho}$ and the chromometric variable $\widehat{A}$.

Can $(\delta t)^2$ be made arbitrarily small? A precise answer to this question would of course constitute a precise statement of the time-energy uncertainty principle. To answer this question, we first use the Heisenberg inequality

$$\delta D \, \delta B \geq \tfrac{1}{2} \left| \mathrm{tr}\,\widehat{\rho}\,[\widehat{D}, \widehat{B}] \right| \, , \tag{2.5}$$

and the identifications $\widehat{D} = t(\widehat{A})$, $\widehat{B} = \widehat{H}$, to write

$$2\delta t \, \delta H \geq \left| \mathrm{tr}\,\widehat{\rho}\,[\widehat{H}, t(\widehat{A}) - \bar{t}] \right| \, . \tag{2.6}$$

Next, using X to denote the right-hand side of (2.6), we seek to minimize it by requiring that its first-order variation vanish. This requirement leads to the condition $[\widehat{H}, dt(\widehat{A}_0)/d\widehat{A}_0] = 0$ for the optimal variable $\widehat{A}_0$. This last condition essentially requires that either $\widehat{A}_0$ commute with $\widehat{H}$ or that $dt(\widehat{A}_0)/d\widehat{A}_0$ vanish. The first possibility actually maximizes X, since in that case $\widehat{A}_0$ would be a constant of the motion, resulting in a clock that is stuck on a fixed value! The second possibility forces $t(\overline{A}_0)$, or $\overline{A}_0(t)$, to be a linear function, so that $\mathrm{tr}\,\widehat{\rho}\,\widehat{A}_0$ is required to be a linear function of t. An immediate consequence of this linearity is that $X = 1$, and (Partovi and Blankenbecler, 1986a)

$$\delta t \, \delta H \geq \tfrac{1}{2} \, . \tag{2.7}$$

In other words, when a quantum system is used as a clock, the dispersion in the measured values of the time of an event cannot be reduced below $1/2\delta H$. In particular, a quantum system which is almost in a stationery state would make a very poor clock. Conversely, a system required to measure time accurately must have a correspondingly large uncertainty in the value of its energy.

Note that the optimal chromometric variable $\widehat{A}_0$ is precisely what we designated as $\widehat{C}$ earlier and characterized as an ideal standard of time. We also stated that such an operator does not in general exist, a well-known fact that follows from the non-existence of a well-behaved canonical conjugate to the Hamiltonian operator. Strictly speaking then, the lower limit in (2.7) is not realizable (even in principle) in actual measurements.

3 MULTITIME MEASUREMENTS

In Section 2 we described how quantum systems may be used as clocks, and established the fact that one can in principle construct clocks of arbitrary accuracy by allowing δH to be sufficiently large. Our next task is to generalize the formalism described in Section 1 to multitime measurements assuming the existence of such clocks of arbitrary accuracy.

Let us consider a measurement of a quantum system involving the observables $\widehat{A}^\nu(t_r^\nu)$, where ν labels different observables as in Section 1, and where the additional label t_r^ν denotes the time at which the measurement was carried out. As before, the results of these measurements are summarized in a set of frequencies $\mathcal{P}_{ir}^\nu$, where

$$\mathcal{P}_{ir}^\nu = tr\left[\widehat{\rho}\,\widehat{\pi}_i^\nu(t_r^\nu)\right] . \tag{3.1}$$

Note that $\widehat{\rho}$ does not carry a time label as it corresponds to the reference time $t = 0$. Comparing (3.1) with (1.1), we see that the multitime measurement is in essence not different from the single time case, once the constraint conditions (3.1) are rewritten in terms of $\widehat{\pi}_i^\nu \equiv \widehat{\pi}_i^\nu(0)$ so that all projection operators refer to $t = 0$. In terms of the evolution operator $\widehat{U}(t)$, defined by

$$i\,\frac{d}{dt}\,\widehat{U}(t) = \widehat{H}\widehat{U}(t)\,, \quad \widehat{U}(0) = 1\,, \tag{3.2}$$

we have

$$\widehat{\pi}_i^\nu(t) = \widehat{U}^\dagger(t)\,\widehat{\pi}_i^\nu\widehat{U}(t)\,, \tag{3.3}$$

so that the constraint equations (3.1) now read

$$\mathcal{P}_{ir}^\nu = tr\left[\widehat{\rho}\,\widehat{U}^\dagger(t_r^\nu)\widehat{\pi}_i^\nu\widehat{U}(t_r^\nu)\right] . \tag{3.4}$$

We can now write the solution for the multitime case as a simple generalization of (1.2):

$$\widehat{\rho} = Z^{-1}\exp\left[-\sum_{\nu ir}\lambda_{ir}^\nu\widehat{\pi}_i^\nu(t_r^\nu)\right]\,, \tag{3.5}$$

with constraint equations similar to (1.3). Equation (3.5) gives the density matrix of a quantum system subsequent to a general, multitime measurement according to the maximum uncertainty/entropy principle (Partovi and Blankenbecler, 1986a).

To illustrate the use of (3.5), we shall apply it to another long-standing problem in time-energy uncertainty relations: How accurately can the energy of a quantum system be determined if the measurement is to last no longer than T seconds? To answer this question, we shall consider a "canonical" measurement where the (x-component of the) position of a free particle is measured at two different times, say $-T/2$ and $+T/2$. Thus the device is a position measurement apparatus which we take to have bins of uniform size Δ arranged symmetrically along the x-axis so that $\alpha_i^x = [(i - \frac{1}{2})\Delta, (i + \frac{1}{2})\Delta]$, $i = 0, \pm 1, \ldots$ The density matrix resulting from this measurement is, according to (3.5),

$$\widehat{\rho} = Z^{-1} \exp \left\{ -\sum_i \lambda_i^- \widehat{\pi}_i^x \left(-\frac{T}{2} \right) + \lambda_i^+ \widehat{\pi}_i^x \left(+\frac{T}{2} \right) \right\} , \qquad (3.6)$$

where $\lambda_i^\pm$ are parameters related to the measured frequencies $\{\mathcal{P}_i^\pm\}$ in the standard way.

We must now study the dispersion

$$(\delta H)^2 = tr \left(\widehat{\rho} \, \widehat{H}^2 \right) - \left(tr \, \widehat{\rho} \, \widehat{H} \right)^2 , \qquad (3.7)$$

and determine how it is related to T. More specifically, we will determine how small δH can be made when T is considered fixed and the $\{\mathcal{P}_i^\pm\}$, and consequently the $\{\lambda_i^\pm\}$ are varied so as to produce the state of lowest possible δH. To avoid technical complications, we shall outline the main points of the argument and leave the details to the literature (Partovi and Blankenbecler, 1986a).

It can be shown that the state $\widehat{\rho}_0$ corresponding to the lowest possible δH has certain symmetry properties which imply the conditions $\lambda_i^+ = \lambda_i^-$ and $\lambda_i^\mp = \lambda_{-i}^\mp$. These conditions in turn imply that $\widehat{\rho}_0$ is self-conjugate under a Fourier transformation that sends $\widehat{x}$ into $(T/2m)\widehat{p}$ and $\widehat{p}$ into $(-2m/T)\widehat{x}$; here $\widehat{x}$ and $\widehat{p}$ are the position and momentum operators, and m is the mass of the particle. This interesting invariance in turn can be used to show that the optimal state $\widehat{\rho}_0$ has an energy dispersion which is no less than $1/2T$. In other words,

$$T\delta H \geq \tfrac{1}{2} . \qquad (3.8)$$

This result is a precise statement of the time-energy uncertainty relation in the form that was often used by Bohr. Note that T in Eq. (3.8) is not a dispersion or uncertainty but the time elapsed between the two position measurements, so that it is the duration of the canonical measurement which was performed to determine the state of the quantum system.

We conclude by briefly discussing the result of applying the above formalism to the derivation of a limit known as the *standard quantum limit* (SQL). This limit naturally arises in connection with the use of laser interforometry in gravitational wave detection (Partovi and Blankenbecler, 1986b). The problem is this: suppose one is trying to detect the presence of a very weak force (the gravitational force resulting from the passage of a gravitational wave in the actual situation) by successively measuring the position of an otherwise free mass m and thereby the acceleration caused by that force. Under these circumstances, optimal sensitivity is obtained when the displacement (i.e., the change in the position) of the mass is measured with the highest possible accuracy. The accuracy, ℓ, with which the position of the mass can be measured for optimal sensitivity in the detection of acceleration (or force) is the limit referred to above, the SQL. Since this type of measurement is essentially the canonical measurement discussed above, the results of our analysis can be applied here. Using these, we have shown that

$$\ell \geq \ell_0 \left\{ 2U_{\min} \left[\frac{m}{2\pi T} (\Delta x)^2 \right] \right\}^{1/2} , \qquad (3.9)$$

where T is the time elapsed between the two measurement, Δx is the resolution of the device used to measure position, $\ell_0 = (T/2m)^{1/2}$, and $U_{\min}[(\Delta x)(\Delta p)/2\pi]$ is the minimum possible value of the dispersion product $\delta x \delta p$ for a measuring device whose position and momentum resolutions are, respectively, Δx and Δp.

Clearly, the minimum value of ℓ in Eq. (3.9) is achieved for the lowest possible value of $U_{\min}$. But the latter is the standard Heisenberg result $\frac{1}{2}$, which is possible when the resolutions Δx and Δp are essentially equal to zero. This gives us the absolute lower bound $\ell \geq \ell_0 = (T/2m)^{1/2}$. For comparison, we note that SQL, the result previously quoted in the literature (and the subject of controversy previous to our work), gives an absolute lower bound equal to $\sqrt{2}\,\ell_0$ (Caves, 1985).

This work was supported in part by a grant from the California State University, Sacramento, NSF Grant No. PHY-8513367, and the Department of Energy under Contract No. DE-AC03-76SF00515.

REFERENCES

Blankenbecler, R. & Partovi, M. H. (1985). Phys. Rev. Lett. <u>54</u>, 373.

Caves, C. (1985). Phys. Rev. Lett. <u>54</u>, 2465.

Deutsch, D. (1983). Phys. Rev. Lett. <u>50</u>, 631.

Jaynes, E. T. (1957). Phys. Rev. <u>106</u>, 620.

Partovi, M. H. (1983). Phys. Rev. Lett. <u>50</u>, 1883.

Partovi, M. H. & Blankenbecler, R. (1986a). Phys. Rev. Lett. <u>57</u>, 2887.

Partovi, M. H. & Blankenbecler, R. (1986b). Phys. Rev. Lett. <u>57</u>, 2891.

ON A DETECTION ESTIMATOR RELATED TO ENTROPY

R.N. Madan
Office of Naval Research, Arlington, VA 22217

Abstract. The paper discusses the problem of estimation of parameters in a Rayleigh distribution modified to take into account the additional information. Madan and Guild [1981] have already given the maximum likelihood estimator (MLE) and the minimum mean squared estimator (MMSE) for the problem. Here we propose a new type of estimator called the entropy estimator for finding the mean of the samples from a small number of observations. The entropy estimator is the ratio of the arithmetic mean to the geometric mean multiplied by a normalizing constant. After normalizing the three estimates appropriately, the tightness of the entropy estimator is demonstrated numerically.

INTRODUCTION

In an earlier presentation, Madan and Guild [1981] had investigated two estimators and evaluated their performance in establishing a detection threshold according to the Neyman-Pearson criterion that maximizes the detection probability for a given probability of false alarm. The estimators investigated were a minimum mean squared estimator (MMSE) and a maximum likelihood estimator (MLE) for N i.i.d. samples of a Rayleigh density function. An average probability of false alarm for various sample sizes was employed as a measure to evaluate the performance differences of the two estimators. In this work we discuss an estimator which is related to the expression for information entropy. We propose and analytically develop a normalization constraint that it must satisfy in order to be considered an estimator. We call this estimator the entropy estimator. It is biased like the MLE, but is shown to be more efficient than the MMSE or the MLE, significantly so for small sample sizes. Some quantitative results for samples of Rayleigh distribution, as would be the case for a linear detector, are reported in this paper.

MMSE AND MLE

Assuming that n_i and n_q are the Gaussian distributed $N(0,\sigma^2)$ random noise variables of the in-phase and the quadrature channels in a detection

257

G. J. Erickson and C. R. Smith (eds.),
Maximum-Entropy and Bayesian Methods in Science and Engineering (Vol. 1), 257–265.

scheme, then the density function of the envelope $x = \sqrt{n_i^2 + n_q^2}$ is given by

$$p(x) = \frac{x}{\sigma^2}\, e^{-x^2/2\sigma^2}\, u(x), \quad \text{Rayleigh} \tag{1}$$

conditioned on σ, related to the noise power σ^2, in each channel. The parameter σ is related to the mean $\bar{x}$ by the relation

$$\bar{x} = \int_0^\sigma x\, p(x)\, dx = \sqrt{\frac{\pi}{2}}\, \sigma. \tag{2}$$

We require a definition of the probability of false alarm to employ it as a measure of performance for various estimators. In the detection of signals, for example, sinusoidal signals in the presence of random noise, probability of false alarm is defined by the probability that a noise signal exceeds detection threshold T and is evaluated by

$$pfa = \int_T^\infty p(x)\, dx = e^{-T^2/2\sigma^2}. \tag{3}$$

Employing (2), we get

$$p(x) = e^{-\frac{\pi}{4}\frac{x^2}{\bar{x}^2}} \quad \text{and} \quad pfa = \int_T^\infty p(x)\, dx = e^{-\pi T^2/4\bar{x}^2} \tag{4}$$

For numerical work, it is convenient to define T by $T = K\bar{x}$. If $\bar{x}$ is normalized to 1, (implying $\sigma = \sqrt{2/\pi}$), the scaling parameter K defines the threshold. For sampled systems, $\bar{x}$ is estimated by $\hat{\bar{x}}(N)$ from N samples employing any estimator, thus the threshold T and pfa become variables of estimation given by

$$\hat{T}(N) = K\hat{\bar{x}}(N) \quad \text{and} \quad pfa(\hat{T}(N)) = e^{-\hat{T}^2(N)/2\sigma^2} \tag{5}$$

A reasonable measure for performance evaluation of various estimators is seen to be the average

$$\overline{pfa} = \int pfa(\hat{T}(N))\, p(\hat{T}(N))\, d\hat{T}(N) \tag{6}$$

where $p(\hat{T}(N))$ is the probability density function of $\hat{T}(N)$ whose analytical expression may not be available for most situations. We have found the exact relationship for one case. Assuming that the conditions for the application of the central limit theorem have been met, we can approximate $p(\hat{T}(N))$ by the Gaussian $N(\mu(\hat{T}), \sigma^2(\hat{T}))$. With this approximation, the integral in equation (6) can be evaluated as

$$\overline{pfa} = \exp\left[-\frac{\mu(\hat{T})^2}{2(\sigma^2(\hat{T}) + \sigma^2)}\right]\sqrt{\frac{\sigma^2}{\sigma^2 + \sigma^2(\hat{T})}}. \tag{7}$$

We briefly reproduce the results of the MMSE and the MLE from Madan and Guild [1981] to establish the background for notation and comparative evaluations. The MMSE of x_i ($i = 1,2,...,N$) i.i.d. Rayleigh samples is known to be

$$\hat{\overline{x}}_M(N) = \frac{1}{N} \sum_{i=1}^{N} x_i \tag{8}$$

The expectation value or the first moment is given by

$$E\{\hat{\overline{x}}_M(N)\} = \sqrt{\pi/2} \, \sigma \tag{9}$$

and the variance by

$$\text{var}(\hat{\overline{x}}_M(N)) = \frac{1}{N} \left(2 - \frac{\pi}{2}\right) \sigma^2 . \tag{10}$$

The threshold $\hat{T}_M = K\hat{\overline{x}}_M(N)$. $\mu_M(\hat{T})$ and $\sigma_M^2(\hat{T})$ will be scaled by K and K^2 respectively. The expression for $\overline{\text{pfa}}$ in the MMSE case is then

$$\overline{\text{pfa}}_M = \frac{1}{1 + \frac{K^2}{N}(2 - \frac{\pi}{2})} \exp\left[-\frac{K^2\pi}{4\{1 + \frac{K^2}{N}(2 - \frac{\pi}{2})\}}\right] \tag{11}$$

The MLE of x_i ($i=1,2,...,N$) i.i.d. Rayleigh samples was evaluated in Madan and Guild [1981] as

$$\hat{\overline{x}}_L(N) = \frac{\pi}{4N} \sqrt{\sum_{i=1}^{N} x_i^2} \tag{12}$$

The expectation value was evaluated as

$$E\{\hat{\overline{x}}_L(N)\} = \frac{1}{\sqrt{N}} \frac{\Gamma(N + 1/2)}{\Gamma(N)} = GR \tag{13}$$

and the variance as

$$\text{var}(\hat{\overline{x}}_L(N)) = 1 - \frac{1}{N} \frac{\Gamma^2(N + 1/2)}{\Gamma^2(N)} = 1 - GR^2 \tag{14}$$

As before, $\hat{T}_L(N) = K\hat{\overline{x}}_L(N)$. $\mu_L(\hat{T})$ and $\sigma_L^2(\hat{T})$ will be scaled by K and K^2 respectively. The expression for $\overline{\text{pfa}}$ in the MLE is then given by

$$\overline{\text{pfa}}_L = \frac{1}{\sqrt{1 + K^2\frac{\pi}{2}(1 - GR^2)}} \exp\left[1 - \frac{K^2}{2} \frac{GR^2}{\{\frac{2}{\pi} + K^2(1 - GR^2)\}}\right] \tag{15}$$

It is easy to realize the limiting values of the quantities in Eq. (9) to (15). In the limit N $\rightarrow$ ∞, E{...} in each case tends to one. With $\sigma = \sqrt{2/\pi}$ the variance tends to zero and $\overline{\text{pfa}}$ tends to exp $\left[-\frac{K^2\pi}{4}\right]$. The calculations were given in Madan and Guild [1981], and the comparative performance of pfa is shown in figure 1.

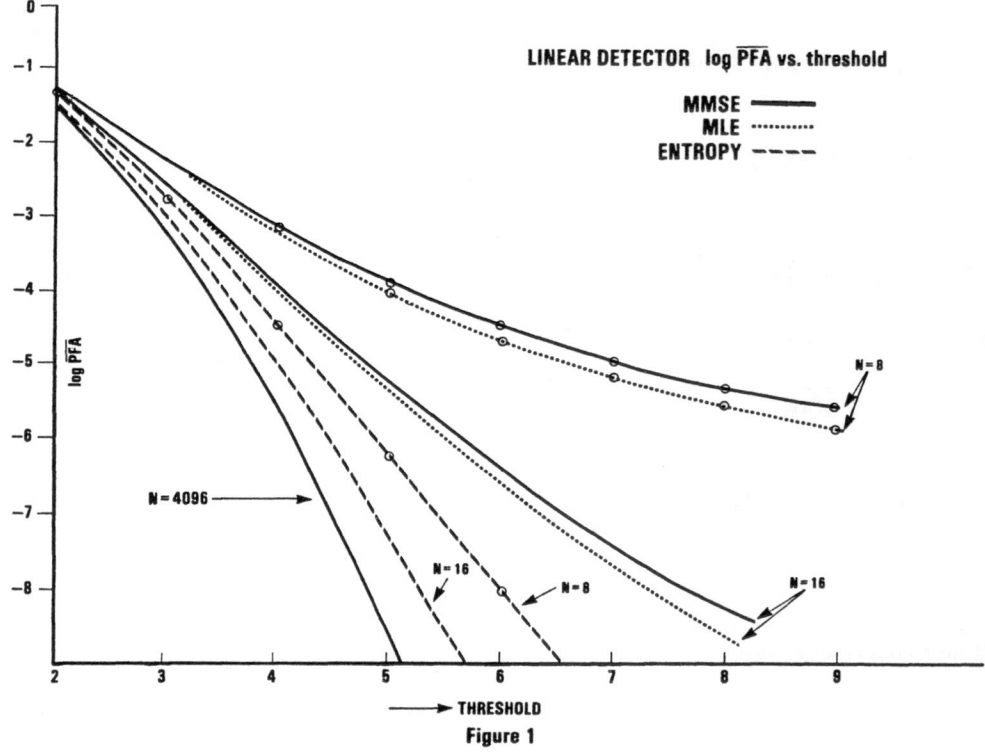

Figure 1

ENTROPY ESTIMATOR

In this paper we start with an ansatz for an estimator, modify it to satisfy the same boundary conditions that the MMSE and MLE satisfy and if found efficient, hope to investigate it further for its intuitive and mathematical relation to physical phenomenon.

Consider the ansatz

$$\hat{\rho}(N) = \frac{\dfrac{1}{N} \sum\limits_{i=1}^{N} x_i}{\prod\limits_{i=1}^{N} (x_i)^{1/N}} \tag{16}$$

for an estimator, where x_i's $(i=1,2,...,N)$ are N i.i.d. samples of any density function, though we shall again confine ourselves to reporting only on the Rayleigh density in this paper. It has been shown in Gray and Markel [1974]* that $\hat{\rho}(N)$ is the discretized version of $\rho(x)$ related to entropy $\epsilon(x)$ by

$$\rho(x) = \exp\left(-\epsilon(x)\right)$$

where

$$\epsilon(x) = \frac{1}{2\pi} \int\limits_{-\pi}^{\pi} \ln\left(S_x(w)/r_x(0)\right)dw \tag{17}$$

As a minimum condition to proceed further, we require that the expectation value of $\hat{\rho}(N)$ tend to the limit 1 in the limit $N \rightarrow \infty$ for i.i.d. Rayleigh samples normalized to unit mean. Evaluating

$$E\{\hat{\rho}(N)\} = \int\limits_0^\infty \int\limits_0^\infty \cdots \int\limits_0^\infty \hat{\rho}(N) \prod_{k=1}^{N} \frac{x_k}{\sigma^2} e^{-x_k^2/2\sigma^2} dx_k$$

with an appropriate change of variables and recognizing that it is a sum of N similar integrals, we get

$$E\{\hat{\rho}(N)\} = \int\limits_0^\infty y^{2-\frac{1}{N}} e^{-y^2/2} dy \left[\int\limits_0^\infty z^{1-\frac{1}{N}} e^{-z^2/2} dz\right]^{N-1}$$

which is easily evaluated employing the Gamma function integral representation

$$E\{\hat{\rho}(N)\} = \frac{\sqrt{\pi}}{2} \Gamma(2 - \frac{1}{N})2^{1/N} \Gamma^{N-2}\left(1 - \frac{1}{2N}\right). \tag{18}$$

To arrive at the limiting value of Eq. (18) as $N \rightarrow \infty$ we require the limiting value of the non-trivial term like

$$\Gamma^{N-2}\left(1 - \frac{a}{N}\right)$$

where $\dfrac{a}{N} \ll 1$.

Consider the expansion in the region $x \ll 1$ for

$$\Gamma(1-x) = \frac{(1-x)\Gamma(1-x)}{1-x} = \frac{\Gamma(1+(1-x))}{1-x}$$

where $\Gamma(1+(1-x))$ can be expanded, employing the series, 6.1.36, Abromowitz and Stegun [1965],

$$\Gamma(1+y) = 1 + \sum_{i=1}^{\infty} b_i y^i, \quad 0 \leq y \leq 1. \tag{19}$$

---

* The possibility of $\hat{\rho}(x)$ leading to a detection estimator arose in discussions with Professor Lloyd J. Griffiths, who later pointed out the relationship between the entropy $\epsilon(x)$ and the spectral power density $S_x(w)$ to the author.

Also expanding $(1-x)^{-1}$ in its series and collecting terms, we get

$$\Gamma(1-x) = 1 + x(1 - \sum_{i=1}^{\infty} i\, b_i) + \text{terms of order } (x^2) \, .$$

$\sum_{i=1}^{\infty} i\, b_i$ can be determined by differentiating Eq. (19) and noticing that it is related to the $\psi(n)$ function as $\sum_{i=1}^{\infty} i\, b_i = \psi(2) = -\gamma + 1$ where $\gamma = $ Euler's constant $= 0.5772156649$. Thus, $\Gamma(1-x) = 1 + \gamma x + \text{terms of order } (x^2)$ and

$$\Gamma^{N-2} \left(1 - \frac{a}{N} \right) = \left(1 + \frac{\gamma a}{N} \right)^{N-2}$$

$$= 1 + \gamma a \left(1 - \frac{2}{N} \right) + \frac{\gamma^2 a^2}{2} \left(1 - \frac{2}{N} \right) \left(1 - \frac{3}{N} \right) + \cdots$$

which in the limit $N \to \infty$ leads to

$$\underset{N \to \infty}{\text{Lt.}} \; \Gamma^{N-2} \left(1 - \frac{a}{N} \right) = e^{a\gamma}. \tag{20}$$

It is now possible to write down the limiting value of Eq. (18) as

$$\underset{N \to \infty}{\text{Lt.}} \; E\{\hat{\rho}(N)\} = \frac{\sqrt{\pi}}{2} \, e^{\gamma/2} \, . \tag{21}$$

Clearly the ansatz of Eq. (16) to be considered as an estimator in the case of the Rayleigh distributed samples should be modified by a multiplying constant or a normalizing constant A that equals the inverse of the result in Eq. (21). Thus we define

$$\hat{\bar{x}}_E(N) = A \, \hat{\rho}(N) \tag{22}$$

as the correct estimator and call it the entropy estimator due to its relationship to the entropy expression referred to earlier. The limit of $E\{\hat{\bar{x}}_E(N)\}$ as $N \to \infty$ tends to 1.

The value

$$A = \frac{2}{\sqrt{\pi}} \, e^{-\gamma/2} \tag{23}$$

is further confirmed to be correct as one evaluates the second moment of $\hat{\rho}(N)$ to form the expression for the variance of the ansatz. Following the procedure in Eq. (18), the integrations in the expression for the second moment are easily carried out and the result is

$$E\{\hat{\rho}^2(N)\} = \int_0^\infty \cdots \int_0^\infty \hat{\rho}^2(N) \prod_{k=1}^N \frac{x_k}{\sigma^2} e^{-x_k^2/2\sigma^2} dx_k$$

$$= \frac{1}{N}\left[1 - \frac{1}{N}\right]\Gamma^N\left[1 - \frac{1}{N}\right] + \left[1 - \frac{1}{N}\right]\Gamma^2\left[\frac{3}{2} - \frac{1}{N}\right]\Gamma^{N-2}\left[1 - \frac{1}{N}\right]. \quad (24)$$

The limiting values of Eq. (24) can be carried out employing Eq. (20) and the result is

$$\underset{N \to \infty}{Lt.} E\{\hat{\rho}^2(N)\} = \frac{\pi}{4} e^\gamma . \quad (25)$$

This implies that the estimator formed out of $\hat{\rho}(N)$ satisfies the second of the conditions that the MMSE and MLE are seen to satisfy. That is,

$$\underset{N \to \infty}{Lt.} var(\hat{\overline{x}}_E(N)) = 0 . \quad (26)$$

In order to form $\overline{pfa_E}$ for the entropy the expression for $\mu_E(\hat{T})$ and $\sigma_E^2(\hat{T})$ are to be obtained and substituted in Eq. (7). The expressions are unnecessarily lengthy so we merely write down the steps involved.

$$\mu_E(\hat{T}) = K\ E\{\overline{x}_E(N)\}$$

$$= K\frac{2}{\sqrt{\pi}} e^{-\gamma/2} E\{\hat{\rho}(N)\} \quad (27)$$

and

$$\sigma_E^2(\hat{T}) = K^2\frac{4}{\pi} e^{-\gamma}[E\{\hat{\rho}^2(N)\} - E^2\{\hat{\rho}(N)\}] . \quad (28)$$

The values of $E\{\hat{\rho}^2(N)\}$ and $E\{\hat{\rho}(N)\}$ are in Eqs. (18) and (24). Substituting these values, we have determined the limit $N \to \infty$ of pfa_E and found it to be $e^{-K^2\pi/4}$ consistent with the limiting values established for the MMSE and MLE estimators. The limiting values for all the three estimators MMSE, MLE and entropy are seen to hold in results of numerical calculations as shown in Figure 1 and Table 1. For large values of N like 4096, all the three estimators seem to converge. They diverge at small values of N where interesting things happen to entropy.

RESULTS AND CONCLUSION

In Table 1, we give results for selected values of N. The expectation value for each estimator or its mean and the variance value is given. The mean in the case of the entropy estimator shows a bias much like the MLE. The magnitude of the bias in each case is almost identical, diminishes with large N, tending to zero as N tends to infinity. The variance of the entropy estimator is less than the MLE variance, which in turn is less than the MMSE variance. The difference is significant for small N, especially in the case of the entropy estimator, which is much tighter at small sample sizes.

Table 1: Rayleigh Distribution

N	MEAN			VARIANCE		
	MMSE	MLE	ENTROPY	MMSE	MLE	ENTROPY
4	1	.9693107	.972383964	.0683098862	.0604367668	.0181720905
8	1	.984506406	.986796404	.0341549431	.0307471368	8.97250425E-03
16	1	.992219199	.993535778	.0170774716	.0155010626	4.46020371E-03
32	1	.996101528	.996797286	8.53873578E-03	7.78174563E-03	2.2251767E-03
64	1	.998048802	.998402053	4.269789E-03	3.89858801E-03	1.11167523E-03
128	1	.99902392	.99919816	2.1346894E-03	1.95120834E-03	5.55547897E-04
256	1	.99951184	.999594723	1.06734197E-03	9.76082636E-04	2.77649721E-04
512	1	.999755888	.999793128	5.33670986E-04	4.88164136E-04	1.37865761E-04
1024	1	.999877936	.999892075	2.66835493E-04	2.44114082E-04	6.9329218E-05
2048	1	.999938962	.999941487	1.33417747E-04	1.22072641E-04	3.48149029E-05
4096	1	.999969469	.999970142	6.67088733E-05	6.10612333E-05	9.57664041E-06

We would like to point out that the estimates in Eqs. (8) and (12) scale linearly with a multiplier of the random variable x_i whereas the function $\hat{\rho}(N)$ in the procedure arrived at in this paper converts $\hat{\rho}(N)$ to $\overline{x}_E(N)$, in Eq. (22), a scale variant quantity. The scaling number is the mean μ_o of the probability density function and is contained in the constant A. Clearly A should be

$$A = \mu_o \ \frac{2}{\sqrt{\pi}} \ e^{-\gamma/2} \tag{29}$$

In our evaluations we employed $\overline{x} = \mu_o = 1$ and included the scaling factor implicitly in Eq. (23).

In Figure 1 we provide a view of the relationship between the threshold (multiplier) K and the average probability of false alarm pfa, for a few values of the sample size N. For N = 4096, a large and limiting value of the sample size, pfa in all three cases MMSE, MLE and the entropy converges to the same curve. For small sample sizes like N = 8 and N = 16, the entropy estimator outperforms the MMSE and MLE by orders of magnitude. In usual estimation situations assuming no knowledge of μ_o, one wants to estimate μ_o from a finite number of samples employing MMSE as the MLE procedure. The procedure of the entropy estimator is useful where one has prior knowledge of μ_0 or a long term mean of the process and then one wants to estimate a short term mean from a few known samples. Applications of this procedure are expected in radar and communication signal processing.

REFERENCES

Abromowitz, A. and Stegun, I.A. (1965). *Handbook of Mathematica! Functions,* pp. 255-293, Dover Publications, Inc., New York.

Gray, A.H. and Markel, J.D. (1974). "A Spectral-Flatness Measure for Studying the Autocorrelation Method of Linear Prediction of Speech Analysis," *IEEE Transactions on Acoustics, Speech and Signal Processing,* ASSP-22, No. 3, pp. 207-217.

Madan, R.N. and Guild, J. (1981). "Maximum Likelihood Estimation in Radar Signals," *International Symposium on Information Theory,* IEEE, February 9-12, 1981, Santa Monica, CA.

THE EVOLUTION OF CARNOT'S PRINCIPLE*

E. T. Jaynes**
St. John's College
Cambridge CB2 1TP, U. K.

Abstract: We trace the development of the technical ideas showing that the Second Law of Thermodynamics became, over a Century ago, a general principle of reasoning, applicable to scientific inference in other fields than thermodynamics. Both the logic and the procedure of our present maximum entropy applications are easily recognized in the methods for predicting equilibrium conditions introduced by Gibbs in 1875. Chemical thermodynamics has been based on them ever since. What is new in this field is not the method, but the recognition of its generality.

*The opening talk at the EMBO Workshop on Maximum-Entropy Methods in x-ray crystallographic and biological macromolecule structure determination, Orsay, France, April 24-28, 1984.
**Visiting Fellow, 1983-84. Permanent Address: Department of Physics, Washington University, St. Louis MO 63130, USA.

G. J. Erickson and C. R. Smith (eds.),
Maximum-Entropy and Bayesian Methods in Science and Engineering (Vol. 1), 267–281.
© 1988 by Kluwer Academic Publishers.

1. INTRODUCTION

The first reaction of nearly everybody, on hearing of a mysterious principle called "maximum entropy" with a seemingly magical power of extracting more information from incomplete data than they contain, is disbelief.

The second reaction, on sensing that there does seem to be something in it, is puzzlement. How is it possible that a quantity belonging to thermodynamics could escape from that setting and metamorphose itself into a principle of reasoning, able to resolve logical ambiguities in situations that have nothing to do with thermodynamics?

Newcomers to this field usually start by asking, not how to apply the method, or even what numerical results it gives; but "I don't see what this has to do with the entropy of thermodynamics -- is there a connection?" Therefore it might be useful, before seeing details of present applications, to explain that connection.

We are taught to think of the First Law of Thermodynamics as a basic law of physics, true of necessity in every case. But attempts to see the Second Law in this way (Kelvin, Clausius, Planck, Boltzmann and many others) never quite succeeded; and Gibbs (1875) recognized that its logic is different. He concluded that "the impossibility of an uncompensated decrease seems reduced to improbability", a remark that Boltzmann quoted 20 years later in the Introduction to his *Gastheorie*.

Clausius saw the second law as a law of physics, but only a qualitative one -- a kind of arrow to tell us in which general direction a process will go. Gibbs, while depriving it of that logical certainty, extended its practical application to serve the stronger purpose of quantitative prediction; to fill the logical void left by the great incompleteness of thermodynamic data. Out of all the different macroscopic behaviors permitted by the data and the laws of physics, which should we choose as, not what must happen, but only what will most likely happen?

Since Gibbs' *Heterogeneous Equilibrium* (1875-78) the second law has been used in practice, not as a "law of physics", but as a principle of human inference; a criterion for resolving the ambiguities of incomplete data. In this service it does indeed extract more information than could have been obtained from the data alone; not by magic but by combining the evidence of the data with the additional information contained in the entropy function.

In other words, Gibbs' use of the second law to predict equilibrium states was virtually identical in rationale with our present maximum-entropy inference. The experimental confirmation of Gibbs' thermodynamic predictions, and the success of maximum-entropy predictions outside thermodynamics, are just two illustrations of the power of that rationale.

The above summarizes our general philosophical viewpoint; now we must justify it by examining those mysterious technical details, to show that there is not just a similarity of philosophy, but an identity of mathematical method.

But we must be prepared for the same disappointment that James Clerk Maxwell felt in 1878 when he examined that mysterious new American invention called a "telephone", and found that it had no ultrafine machinery and ran on principles readily understood by every schoolboy. He reported that his disappointment at its simplicity "was only partially relieved on finding that it actually did work".

We hope that the disappointment of some at the simplicity of the maximum entropy principle, will be partially relieved on finding that it actually does work; and perhaps even fully compensated on seeing the connection with thermodynamics, and the logical unity of these seemingly different fields. This may be hoped for particularly in a lecture given in Paris; for the general viewpoint and the specific principle, from which it all follows, were given by Laplace and Carnot, not far from where we are now. After more than 1½ Centuries, we are beginning to understand them.

In the following we survey the reasoning that follows from Carnot's principle, not in the confusion of the actual historical development, but as today's hindsight shows us it could have been done. Of course, if our object were merely to explain maximum-entropy inference, that can be done by much shorter arguments without reference to thermodynamics, as will doubtless be shown by other speakers here. Indeed, we think that in the future thermodynamics itself will be approached by that shorter route and will be seen as only one particular application of maximum-entropy inference. But our present object is to make the logical connection to what is familiar to scientists today; the historical approach to thermodynamics.

Part 2 recalls the background of Carnot's work. Parts 3–6 offer a short course in "Classical Thermodynamics Made Easy". All the results will be familiar from textbooks, but the reasoning is simpler; even in the initial period when it was considered a new physical law, we can now see that Carnot's principle was actually used only to resolve ambiguities. Then in Part 7 the "new" maximum entropy methods will be seen as just one more step in the natural development of the subject.

Since entropy maximizers are sometimes accused of trying to "get something for nothing", we note that the method expresses, and has evolved from, an explicit statement of the opposite; that you *cannot* get something for nothing.

2. CARNOT'S PRINCIPLE

In the revolutionary years 1791-1797 an army engineer named Lazare Carnot found himself suddenly catapulted into the ruling bodies of France. As a member of the Legislative Assembly, the Committee of Public Safety, and the Directory, he participated in voting for the execution of Louis XVI, the appointment of Napoleon to his Italian command; and in so many other things that his biography (Reinhard. 1950) fills two volumes.

But before and after that period, Lazare Carnot published articles (Gillispie, 1971) on his true lifelong interest, the most general statement of the principles of mechanics and inferring from them the impossibility of a perpetual motion machine. His son Sadi then turned that idea neatly around, inferring a new principle from the assumed impossibility of a perpetual motion machine. (It was Lazare's grandson, another Sadi Carnot, the nephew of our Sadi, who later became President of the French Republic).

An important technical problem of the time was the design of steam engines. How much work can be extracted from a kilo of coal? Can the efficiency be improved by different temperatures or pressures, a different working substance than water; or some different mode of operation than pistons and cylinders?

In the absence of any sound understanding of the nature of heat, misconceptions flourished. In 1818 Petit suggested that an air engine would be far more efficient than a steam engine, because no heat of vaporization need be supplied; an erroneous argument that must have inspired much wasted effort.

Sadi Carnot (1824) resolved this confusion by enunciating a single, intuitively compelling principle that answered all these questions, and a hundred others not yet asked. He envisaged a generalized *heat* engine (and the qualification "heat" is essential — see Appendix B) which operates by drawing heat q_1 from a source which is at thermal equilibrium at temperature t_1, and delivering useful work W. He saw that, in order to operate continuously, the engine requires also a cold reservoir, $t_2 < t_1$, to which some heat q_2 can be discharged. The temperature scale t was any convenient thermometer calibration.

Now Carnot had the happy idea of a reversible engine; one can turn the shaft backwards, delivering the same work W back to the engine, which then delivers the same heat q_1 back to the high-temperature reservoir. One does not think about this very long before perceiving:

Carnot's Principle: No heat engine E can be more efficient than a reversible one E_r operating between the same temperatures.

For suppose that some engine E, given heat q_1, can deliver a greater amount of work $W > W_r$ than does E_r. Then we need only connect them mechanically so that E runs E_r backwards, delivering the work W_r to it, and thus pumping the heat q_1 back into the source reservoir, ready for re–use. The excess $W - W_r$ can then be used to drive our ships, locomotives, and factories. Once started, this would run forever, delivering an infinite amount of useful work without any further expenditure of fuel. We would have a new kind of perpetual motion machine.

We can imagine the economic impact that invention of such a machine would have today. As Max Planck put it, "we expect to make a most serviceable application" of any physical phenomenon that is found to deviate from Carnot's principle, or from any other principle that can be deduced from it. But an astonishing number of things can be deduced from Carnot's principle, as we shall now see.

For example, it follows at once that all reversible heat engines have the same efficiency, independent of the working substance or mode of operation; *i.e.*, the reversible efficiency is a universal function $e(t_1, t_2)$ of the two temperatures. This answered in one stroke all those questions about improvements in steam engines; already Carnot's principle had resolved a mass of ambiguities.

Carnot's reasoning is outstandingly beautiful, because it deduces so much from so little -- and with such a sweeping generality that rises above all tedious details -- but at the same time with such a compelling logical force. In this respect, I think that Carnot's principle ranks with Einstein's principle of relativity.

But Carnot solved the problem only implicitly; while he made it clear that one should strive to make an engine more nearly reversible, he did not find the explicit formula for the reversible efficiency that would result.

3. FIRST METAMORPHOSIS: KELVIN

Wm. Thomson (later Lord Kelvin) was collaborating with James Prescott Joule, who in a private laboratory in Manchester was doing the quantitative measurements that established the validity of the First Law. This gave him the essential fact that Carnot had lacked. Recognizing a universal "mechanical equivalent of heat" h, we can express heat $Q = hq$ and work W in the same units; then we have the relation

$$W = Q_1 - Q'_2 \tag{1}$$

and deduction of the quantitative consequences of Carnot's principle can begin (the prime will be explained presently).

Consider two reversible engines connected in series; engine A receives the heat Q_1 at temperature t_1, delivers work W_A, and discharges heat Q'_2. Engine B receives the heat Q'_2, delivers work W_B, and discharges heat Q_3 to a reservoir at temperature $t_3 < t_2$. Their efficiencies are

$$e(t_1, t_2) = W_A/Q_1 \quad , \quad e(t_2, t_3) = W_B/Q'_2 \quad . \tag{2}$$

But by a simple mechanical linkage, A and B can be combined into a single reversible engine C, which receives heat Q_1 and delivers work $W_C = W_A + W_B$. So we must have also

$$e(t_1, t_3) = (W_A + W_B)/Q_1 \tag{3}$$

From (1), (2), (3) we find that Carnot's principle requires that (using the abbreviation $e_{13} = e(t_1, t_3)$, etc.), the reversible efficiency must satisfy the functional equation

$$e_{13} = e_{12} + e_{23} - e_{12}e_{23} \quad , \quad t_1 > t_2 > t_3 \quad . \tag{4}$$

This is a condition of consistency, and it reminds us of a relation of probability theory. The change of variables $x = \log(1-e)$ makes the general solution obvious: the reversible efficiency must have the functional form

$$e_r(t_1, t_2) = 1 - f(t_2)/f(t_1) \quad . \tag{5}$$

From (2a) and (3), $f(t)$ is a monotonic increasing function. By Carnot's principle the ratio $f(t_1)/f(t_2)$ must be the same function of t_1, t_2 for all reversible engines.

Now the temperature scale t was basically arbitrary (uniformly spaced marks on a mercury thermometer and on a gas thermometer do not agree), long a troublesome problem. Turning the argument around, Kelvin perceived that Carnot's principle resolves another ambiguity; if the reversible efficiency is a universal function of the temperatures, then it in effect defines a universal temperature scale that is independent of the properties of any particular substance like mercury.

We define the *Kelvin temperature scale* by

$$T(t) = Cf(t) \tag{6}$$

and now only one free choice is left to us; we may choose the arbitrary multiplicative factor C, as a convention, to indicate the size of the units in which we measure temperature.

Carnot's reversible efficiency is then

$$e_r = 1 - T_2/T_1 \tag{7}$$

from which one readily determines the measurements by which one can calibrate his thermometer to read T. All such thermometers, however constructed, and independently calibrated, will then agree in their readings at all points, if the units are chosen so that they agree at one point (at least, if a thermometer is ever found that fails to do this, then we shall have the means to realize Carnot's perpetual motion machine after all).

With this, Carnot's principle starts its metamorphosis. It now says that the efficiency e of a real heat engine must satisfy the inequality $e \leq e_r$; or from (1), (2), (7),

$$e = 1 - Q'_2/Q_1 \leq 1 - T_2/T_1 \quad . \tag{8}$$

Writing now $Q_2 = -Q'_2$ so that Q_1, Q_2 are both quantities of heat delivered from a heat reservoir to the engine, it takes the more suggestive form

$$Q_1/T_1 + Q_2/T_2 \leq 0 \tag{9}$$

with equality if and only if the engine is reversible.

A simple generalization is then obvious; we may consider a more complicated heat engine that runs cyclically, making contact successively with n reservoirs at temperatures $(T_1 \cdots T_n)$. Then the first law is $W = \Sigma Q_i$, and Carnot's principle becomes

$$\Sigma Q_i/T_i \leq 0 \quad . \tag{10}$$

This is the form that Kelvin used in his work of 1854 on the thermocouple, in which the Seebeck potential and the Peltier, Joule, and Thomson heat effects were all analyzed correctly, leading to the thermoelectric equations still in use.

Equation (10) is pivotal for all further developments, forming the starting point for two very different extensions. Some comments on its meaning and generality are given in Appendix A. In Appendix B we note some important limitations on the applicability of the reversible efficiency formula (7), and speculate about further generalizations.

4. SECOND METAMORPHOSIS: CLAUSIUS

It is curious that, having perceived such an important consequence of Carnot's principle as the temperature scale (6), Kelvin does not seem to have perceived the still more important fact that was now staring him in the face in (10). This was left for Rudolph Clausius. Imagine (10) extended to arbitrarily large n, the sum going into a cyclic integral:

$$\oint dQ/T \leq 0 \quad . \tag{11}$$

In the limit of a reversible process, where the equality applies, T is also the temperature of the system. But then (11), holding for any cycle, is the condition that the line integral over any part of a cycle is independent of the path.

Thus was discovered a new function S of the thermodynamic state of the system, defined to within an additive constant by the difference

$$S_A - S_B = \int_B^A dQ/T \tag{12}$$

where we integrate over a reversible path R; i.e., a locus of equilibrium states. As we know, Clausius coined the name "entropy" for this quantity.

The integral (12) is over a reversible path, but only part of a cycle. Then complete the cycle by adding to it a return from state A to state B over any path P, reversible or irreversible. With this choice of cycle, (11) becomes

$$\int_A^B dQ/T \leq S_B - S_A \tag{13}$$

with equality if and only if the process A → B is reversible (the meaning of the word "reversible" may now be extended beyond what Carnot had in mind, as explained in Appendix C).

Note, from its origin in (10), that in (13) T denotes the temperature of a heat bath with which the system is momentarily in contact. This may or may not be the temperature of the system. Therefore, in all cases, reversible or irreversible, the negative of the left-hand side of (13) is the entropy gained by the heat-reservoirs which constitute, for the system, the "rest of the universe". So Carnot's principle has now become: in the change from one thermal equilibrium state to another, the total entropy of all bodies involved cannot decrease; if it increases, the process is therefore irreversible:

$$S(\text{final}) \geq S(\text{initial}) \quad . \tag{14}$$

Note also that (14) describes only the net result of a process that begins and ends in thermal equilibrium. Carnot's principle does not permit us to draw any such conclusion as $dS/dt \geq 0$ at intermediate times. Indeed, entropy has been defined only for equilibrium states, in which there is no time dependence.

We shall take (14) as the fundamental Clausius statement of the second law. Other statements have been proposed, but (14) is logically simpler and it has stood the test of time, remaining valid in situations such as negative spin temperatures where some others failed. However, from the standpoint of logic (14) only restates Carnot's principle in a more useful form; it adds nothing to its actual content. In particular, Clausius could still see it as a law of physics.

5. THIRD METAMORPHOSIS: GIBBS

The above statements of the Second Law are the ones traditionally taught to physicists, although they have severe limitations. Equation (14) gives us one piece of information about the general direction in which an irreversible process will go; but it does not tell us how fast it will go, how far, or along what specific path. And it refers only to a closed system (no particles enter or leave).

Gibbs showed how to remove two of those limitations. He generalized the definition of entropy to open systems, as needed for many applications. More important for our purposes, he perceived the correct logical status of Carnot's principle, which enabled him to extend its application to quantitative prediction, thus answering the question: "How far?".

Instead of Clausius' weak statement that the total entropy of all bodies involved "tends" to increase, Gibbs made the strong prediction that it *will* increase, up to the maximum value permitted by whatever constraints (conservation of energy, volume, mole numbers, etc.) are imposed by the experimental arrangement and the known laws of physics. Furthermore, the systems for which this is predicted can be more complicated than those envisaged by Clausius; they may consist of many different chemical components, free to distribute themselves over many phases.

Gibbs' variational principle resolved the ambiguity: "Given the initial macroscopic data defining a nonequilibrium state, there are millions of conceivable final equilibrium states to which our system might go, all permitted by the conservation laws. Which shall we choose as the most likely to be realized?"

Although he gave a definite answer to this question, Gibbs noted that his answer was not found by deductive reasoning. Indeed, the problem had no deductive solution because it was ill-posed. There are initial microstates, allowed by the data and the laws of physics, for which the system will not go to the macrostate of maximum entropy. There may be additional constraints, unknown to us, which make it impossible for the system to go to that state; for example new "constants of the motion". So on what grounds could he justify making that choice in preference to all others?

At this point thermodynamics takes on a fundamentally new character. We have to recognize the distinction between two different kinds of reasoning; *deduction* and *inference*. Instead of asking, "What do the laws of physics require the system to do?", which cannot be answered without knowledge of the exact microstate, Gibbs asked a more modest question, which can be answered: "What is the best guess we can make, from the partial information that we have?" Of course, this implies some statement of what we mean by "best".

At first glance, this does not seem a very radical move. It must be clear to every child that virtually all human reasoning, in or out of science, is of necessity inference rather than deduction. Yet to sophisticated scientists this change of thinking has been very difficult conceptually, and it can arouse bitter controversy. To recognize "officially" that we are only doing inference rather than deduction, is very foreign to the attitude that scientists are taught.

The conventional attitude is exhibited by those who would object to Gibbs' answer on the grounds that there may be unknown constraints that prevent the system from getting to the state of maximum entropy; and so Gibbs' answer might be wrong. But the same kind of objection would apply whatever answer he gave. If such an objection were sustained, Gibbs would be prohibited from giving any answer at all. Science does not advance on that kind of timidity; let us note how much more realistic and constructive is the opposite attitude. To one who raised that objection, Gibbs might reply as follows:

"Of course, my answer might be wrong. You seem to think that would be a calamity that we must avoid; but you are like a chess player who thinks only one move ahead. If you will think ahead two moves, you will see that, on the contrary, getting a wrong answer would be even more valuable than getting a right one. As you note, at present we do not know whether there may exist unknown constraints that would prevent the system from getting to the maximum entropy state. But I choose to ignore that warning, go ahead with my calculation, and then ask an experimentalist to compare my prediction with observation. What conclusions will we be able to draw from his verdict?

"Suppose my prediction turns out to be right. That does not prove that no unknown constraints exist; but it does prove that there are none which prevent the system from getting to the state of maximum entropy. So the calculation has served a useful predictive purpose, and its success gives us more confidence in future predictions.

"But suppose my prediction turns out to be wrong; the experiment repeatedly gives a different result. Then we have learned far more; we know that there *is* some new (*i.e.*, previously unknown) constraint affecting the macroscopic behavior, and the nature of the error gives us a clue as to what that new constraint is. We would have a start toward learning a fundamental new physical fact. I do not see this as a calamity; how else can we advance to a new state of knowledge about physical law, but by having the courage to go ahead with the best inferences we can make on our present state of knowledge?"

The words we have just put into Gibbs' mouth are not fanciful. Gibbs' classical statistical mechanics did make incorrect predictions of specific heats and vapor pressures. These were the first clues, indicating the new constraints of discrete energy levels, pointing to the quantum theory. Nobody would have realized that such things were relevant to the question, had Gibbs lacked the courage to make an inference because he might be wrong.

After development of the Schroedinger equation, the Gibbs formalism based on maximizing the new quantum expression for entropy has yielded so many thousands of quantitatively correct equilibrium predictions that there seems to be almost no chance that it will ever fail in that problem. Whenever it did seem to fail -- as in the case of ortho and para hydrogen -- it was seen quickly that it was only performing its second function, revealing an unexpected constraint.

Today, we are only in the initial stages of extensions to predict the details of nonequilibrium behavior; these put our entropy expressions to a more severe test. We can by no means rule out the possibility that nonequilibrium statistical mechanics may lead to incorrect predictions, which would then point the way to the next higher level of understanding of physical law, beyond our present quantum theory. We may be seeing the incipient beginnings of this in the lore of "strange attractors".

We think that this scenario will be repeated many times in the future outside thermodynamics, particularly as the method moves into biology. Most maximum entropy inferences will be correct, serving a useful predictive purpose. But some of the predictions will be wrong; those instances, far from being calamities, will open the doors to new basic knowledge.

Another of the curiosities of this field is that, having done so much with entropy and demonstrated such a deep understanding of the logic underlying the second law, giving thermodynamics an entirely different character, Gibbs said almost nothing about what entropy really means. He showed, far more than anyone else, how much we can accomplish by maximizing entropy. Yet we cannot learn from Gibbs: "What are we actually doing when we maximize entropy?" For this we must turn to Boltzmann.

6. FOURTH METAMORPHOSIS: BOLTZMANN

Entropy first appeared, unanticipated and without warning, merely as a mathematical construct in equation (12). Even after its fundamental nature and usefulness were recognized and exploited, the question "What is it?" continued to mystify and confuse. It appears that the answer was first revealed to Ludwig Boltzmann, when he calculated the phase volume of an ideal gas of N atoms in volume V, for which the energy lies in (E, E + dE):

$$W = \int d^3x_1 \dots d^3x_N \, d^3p_1 \dots d^3p_N = CV^N E^{3N/2-1} \, dE \tag{15}$$

where the region of integration is those points for which all coordinates are within a volume V, and the momenta satisfy

$$E < \Sigma \, p^2/2m < E + dE \quad . \tag{16}$$

The constant C is independent of V and E.

It was evident that log W has the same volume and energy dependence as the entropy of that gas, calculated from (12), so to within an additive constant it was true that

$$S = k \log W \quad . \tag{17}$$

This is such a strikingly simple relation that one can hardly avoid jumping to the conclusion that it must be true in general; *i.e.*, the entropy of any macroscopic thermodynamic state A is a measure of the phase volume W_A occupied by all microstates compatible with A.

It is convenient verbally to say that S measures the "number of ways" in which the macrostate A can be realized. This is justified in quantum theory, where we learn that a classical phase volume W does correspond to a number of global quantum states $n = W/h^{3N}$. So if we agree, as a convention, that we shall measure classical phase volume in units of h^{3N}, then this manner of speaking will be appropriate in either classical or quantum theory.

We feel quickly that the conjectured generalization of (17) must be correct, because of the light that this throws on our problem. Suddenly, the mysteries evaporate; the meaning of Carnot's principle, the reason for the second law, and the justification for Gibbs' variational principle, all become obvious. Let us survey quickly the many things that we can learn from this remarkable discovery.

Given a "choice" between going into two macrostates A and B, if $S_A < S_B$, a system will appear to show an overwhelmingly strong preference for state B, not because it prefers any particular microstate in B, but only because there are so many more of them. As noted in Appendix C, an entropy difference $(S_B - S_A)$ corresponding to one microcalorie at room temperature indicates a ratio $W_B/W_A > \exp(10^{15})$. Thus violations are so improbable that Carnot's principle, or the equivalent Clausius statement (14), appear in the laboratory as absolutely rigid constraints suggesting a law of physics rather than a matter of probability.

Let us see the light that this casts on Gibbs' method, by examining a simple application. We have two systems of one degree of freedom (*i.e.*, their energy and temperature can vary when in contact with other systems). Then their entropy functions are

$$S_1(E_1) = k \log W_1(E_1) \quad , \qquad S_2(E_2) = k \log W_2(E_2) \quad , \tag{18}$$

The systems start out in thermal equilibrium with arbitrary initial energies E_{1i}, E_{2i}. Then they are placed in contact so they can exchange energy in such a way that the total amount is conserved:

$$E = E_1 + E_2 = const. \quad , \quad E_1 > 0 \quad , \quad E_2 > 0 \quad . \tag{19}$$

Required: to predict the final energies E_{1f}, E_{2f} that they will reach when they come into equilibrium with each other.

This is manifestly an ill-posed problem; for the final energies must depend on the initial microstates which are unknown; and all values compatible with (19) are possible without violating any known laws of physics. We are thus obliged to use inference rather than deduction. Gibbs' algorithm was: predict that energy distribution that maximizes the total entropy $S_1 + S_2$ subject to the constraint (19). At first this seems arbitrary; but now if (17) is correct we can see why this guess is "best". We are maximizing the product

$$M(E_1) = W_1(E_1)W_2(E - E_1) \tag{20}$$

with respect to E_1; but that product is just the multiplicity, or number of ways in which the energy distribution (E_1, E_2) can be realized. So in the light of (17) Gibbs' rule now says, in effect: "Predict that energy distribution that can happen in the greatest number of ways, subject to the information you have".

Experimentally, one says that equilibrium is reached when the systems have equal temperature. Differentiating (20), we find that the maximum is reached when d log W_1/dE_1 = d log W_2/dE_2. But the general thermodynamic relation $T^{-1} = dS/dE$ that follows from (12) becomes, in the light of (17)

$$(kT)^{-1} = d \log W/dE \quad . \tag{21}$$

So the general interpretation of entropy by (17) not only predicts equal temperature as the condition for equilibrium; it gives a simple explanation of why this is true.

The above explains why Gibbs' method gives, in a sense, the best guess one could have made in view of our great ignorance as to the microstate; but does not explain why it is so uniformly successful. If the multiplicity (20) had a broad maximum, or many local maxima, one would not expect Gibbs' rule to be very reliable in practice. This raises the question: How sharp is the maximum in the multiplicity (20)? Note that differentiating (21) once more gives the heat capacity:

$$d^2 \log W/dE^2 = -(kT^2C_v)^{-1} \quad . \tag{22}$$

But, as (15) shows for an ideal gas and is true in general, C_v may be interpreted as $C_v = nk/2$, where n is the effective number of degrees of freedom of the system (in quantum theory, the number excited at the temperature T), of the order of Avogadro's number for a macroscopic system. Therefore, expanding log $M(E_1)$ about its peak at E' we have

$$M(E_1) = M(E')\exp[-(E_1 - E')^2/2D^2] \tag{23}$$

with the RMS deviation

$$D = kT[n_1n_2/(n_1 + n_2)]^{1/2} \quad . \tag{24}$$

which is of the order of $E'/n^{1/2} = 10^{-12} E'$. Therefore, not only is E' the value of E_1 that can happen in the greatest number of ways for given total energy E; the vast majority of all possible microstates with total energy E have E_1 very close to E'. Less than 1 in 10^8 of all possible states have E_1 outside the interval $(E' \pm 6D)$, far too narrow to measure experimentally. From (17), then, we understand also why Gibbs' method succeeds.

But there is still more to be learned from (17). Imagine system 2 (i.e., n_2) to become very large; then we may expand using (21):

$$\log W_2(E - E_1) = \log W_2(E) - E_1/kT + \cdots \tag{25}$$

and from (22) the next term is negligible. But then the fraction of the multiplicity (23) in the interval $(E_1, E_1 + dE_1)$ becomes

$$f(E_1)dE_1 = Z^{-1} W_1(E_1)\exp(-E_1/kT) dE_1 \tag{26}$$

which is the distribution of Gibbs' "Canonical Ensemble", the basis of his later work on Statistical Mechanics.

The normalization constant

$$Z(\beta) = \int W_1(E) \exp(-\beta E_1)dE \quad , \qquad \beta = 1/kT \tag{27}$$

is Gibbs' partition function, and if we refine the inference procedure by taking as our prediction the mean value over the distribution (26) instead of the peak E', our prediction reduces to

$$<E_1> = -d \log Z/d\beta \tag{28}$$

the basic predictive rule of statistical mechanics.

All these relations generalize effortlessly to systems with more macroscopic degrees of freedom (volume, magnetization, angular momentum, mole numbers, etc.) corresponding to Gibbs' grand canonical ensemble and its generalizations. So the interpretation (17) of entropy has given us the key to essentially everything that has happened since in the field of equilibrium thermodynamics and statistical mechanics. This was recognized, and exploited in their fundamental research, by both Planck and Einstein.

7. CONCLUSION

We have followed the evolution of Carnot's principle, via Kelvin's perception that it defines a universal temperature scale, Clausius' discovery that it implied the existence of the entropy function, Gibbs' perception of its logical status, and Boltzmann's interpretation of entropy in terms of phase volume, into the general formalism of statistical mechanics. But now, we can see how utterly simple it all is, and that the reasoning had nothing fundamentally to do with thermodynamics.

From our present vantage point, everything we have done could have been found as a trivial consequence of (20). But, as a principle of reasoning, that had been given in far greater generality by James Bernoulli (1713) and Laplace (1812). Had their works been better understood and applied, we might have passed directly from first principles of inference to the canonical ensemble, with the second law as a straightforward predicted consequence rather than a puzzling empirical fact. The genius of Carnot was to have seen the one case where that fact is not puzzling, but intuitively compelling.

In our comments on Gibbs' work we have noted two fundamentally different attitudes, epitomized by the words *deduction* and *inference*. But the inference attitude that Gibbs introduced into thermodynamics is just what Bernoulli had recognized and expounded, much earlier. With an accuracy and honesty almost unique in probability theory, Bernoulli called his work, simply, *Ars Conjectandi*, the "Art of Conjecture". Seeing correctly that a major intellectual problem of both science and everyday life is the necessity of reasoning somehow from incomplete information, he asked whether there are any general principles, of consistency and rationality, which would help us in this.

A single desideratum of consistency; that propositions about which we are in the same state of ignorance should be given equal weight in our reasoning, implies all else. This desideratum led Bernoulli to write down the general equations of probability theory.

In the hands of Bernoulli those equations were seen, not merely as rules for calculating frequencies; but the consistent rules for conducting inference; a probability distribution is used as a means of describing our state of incomplete knowledge, and the equations show how probabilities of different propositions must be related for consistency. Laplace adopted this viewpoint and applied it with great success to a mass of problems of scientific inference. In the course of this, he developed the analytical theory relevant to applications to a level that is not often surpassed today. The history is recounted in more detail in Jaynes (1983).

Now by Bernoulli's principle the multiplicity function $M(E_1)$ in (20) expresses just the relative probabilities that we should assign to different values of E_1 to represent a state of knowledge about E_1. "*Whose* state of knowledge?" everyone asks. *Answer*: a person whose relevant information consists of the phase volume functions W_1, W_2, the total energy E; and nothing else. These equations represent the "best" inferences that can be made by a person in that state of knowledge. Aware of the microstates and their relative numbers but, having no grounds for preferring any particular microstate consistent with our knowledge to any other, an honest description of what we know requires us to assign equal probabilities to them, resulting in (20).

A person with greater knowledge would have a smaller set of possible states, and would be able to make better predictions of some things. But he would seldom do better in prediction of reproducible phenomena, because those are the same for virtually all microstates in an enormously large class C; and therefore also in virtually any subset of C. Indeed, as Gibbs showed, in almost every case the knowledge supposed above is already sufficient to predict equilibrium states correctly. Still greater knowledge (such as, perhaps, that the real system stays in some complicated fractal subset of C) might be very interesting and important for future purposes; but it would not have helped for the predictions that Gibbs was making.

Knowledge of the "data" E alone would not enable us to choose among the different values of E_1 allowed by (19); the additional information contained in the entropy functions, nevertheless leads us to make one definite choice as far more likely than any other, *on the information supposed.*

On the other hand, we need not know the entropy functions in advance; nothing prohibits us from trying out some guess, as a working hypothesis, to see what predictions it would make. Good predictions give us confidence that our guess was right; bad predictions can, from the particular way they fail, give us clues to a better guess. We expect that many applications will take this form.

The sharpness of the thermodynamic predictions, and the resulting "stone wall" character of the second law, arise from the extremely high dimensionality of the space of microstates, with the result that W(E) is an enormously rapidly increasing function, of the order of E to the N'th power, where N is Avogadro's number.

The same reasoning will apply to many situations outside thermodynamics, in which we can enumerate a set C of conceivable situations (corresponding to the global quantum states in thermodynamics), and have some data that, although incomplete, restricts them to some subset of C. In thermodynamics we are generally trying to predict only a few quantities, seldom more than three or four; but in image reconstruction it is not unheard of to estimate over a million pixel intensities. This is not a difference of principle, although it is a major difference to a computer programmer.

Our numerical values will not in general be as extravagant as in thermodynamics, but may still be enormous by ordinary standards. The preferred choice may be indicated over others by a factor of only a million, instead of the fantastic numbers of statistical mechanics. But the general conclusions and usefulness of the reasoning will be the same; there will still appear to be a "second law" favoring situations of high entropy, because they can be realized in more ways.

In the above we have tried to show the basic unity of thermodynamics and the "new" maximum entropy methods, by expounding a general philosophy and rationale that pertains equally to both. But it may not yet be clear where the mathematical connection lies. The mathematical appearance of our solution depends on how we choose the basic "hypothesis space" that corresponds to enumerating the global quantum states of thermodynamics.

As a highly oversimplified example, if in thermodynamics our system consisted of N molecules with r possible energy levels each, there would be r^N possible global states, each one defined by specifying the state of each molecule. If we wanted to predict the number n_k of molecules in the k'th energy level, subject to some incomplete data D that partially restricts the possibilites, we would want to maximize the number of ways a certain set $(n_1 ...n_r)$ of occupation numbers could be realized:

$$W(n_1 \cdots n_r) = \frac{N!}{n_1! \cdots n_r!} \quad . \tag{29}$$

But when these numbers become large we may use the Stirling approximation, and the entropy corresponding to W is given by

$$(1/N) \log W \to -\Sigma f_i \log f_i \tag{30}$$

where $f_i = n_i/N$ is the fraction of molecules in the i'th energy level. Maximization of the expression (30) gives the familiar mathematical form of the principle.

Now there are many other applications where enumeration of the possibilities leads us to write the same combinatorial factor (29); and therefore the solution will proceed in the same mathematical way. For example, we may think of an image created by strewing N little "elements of luminance" over r pixels, the k'th one receiving $n_k = Nf(k)$ of them. The set of proportions $\{f(k)\}$ constitutes the "true" image, and to estimate the $f(k)$ from incomplete data we would maximize (30) subject to the constraints of the data. This would give us the reconstructed image that is "most likely" on the supposed information; i.e., it is the one that can be realized in more ways than any other that agrees with the data. The situation is logically equivalent to that of thermodynamics.

But suppose we are dubious as to whether our image has been formed by anything like this strewing process. Is there now any reason for using the same maximum entropy algorithm? It appears that there may be several.

One is the argument that we put, rather presumptuously, into Gibbs' mouth above: if there is anything seriously wrong with this hypothesis space, then reconstructions based on it ought to show systematic deviations from the true scene. If such deviations are found, then they give us a clue to a better hypothesis space; if not, there is no need for a different one. Therefore, if those doubts are not specific enough to suggest a definite alternative hypothesis space, we have nothing to lose and something to gain by continuing to use the above one.

But the maximum-entropy procedure may be supported by more than one rationale. There are other arguments, based on logical consistency or on information theory, rather than multiplicity, which also point uniquely to the maximum entropy rule.

Finally, we may invoke entropy not as an element in a probabilistic argument, but as a value judgment; out of all reconstructions that agree with the data, the maximum entropy one is preferred, simply because it is cleaner, free of artifacts, and therefore more informative and safer to use. This comes about because entropy is not only a measure of multiplicity; the expression (30) is also a measure of smoothness.

Any change in the direction of equalizing any two elements f_i and f_j, increases the entropy (30). When we maximize (30), f_i and f_j will therefore be equalized unless this is prevented by some constraints coming from the data; the variational principle that generates it guarantees that the maximum entropy reconstruction cannot show any detail for which there is no evidence in the data. To the best of our knowledge, no other algorithm that has been proposed for this problem has this safety feature.

Indeed a procedure is something that exists in its own right, and is not tied down to any particular theoretical rationale. In the last analysis, the methods of science are determined by their pragmatic success; a procedure that yields useful results will be used even if it has no theoretical basis at all. If it does not yield useful results, it will not be used however strong the theoretical arguments for it. So, now let us hear from others how the method is working on some real, important, and highly nontrivial problems.

APPENDIX A. COMMENTS ON KELVIN'S RELATION

Some have thought it inelegant to base a scientific theory on such vulgar things as heat engines; but in Kelvin's form (10) of Carnot's principle:

$$\Sigma Q_i / T_i \leq 0 \tag{A1}$$

the vulgarity and inelegance are gone. (A1) refers to any cyclic process involving coupling a system of interest to n heat reservoirs, whether or not any work is actually done. The engines are only, so to speak, held in abeyance. We infer a number of general inequalities which make no reference to any engine; but if ever we find a physical phenomenon that violates one of these inequalities, then some marvelous engines can be built.

Equation (A1) is more general in another respect. With engines in mind we have been making the tacit assumption that contacts with the different heat reservoirs are to be made successively in time. But (A1) makes no explicit reference to time either, and a moment's thought persuades us that this assumption was not necessary for our argument. Kelvin applied (A1) to the continuous flow of electric current around his thermocouple, which was in continuous contact with the heat reservoirs.

Although it goes beyond our present topic, we note that Kelvin in using (A1) launched another new field by introducing a new assumption; that we may apply Carnot's principle in the form (A1) with the equality sign, to the reversible Peltier and Thomson heat effects, even though irreversible heat conduction is also present.

But a heat current presumably "drags" some electric current with it; i.e. conduction electrons ought to be carried along a little by a stream of phonons, like sand grains carried downstream by the current of a river. Conversely, a stream of moving electrons ought to drag along some phonons, generating a heat current. The reversible effects cannot be physically independent of the irreversible ones; and this should have introduced an error into Kelvin's analysis.

The fact that Kelvin's equations were nevertheless verified means that these two errors must have cancelled each other. This was recognized later as implying a proportionality relation between the two dragging coefficients; the first example of an "Onsager Reciprocal Relation". Thus both the Clausius-Gibbs entropy developments in equilibrium theory, and Onsager's irreversible thermodynamics (1931), had their origins in Kelvin's use of (A1).

APPENDIX B: ANTI-CARNOT ENGINES

It is important that we understand some limitations on the applicability of the reversible efficiency formula (7):

$$e_r = 1 - T_2/T_1 \quad . \tag{B1}$$

This applies, not to every type of energy converter but only to *heat* engines -- *i.e.*, engines which operate by extracting heat from one reservoir which is at thermal equilibrium at some temperature T_1 and delivering heat to a similar reservoir at a lower temperature T_2. But there is no reason why (B1) should apply to engines that deliver work by a different mode of operation.

Indeed, the world's most universally available source of work -- the animal muscle -- presents us with a seemingly flagrant violation of that formula. Our muscles deliver useful work when there is no cold reservoir at hand (on a hot day the ambient temperature is at or above body temperature) and a naive application of (B1) would lead us to predict zero, or even negative, efficiency. But according to Lehninger (1965), under these conditions they still deliver an efficiency of about 20 percent. According to Alberts *et al.* (1983), under favorable conditions the efficiency of a muscle can be as high as 70%.

The answer, of course, is that a muscle is not a heat engine. It draws its energy, not from any heat reservoir, but from the activated molecules produced by a chemical reaction. This is why we stressed the word "heat" when we introduced Carnot's principle.

Only when we first allow that activation energy to degrade itself entirely into heat -- and then extract only that heat for our engine -- does Kelvin's formula apply. If we can learn how to capture the activation energy before it has a chance to degrade, as our muscles have already learned how to do, then we shall be able to achieve higher efficiency than (B1) in an engine. Such an anti-Carnot engine will, of course, not violate the second law. Rather, to achieve it will require a very clear understanding of what the second law really says.

What efficiency might one hope for in such an engine? There is no reason to doubt that, with proper understanding of these matters, the performance of our muscles could be at least equalled in vitro. Now, whatever the theoretical maximum efficiency, it can always be written in the Kelvin form (B1) if we wish to do so; the question then becomes: "What are the effective upper and lower temperatures?"

As a partial answer we imagine that our engine will, like our muscles, eventually discharge some heat to the outside world; then let us take T_2 as the ambient temperature; for our muscles, body temperature. What is the effective upper temperature? It appears to us that this was answered already by Gibbs; it is the highest temperature to which the activated molecules could deliver heat, which would make the maximum efficiency close to 100%.

To see this more specifically, note that at room temperature the average thermal energy ($kT/2$) per degree of freedom is about 1/80 ev. A chemical reaction might leave a product molecule in an excited state with perhaps $E_a = 0.5$ ev of activation energy. If this is concentrated in only $N = 2$ or 3 degrees of freedom, it thus represents a tiny "hot spot" with an effective temperature $T_{eff} = 2E_a/N_k$, of the order of 20 times room temperature. This, we conjecture, is the temperature T_1 that we should use in Kelvin's formula.

If we can convert that little bubble of concentrated energy into useful work before it has a chance to thermalize by spreading out over 100 vibrational degrees of freedom, we should in principle be able to realize something like 95% conversion efficiency. Thus the 20% to 70% actually realized by our muscles ceases to be puzzling.

APPENDIX C: REVERSIBILITY

Thermodynamics is notoriously a field which encourages confusion and illogic by a terminology which may use a common technical term with several different meanings, and fails to distinguish between them. We have noted before (Jaynes, 1980) some of the many different, mutually inconsistent meanings that have been attached to the word "entropy". An equally serious confusion arises from the fact that the word "reversible" is used with different meanings; and few writers since Gibbs and Planck have taken sufficient note of this.

Let A and B stand for two different macrostates, defined by specifying (*i.e.*, controlling or observing) a few macroscopic quantities like temperature, volume, pressure, magnetization, such that the change A → B can be carried out in the laboratory. What do we mean by saying that it is reversible? In the literature, we find three different meanings:

1. *Mechanical Reversibility.* Reversing all molecular velocities in B, the equations of motion carry the system back along exactly its previous path to A. In the end this would restore every individual molecule to its original position.

But this is manifestly not what Carnot had in mind. In his reversible engines he is considering instead:

2. *Carnot Reversibility.* The macroscopic physical process can be made to proceed in the opposite direction B → A, restoring the original macrostate.

This is an enormously weaker condition than mechanical reversibility. But it was noted by Clausius, Gibbs, and Planck that thermodynamic reversibility is a still weaker condition (one that Carnot could have used without invalidating his argument):

3. *Thermodynamic Reversibility.* Even if the backward process B → A cannot be made to take place reversibly (for example, because of supercooling at a phase transition), if by any means such as B → C → D → A the original macrostate can be recovered without external change, then all entropies are unchanged and the process A → B is thermodynamically reversible.

From this we see that the common phrase "--- the paradox of how to reconcile the irreversibility of the second law with the reversibility of the equations of motion ---" does not define any real problem at all; it is a nonsense utterance, using the term "reversible" in the totally different meanings (1) and (3) in the same sentence.

These observations are hardly new. The distinction between mechanical and thermodynamic reversibility was stressed by Gibbs (1875) in his discussion of gas diffusion. Confusion of thermodynamic reversibility with Carnot reversibility was called by Planck (1949), "--- an error against which I have fought untiringly all my life, but which seems impossible to eradicate."

Despite the efforts of Gibbs and Planck, these distinctions have been nearly lost today. We have found no recognition of them in current thermodynamics textbooks, or in the current literature of statistical mechanics. In our opinion, recent efforts to "explain irreversibility" by tampering with the equations of motion or the definition of entropy, address themselves to a non-problem, for reasons that Gibbs explained cogently over 100 years ago.

It should not have required the labors of Carnot, Kelvin, and Clausius to convince us that one cannot reverse all molecular velocities with the technology (pistons, stoves, magnets, etc.) available to experimenters. Thermodynamics is concered with macrostates, for the pragmatic reason that those are the things the experimenter can control and observe. And thermodynamics, like (we hope) all of physics, is concerned with reproducible phenomena.

By experimental means of macroscopic coarseness one can generate a class of initial states from which a macroscopic process A → B takes place reproducibly; but in general the reversed process B → A cannot be achieved reproducibly by macroscopic means. That the microscopic equations of motion may nevertheless be "reversible" in the mechanical sense, is quite irrelevant to what the experimenter can actually do.

If the experimenter's apparatus is able to put his system only in some uncontrolled point in W_A, then because of Liouville's theorem (conservation of phase volume) the process $A \rightarrow B$ cannot be reproducible unless $W_B \geq W_A$, or $S_B \geq S_A$. If the inequality holds, then the reverse process is, as Gibbs noted, not impossible, but only improbable; *i.e.*, not reproducible. The probability of success is something like $p = W_A/W_B = \exp(-(S_B - S_A)/k)$. If the entropy difference corresponds to one microcalorie at room temperature, $p < \exp(-10^{15})$. We do not see why any more than this is needed to understand and explain the observed phenomenological irreversibility of thermodynamics.

REFERENCES

B. Alberts, D. Bray, J. Lewis, M. Raff, K. Roberts & J. D. Watson, *Molecular Biology of the Cell*, Garland Publishing Co., New York; pp. 550-609 (1983).

Sadi Carnot, *Reflexions sur la puissance motrice du feu*, Bachelier, Paris, (1824).

J. Willard Gibbs, "On the Equilibrium of Heterogeneous Substances", Trans. Conn. Acac. Sci (1875-78). Reprinted in *The Scientific Papers of J. Willard Gibbs*, Vol. 1; Dover Publications, Inc., N. Y. (1961).

C. C. Gillispie, *Lazare Carnot, Savant*, Princeton University Press (1971). A technical analysis of his work, and its relation to that of his son Sadi. Original manuscripts.

E. T. Jaynes, "The Minimum Entropy Production Principle", in Annual Review of Physical Chemistry, Vol. 31, 579-601 (1980). Reprinted in E. T. Jaynes, *Papers on Probability, Statistics and Statistical Physics*, R. Rosenkrantz, Ed., D. Reidel Publishing Co., Dordrecht-Holland (1983)

A. L. Lehninger, *Bioenergetics*, W. A. Benjamin, N. Y. (1965)

A. L. Lehninger, *Biochemistry, the Molecular Basis of Cell Structure and Function*, Worth Publishers, Inc., 444 Park Ave. South, New York, N. Y. (1975).

L. Onsager, Phys. Rev. **37**, 405; **38**, 2265 (1931)

M. Planck, *Scientific Autobiography*, Philosophical Library, N. Y. (1949); pp.17-18.

M. Reinhard, *Le Grand Carnot*, 2 vols., Paris, 1950-52.

A LOGIC OF INFORMATION SYSTEMS

N.C. Dalkey
UCLA Cognitive Systems Laboratory
Los Angeles, CA. 90024-1596

Abstract. A logic can be formulated with information systems as elements. The calculus of this logic is similar to, but not identical with, Boolean algebra. The logic is inductive--conclusions have more information than premises. Inferences have a strong justification; they are valid for all proper scoring rules.

DOMINANCE.

Information systems (*IS*) are well-known constructs in the knowledge sciences. Examples are: experiments, communication coding schemes, signal systems, pattern recognition techniques, surveillance systems, medical diagnosis, many expert systems, etc. Despite the wide variety of applications *IS* have a common underlying structure:

1. A set of events E (hypotheses, events of interest, target events, states of the world,...)

2. A set of events I (observations, data, signals, messages,...)

3. A joint probability distribution $P(E.I)$ on hypotheses and observations (the period in $P(E.I)$ denotes the logical conjunction "and".)

IS have a significant property from the standpoint of creating a logic, they allow *dominance* --one information system can have a higher expected value than another *for all payoff functions*. This property contrasts sharply with probability distributions. If P and Q are any two non-identical probability distributions, then there is a payoff function (decision matrix) that engenders a higher expected value for P, and another payoff function that engenders a higher expected value for Q.

Representation of expected value is simplified by the notion of *proper scoring rule*. Let P be a probability distribution on the partition of events $E = (e_1, ..., e_n)$ and let $S(P, e)$ be a function which assigns a score (rating, reward, payoff) to P given that the event e occurs. S is called *proper* (admissible, reproducing, honesty promoting) if it fulfills the condition

G. J. Erickson and C. R. Smith (eds.),
Maximum-Entropy and Bayesian Methods in Science and Engineering (Vol. 1), 283–294.
© *1988 by Kluwer Academic Publishers.*

$$\sum_E P(e)S(P,e) \geq \sum_E P(e)S(Q,e) \tag{1}$$

That is, a score rule S is proper if the expectation is a maximum when the score is determined by same distribution as that determining the expectation.

There is an infinite family of functions that fulfill (1). Among them is the logarithmic score $S(P,e) = \log P(e)$, and the set of decisional scores. For the latter, let $U(a,e)$ be the payoff if action a is taken and the event e occurs, and let $a^*(P)$ be the optimal action if P is the probability distribution on E. $S(P,e) = U(a^*(P), e)$ is a proper score. The expectation of the logarithmic score is the negative of the Shannon entropy of P (often called the information in P) and links the theory of proper scores to information theory. Decisional scores tie the theory of proper scores to decision theory.

Abbreviate $\sum_E P(e)S(P,e)$ by $G(P)$ and $\sum_E P(e)S(Q,e)$ by $G(P,Q)$. The expected score of an IS is given by

$$H(P) = \sum_I P(i)G(P(E \mid i))$$

where $P(i) = \sum_E P(e.i)$ is the initial probability of the observation i. $H(P)$ is thus the average over the potential observations of the expected score of the posteriors. The expected relative score is defined analogously,

$$H(P,Q) = \sum_I P(i)G(P(E \mid i), Q(E \mid i))$$

It is readily verified that H fulfills the analogue of (1), that is

$$H(P) \geq H(P,Q) \tag{2}$$

It is also straightforward to demonstrate that $H(P)$ is convex, and $H(P,Q)$ is linear in P. (Dalkey 1987).

An IS P is said to dominate an IS Q, in symbols $P \geq Q$, if $H(P) \geq H(Q)$ for all proper score rules S. It is clear that $\geq$ is a partial order, i.e., it fulfills:

1.   Transitivity. $P \geq Q$ and $Q \geq R \rightarrow P \geq R$

2.   Reflexivity: $P \geq P$.

3.   Antisymmetry: $P \geq Q$ and $Q \geq P \to P = Q$

In addition, for a given set of events E and a given prior distribution (E), $\geq$ has an absolute upper bound $P*$ where for each e, $P*(e \mid i) = 1$ for some i, and 0 otherwise. $P*$ is often called "perfect information" in decision theory, or more jocularly, "the clairvoyant." $\geq$ also has an absolute lower bound P^0, where $P^0(e.i) = P(e)P(i)$. P^0, in effect, consists in implementing the prior distribution. It is readily verified that

$$P* \geq P \geq P^0 \tag{3}$$

The second inequality $P \geq P^0$ is often called the positive value of information principle (PVI), any IS is at least as valuable as the prior IS (assuming that information is free.) (3) is a well-known illustration of the fact that $\geq$ is not an empty relation. (LaValle 1978).

LATTICE STRUCTURE
To proceed further in using the dominance relation as the basis for a logic, it is pertinent to examine the lattice properties of the relation. Lattices have received extensive attention as foundations for logics. (Birkhoff 1940).

A partial order such as $\geq$ is called a lattice if for each pair of elements P, Q there is least upper bound (l.u.b) w.r.t. $\geq$ and a greatest lower bound. Examples can be found of pairs of IS that do not have a l.u.b., and thus $\geq$ is not in general a lattice. However, for an important subclass of IS, namely, those with binary hypotheses, $\geq$ is a lattice.

Theorem 1.   For the set of IS with binary hypotheses, $\geq$ is a lattice.

Proof: Theorems 8 and 8' in (Dalkey 1980).

IS with binary hypotheses are those which address a yes-no question: Does the patient have AIDS? Is there life on Mars? Will a Republican be elected president of the U.S. in 1988? Is the crystal structure of the substance octahedral? In practice, binary hypotheses are part of the stock-in-trade of the analyst. In addition, although the typical IS is not binary, decisional problems often "boil down to" a binary question.

Denote the l.u.b. of P and Q by $P + Q$, and the g.l.b. by $P \cdot Q$. $P \cdot Q$ expresses the information that P and Q have in common. $P + Q$ expresses the "sum" of the information in the two.

THE CALCULUS
Given the operations $+$ and $\cdot$, a calculus can be formulated. Listed below

are a set of postulates for the calculus. They are listed in parallel, one set for + and the analogous set for ·. Although they are listed as postulates, they can be verified by the construction methods described below. The calculus differs from Boolean Algebra in that it does not have complements (negation) and is not distributive. The lack of a negation is partially compensated for by the duality rule described below.

P1. $P+P \geq P$ $P \cdot P \leq P$

P2. $P+Q \geq Q+P$ $P \cdot Q \leq Q \cdot P$

P3. $P+Q \geq P$ $P \cdot Q \leq P$

P4. $P \geq Q$ and $P \geq R \rightarrow P \geq Q+R$ $P \leq Q$ and $P \leq R \rightarrow P \leq Q \cdot R$

P5. $P+Q = P+R \rightarrow P+Q = P+Q \cdot R$ $P \cdot Q = P \cdot R \rightarrow P \cdot Q = P \cdot (Q+R)$

It is noteworthy that the first four postulates for + are homologous to the basic four postulates for the propositional calculus; however, they are not as powerful because of the lack of a negation. The first four are true of any lattice. P5 expresses a property that does not hold for lattices in general and distinguishes *IS* logic. I do not have a proof that P1-5 are complete. The existence of a model--the canonical representation described below--shows that the postulates are consistent.

The basic inference rules for the calculus are the transitivity of $\geq$ and the rule of replacement--in any statement, if $P = Q$, then P can be replaced by Q in any position. In addition, the usual rules of substitution for variables and the inference rules for two-valued logic (e.g., modus ponens for $\rightarrow$) are assumed. A derived rule, the duality principle, is particularly useful. It states that any postulate or theorem remains true if + is replaced by · throughout, and $\geq$ is replaced by $\leq$. Note that with the duality principle, only the + versions of P1-5 are needed. The · versions can be derived immediately with the duality principle.

From P1-5 a variety of theorems can be generated. Among the more familiar:

T1.   Idempotence: $P+P = P, P \cdot P = P$.

T2.   Associativity: $(P+Q)+R = P+(Q+R), P \cdot (Q \cdot R) = (P \cdot Q) \cdot R$.

T3.   Consistency: $P \geq Q, P+Q = P, P \cdot Q = Q$ are mutually equivalent.

T5.   Absorption: $P = P+P \cdot Q = P \cdot (P+Q)$.

T6.   Semi-distributivity: $P \cdot (Q+R) \geq P \cdot Q +P \cdot R, (P+Q) \cdot (P+R) \geq P+Q \cdot R$.

The next three theorems are provided with proofs as an example of using the calculus. Let $P \mid Q$ mean that P does not dominate Q and Q does not dominate P.

T7. $P \mid Q \rightarrow P+Q > P$ and $P+Q > Q$, i.e., $P \mid Q$ implies that $P+Q$ strictly dominates both P and Q.

> *Proof:* Suppose $P \geq P+Q$. From P3, $P+Q \geq Q$, and thus by transitivity of $\geq$, $P \geq Q$ contrary to the hypothesis.

T8. $P = Q \rightarrow P+R = Q+R$.

> *Proof:* From P 1, $P+R = P+R$. Whence, by replacement, $P+R = Q+R$.

T9. $P+Q = P+R \rightarrow P+Q = P+Q+R$.

> *Proof:* From the hypothesis and T8, $P+Q+Q = P+R+Q$. From T 1, $Q+Q = Q$, and thus by replacement, $P+Q+Q = P+Q = P+Q+R$.

T9 is of special interest with regard to the design of information systems. It states that even though P, Q, and R are mutually non- dominating--i.e., each contains information neither of the other two contain--if $P+Q = P+R$, then either Q or R is eliminable. This contrasts with T7, which states that if $P \mid Q$, then the sum is strictly more informative than either alone.

T10. $P+Q = P+R \rightarrow P+Q+R = P+Q \cdot R$.
       Proof: Immediate from P5 and T9.

T10 states that if the sum of P and Q is the same as the sum of P and R, then the sum of all three is just the sum of P and the common part of Q and R.

INFERENCE
         One mode of application of *IS* logic stems from elaborating the set of theorems derivable in the calculus. This body of results appears promising in the design of information systems, e.g., design of experiments. In a sense, this mode is deductive, determining the consequences of the postulates.

A somewhat different mode stems from applying the logic to the problem of combining evidence. This mode is inductive. The basic inference rule in this mode is: If P and Q are known, but the dependencies (correlation) of P and Q are not known, assume $P+Q$. The justification for this rule requires some preliminaries.

Let the observation set for P be I and the observation set for Q be J; i.e., P is the joint distribution $P(E.I)$ and Q is the joint distribution $Q(E.J)$ Let $I.J$ denote the cartesian product of I and J. The composition of P and Q, denoted by $P.Q$ is a joint distribution $R(E.I.J)$. Knowing P and Q is not sufficient to determine R. All that is known is that R must be compatible with both P and Q, i.e.:

$$P(E.I) = \sum_J R(E.I.J) \tag{4}$$

$$Q(E.J) = \sum_I R(E.I.J)$$

Let $K(P,Q)$ be the set of R which fulfills (4). K is the set of compositions of P and Q. It is an immediate consequence of PVI that for any R in K, $R \geq P$ and $R \geq Q$. Since (4) is a set of linear constraints, K is convex and closed. K does not contain all IS which dominate both P and Q; however, for any R which dominates P and Q, there is an R' in K such that $R \geq R'$. (Dalkey 1987).

$P+Q$ is in K, since $P+Q$ dominates both P and Q, and if $P+Q$ were not in K, there is an R' in K, $P+Q \geq R'$. If $P+Q \neq R'$, $P+Q$ would not be the l.u.b. of P and Q. Thus, $P+Q$ is the g.l.b. of K.

The justification for assuming $P+Q$ when P and Q are known is based on the following theorem:

Theorem 2. If R is in $K(P,Q)$ then $H(R,P+Q) \geq H(P+Q) \geq$ max $[H(P), H(Q)]$ for every proper score.

Proof. Since K is convex, $R' = aR + (1-a)(P+Q)$, $0 \leq a \leq 1$, is in K, and since $P+Q$ is the g.l.b. of K, $H(R') \geq H(P+Q)$; thus, $H(R')$ is monotonically decreasing (with decreasing a) between R and $P+Q$. $H(R', P+Q)$ is the line tangent to $H(R')$ at $P+Q$. Thus, since $H(R')$ is convex, $H(R, P+Q) \geq H(P+Q)$ (cf. Figure 1).

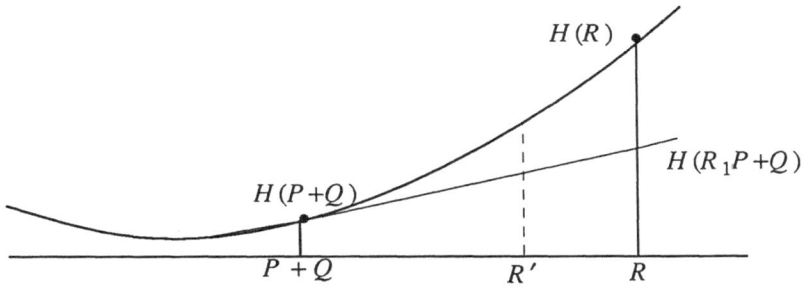

$H(P+Q) \geq \max [H(P), H(Q)]$ from P3.

We can assume that the actual composition R of P and Q (i.e., the R that would be observed in a sufficient set of observations of E.I.J) is in K. The theorem states that whatever R may be, the actual relative expectation of $P+Q$ $H(R, P+Q)$ is greater than the apparent expectation of either P or Q. In other words, the expectation of $P+Q$ is guaranteed and guaranteed to be greater than that of P or Q, no matter what the payoff function of the user. It is this guarantee which justifies the use of the term *logic*.

Note that the inference from P and Q to $P+Q$ is *inductive*. We cannot derive $P+Q$ from P and Q by deductive reasoning. We could derive the actual composition R from P and Q if we knew the dependencies between P and Q, e.g., if we knew they were independent. Without knowing the dependencies, however, we can recommend accepting $P+Q$ on the basis of the strong guarantee.

In the interesting case where P and Q are mutually non-dominating (if P dominates Q for example, then from consistency, T4, $P+Q = P$) T7 assures that $P+Q$ strictly dominates P and Q.

COMPUTATION

For the case of IS with binary hypotheses, the computation of $P+Q$ and $P \cdot Q$ is particularly simple. Let $t(i)$ denote the vector $(P(i \mid e), P(i \mid \bar{e}))$, the supra-bar indicating negation or "non-e." Order these vectors in the decreasing order of the ratio $P(i \mid e)/P(i \mid \bar{e})$, and reindex the observations numerically in the new order. Define $T(i) = \sum_{j \leq i} T(j)$. $T(0) = (0, 0)$. Since $\sum_I P(i \mid e) = 1$, if there are m observations in I, $T(m) = (1, 1)$.

The vectors $T(i)$ can be plotted in the plane, and joining them with straight lines generates a concave, piece-wise linear curve in the unit square lying above the diagonal, as in Figure 2. The convex closure $C(P)$ of this curve--i.e., the points between the curve and the diagonal and including the curve and the diagonal--can be called a canonical representation of the *IS P*. It can be shown that $P \geq Q$ if and only if $C(Q) \subset C(P)$. (Dalkey 1980).

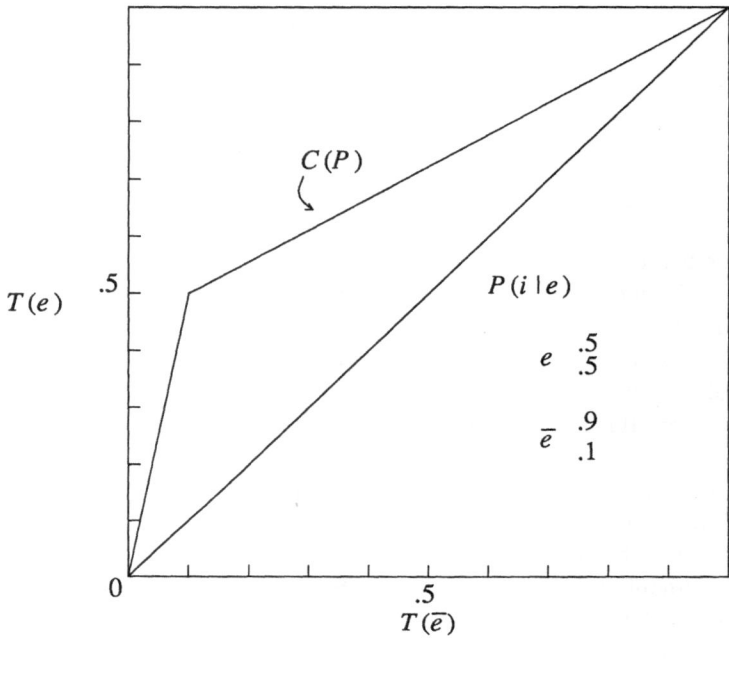

Fig. 2

If two *IS P* and *Q* are plotted, then $C(P+Q) = <C(P), C(Q)>$ the convex closure of $C(P)$ and $C(Q)$. $P \cdot Q = C(P)$. $C(Q)$, the intersection of the two representations. This construction is illustrated in Figure 3. For small *IS*--those with a relatively few observations-- the construction is readily made by hand. For larger *IS*, the construction is easily programmable for a computer.

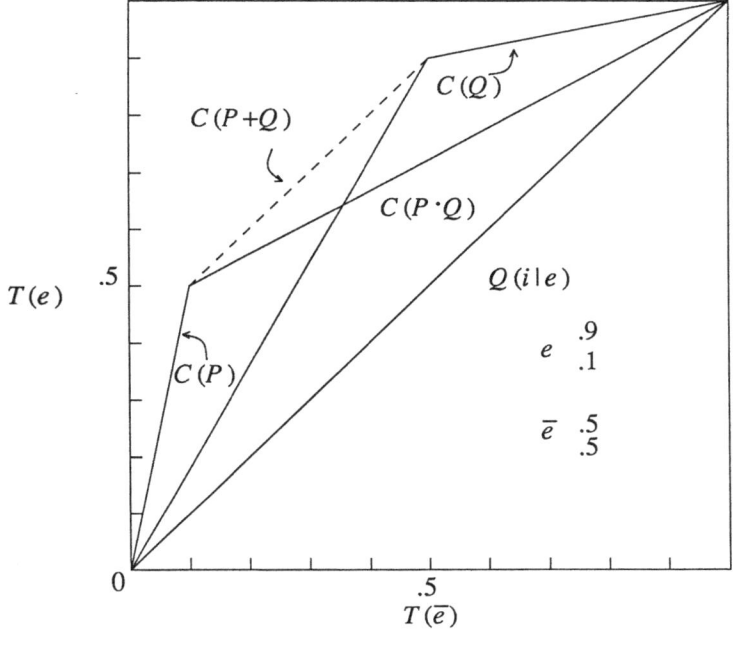

Fig. 3

EXAMPLE

Suppose you are worried about AIDS. You know there are two tests, each developed by a different organization. You know that each has been studied by its developer, and relatively good statistics exist concerning its diagnosticity. However, statistics are insufficient concerning the joint diagnosticity of the two tests taken together. You would prefer the (potentially) greater information of both tests, but you have no way of interpreting the results if the two tests give conflicting results. Let *A* mean "The patient has AIDS," and + mean "The test result is positive." Suppose the likelihoods of test results are: (Any similarity between the numbers in the table and those for actual tests is a miracle. I made the numbers up.)

		T_1	T_2
A	+	.5	.9
	-	.5	.1
$\overline{A}$	+	.1	.5
	-	.9	.5

The diagnosticity of T_1, if the patient has AIDS, is low, but it is high if the patient doesn't have AIDS. These characteristics are reversed for T_2. For the minimal composition T_1+T_2 of the two tests, there are four possible outcomes: both positive, both negative, or the two mixed cases.

	T_1	T_2	$P(\cdot \mid T_1 . T_2)$
A	+	+	.5
	+	-	.0
	-	+	.4
	-	-	.1
$\overline{A}$	+	+	.1
	+	-	.0
	-	+	.4
	-	0	.5

The O's at $+ -$ are a mathematical artifact. In practice, the likelihoods for these cases would be determined by a separate computation.

The table gives the likelihoods of test results given the disease state. To determine the posterior probabilities of the disease states given test results, it is necessary to know the prior probabilities of the disease states for a relevant population. The table below lists the posteriors for prior probabilities of AIDS of .5 and .2.

Posterior Probability

T_1	T_2	$P(A)=.5$	$P(A)=.2$
+	+	.83	.56
+	-	?	?
-	+	.5	.2
-	-	.17	.05

Note that the double negative is a good deal more reassuring if the prior probability is low. The values for this example are those used in Figures 2 and 3. The joint likelihoods were read from the graph in Figure 3.

IS WITH NON-BINARY HYPOTHESES

Although *IS* with binary hypotheses are of practical importance, most *IS* arising in practice have multiple hypotheses. (I recently read of a computerized medical diagnostic service with a list of over 1900 disease states and over 4000 symptoms and test results. It was called "Hypochondriac Heaven.") Examples can be devised with as few as three hypotheses where a pair of *IS* do not have a l.u.b. w.r.t. $\geq$. Thus $\geq$ is not a lattice for non-binary *IS*. On the other hand examples are easily devised for pairs of non-binary *IS* which do have a l.u.b. It should be clear from the preceding analysis that for pairs of *IS* with non-binary hypotheses, if $P + Q$ exists, then the inference from P and Q to $P + Q$ is just as solid as it is for binary *IS*.

At present there is no algorithm for determining whether a pair of non-binary hypothesis *IS* have a l.u.b. This is clearly an area inviting research.

There is a non-trivial fall-back possibility. If a specific payoff function--e.g., the logarithmic score--is considered "adequate," then the H function imposes a complete order on *IS*, and $P + Q$ and $P \cdot Q$ w.r.t. that score always exist. For this case,

$P + Q = \underset{K(P, Q)}{arg\ min} H(R)$. Define the set $L(P, Q) = \{R \mid P \geq R, Q \geq R\}$.

$P \cdot Q = \underset{L(P, Q)}{arg\ max} H(R)$. The guaranteed expectation for $H(P + Q)$ holds for the given score rule. This formalism was the basis for a weaker inductive logic proposed earlier. (Dalkey 1985).

REFERENCES

Birkhoff, G. (1940). Lattice Theory. American Mathematical Society (Colloquium Publications, 25), New York.

Dalkey, N.C. (1980). The Aggregation of Probability Estimates. UCLA-ENG-CSL-8025.

Dalkey, N.C. (1985). Inductive Logic and the Maximum Entropy Principle. *In* Maximum Entropy and Bayesian Methods in Inverse Problems, eds. C. Ray Smith & W.T. Grundy. Boston: D. Reidel Publishing Co.

Dalkey, N.C. (1987). Information Systems. UCLA-ENG-CSL Report, in preparation.

LaValle, . (1978). Fundamentals of Decision Analysis. New York: Rhinehart & Winston.

METHODOLOGICAL PRINCIPLES OF UNCERTAINTY IN
INDUCTIVE MODELLING: A New Perspective

G.J. Klir
Department of Systems Science
T.J. Watson School of Engineering, Applied Science,
and Technology
State University of New York at Binghamton
Binghamton, New York 13901, U.S.A.

Abstract. It is argued that the concept of uncertainty
plays a fundamental role in inductive (data-driven)
systems modelling. In particular, it is essential for
dealing with two broad classes of problems that are
essential to inductive modelling: problems involving
ampliative reasoning (reasoning in which conclusions are
not entailed within the given premises) and problems of
systems simplification. These problem classes are
closely connected with the principles of maximum and
minimum uncertainty. When models are conceptualized in
terms of probability theory, these principles become the
well established principles of maximum and minimum
entropy. However, when the more general framework of
the Dempster-Shafer theory of evidence is employed, four
different types of uncertainty emerge. Well justified
measures of these types of uncertainty are now available
and are described in the paper. The meaning of these
four types of uncertainty is captured by the suggestive
names "nonspecificity", "fuzziness," "dissonance," and
"confusion." Since uncertainty is a multidimensional
entity in evidence theory, the maximum and minimum uncertainty principles lead to optimization problems with
multiple objective criteria.

INDUCTIVE MODELLING

Systems science, which is the field of my current research interests, deals with two large classes of problems:

1. Systems inquiry—the full scope of activities by which we attempt
to construct systems that are adequate models of some aspects of
reality.

2. Systems design—the full scope of activities by which we attempt
to construct systems that are adequate models of desirable man-
made objects.

The purpose of systems inquiry is to understand the phenomenon under
investigation, to make adequate predictions or retrodictions, to
control the phenomenon in any desirable way, and to utilize all
these capabilities for making appropriate decisions. The purpose of

G. J. Erickson and C. R. Smith (eds.),
Maximum-Entropy and Bayesian Methods in Science and Engineering (Vol. 1), 295–304.
© 1988 by Kluwer Academic Publishers.

systems design is to prescribe operations by which a desirable artificial object can be constructed in such a way that desirable objective criteria are satisfied within given constraints.

A system is always an abstraction that characterizes an appropriate
type of relationship among some abstract entities. The term "relationship" is used here as a general term intended to capture not
only the well defined concept of a mathematical relation, but the
whole set of kindred concepts such as constraint, cohesion, inter-
dependence, interaction, coupling, linkage, structure, organization,
and the like.

A given system qualifies as a model of some aspect of reality
(natural or man-made) if a mapping that is homomorphic with respect
to the relationship involved is established from relevant entities
of the real world into the abstract entities of the system. It is
important to realize, however, that this homomorphic mapping cannot
be defined mathematically in many instances, but only in terms of an
appropriate physical device (a measuring instrument).

A wide spectrum of approaches to systems modelling can be recognized.
This spectrum is bounded by two idealized approaches, which are
usually called a postulational approach and a discovery approach.

In the postulational approach, a hypothetical system is postulated
within some mathematical formalism. This postulated system represents basically a frame for a specific type of deductive reasoning:
given a particular set of premises regarding conditions of the
system's environment as well as initial or boundary conditions of
the system itself, the system allows us to derive approriate conclusions. The validity of the system as a model of some specific
aspect of reality depends on the degree of agreement of the derived
conclusions with relevant empirical evidence.

The discovery approach is data-driven. That is, models are derived
by processes that discover patterns in data and utilize them for
making inductive inferences. The criterion of validity of models
derived in this way does not involve the issue of the agreement between the evidence and inferences made from the model (the agreement
is perfect by definition), but rather the issue of the justification
of induction.

In praxis, neither of the two idealized approaches to systems
modelling is actually used in its pure form. Nevertheless, each
actual approach is usually closer to one of the two extremes of the
whole spectrum of approaches.

My own work has recently focused on a particular blend of the discovery and postulational approaches to systems modelling in which
our background knowledge is employed for restricting the class of

possible models while empirical evidence is utilized inductively for determining a particular model or a set of models from the delimited class (Klir, 1985). This approach to systems modelling is usually called an inductive modelling.

The key issue in inductive modelling is the extent to which a model inferred from given data is required to account for the data. Ideally, we should require that the model represent the data completely. This is the same as requiring that the model be based on a deterministic system capable of reproducing the data exactly. However, such a requirement is not realistic since it leads, in general, to excessively complex models. This fact was demonstrated by Gaines (1976,1977), who showed that "the universe becomes incredibly complex and our models of it nonsensical if we assume determinism in the face of even a slight trace of acausal behavior." This important point was further studied by Pearl (1978), who established (on theoretical grounds) that the predictive credibility of deterministic models inferred from data (the likelihood of finding another model of similar complexity but leading to different predictions) increases with the data size and decreases with the descriptive complexity of the model (the shortest description or the model in some standard language). Since the complexity of deterministic models inferred from data tends to be proportional to the data size (Gaines, 1976, 1977), no appreciable predictive credibility of deterministic models inferred from data can be obtained regardless of the size of the data.

We may conclude from these facts that inductive modelling can be meaningful and practical only if it is formulated in terms of nondeterministic systems of some sort. By definition, each nondeterministic systems involves some degree of predictive (or retrodictive) uncertainty. This degree, when appropriately defined, is always one of two key criteria for comparing models. The other criterion is complexity. In general, we try to minimize both uncertainty and complexity of inferred models. Unfortunately, these two criteria conflict with each other. When we reduce complexity, uncertainty increases or, at best, remains the same.

The concept of uncertainty plays thus a fundamental role in inductive modelling.

THE ROLE OF UNCERTAINTY IN INDUCTIVE MODELLING

The concept of uncertainty is closely connected with the concept of information. When our uncertainty in some situation is reduced by an act (such as observation, performing an experiment, or receiving a message), the act may be viewed as a source of information pertaining to the situation under consideration. The amount of information obtained by the act may then be measured by the reduction in uncertainty due to the act.

The capability of measuring information in terms of uncertainty is essential for dealing with two broad classes of problems that are fundamental to inductive modelling:

(a) problems involving reasoning in which conclusions are not en-tailed in the given premises, referred to in this paper as ampliative reasoning;

(b) problems of systems simplification.

On intuitive grounds, we can easily see that the following general principles are fundamental for dealing with these classes of problems.

A general principle of ampliative reasoning may be expressed by the following requirement: in any ampliative inference, use all but no more information than available. This principle thus requires that conclusions resulting from ampliative inferences maximize the relevant uncertainty within the constraints representing the premises. This principle, which may appropriately be called the principle of maximum uncertainty, guarantees that our ignorance be fully recognized when we try to enlarge our claims beyond the given premises and, at the same time, that all information contained in the premises be fully utilized.

A general principle of systems simplification may be expressed as follows: a sound simplification of a system should minimize the loss of relevant information (or increase of relevant uncertainty) while achieving the required reduction of complexity. That is, we should accept only such simplifications at any desirable level of com-plexity for which the loss of relevant information (or the in-crease of relevant uncertainty) is minimal. Let this principle be called the principle of minimum uncertainty.

The principles of maximum and minimum uncertainty were in fact em-bedded in ancient Chinese wisdom and expressed by the Chinese philosopher Lao Tsu as early as the sixth century B.C. in terms of two simple statements of remarkable clarity and beauty (Tsu, 1972):

> Knowing ignorance is strength.
> Ignoring knowledge is sickness.

In order to make these principles operational, a well justified measure of uncertainty (and the associated information) is needed. Such a measure depends, obviously, on how uncertainty is concept-ualized. The key measures of uncertainty, which are now available within several alternative mathematical theories, are overviewed in the next section. For the sake of simplicity, the presentation is restricted to measures of uncertainty defined on a finite set of alternatives.

MEASURES OF UNCERTAINTY

        The classical mathematical apparatus for characterizing
situations under uncertainty has been probability theory. Since the
mid 1960's, however, several alternative mathematical theories became available for this purpose. They are subsumed under the
general concept of a fuzzy measure (Sugeno, 1977). Most notable
among these are possibility theory (Zadeh, 1978) and evidence theory
(also called Dempster-Shafer theory) (Shafer, 1976). I assume that
the reader is familiar with the fundamentals of these theories and I
follow the terminology and notation employed in the book by Klir and
Folger (1987).

In probability theory, uncertainty is measured by the well known
Shannon entropy.

$$H(p(x)|x \varepsilon X) = - \sum_{x \varepsilon X} p(x) \log_2 p(x), \qquad (1)$$

where $(p(x)|x \varepsilon X)$ denotes a probability distribution on a finite set
X. This function is firmly established as the only justifiable
measure of uncertainty in probability theory (Rényi, 1970; Klir &
Folger, 1988), and the associated principles of maximum and minimum
entropy have been applied with great success in inductive modelling
(Christensen, 1980-81, 1983, 1985, 1986).

When generalized to evidence theory, the Shannon entropy bifurcates
into the forms

$$E(m(A)|A \varepsilon P(X)) = - \sum_{A \subset X} m(A) \log_2 Pl(A) \qquad (2)$$

and

$$C(m(A)|A \varepsilon P(X)) = - \sum_{A \subset X} m(A) \log_2 Bel(A), \qquad (3)$$

where Pl and Bel denote the dual plausibility and belief measures
associated with the basic assignment m defined on the power set P(X)
of a finite set X (Shafer, 1976; Klir & Folger, 1988). The following are basic properties of m, Pl, and Bel:

$$m(\emptyset) = 0 \text{ and } \sum_{A \subset X} m(A) = 1, \qquad (4)$$

$$Pl(A) = \sum_{B \cap A \neq 0} m(B), \qquad (5)$$

$$Bel(A) = \sum_{B \subset A} m(B). \qquad (6)$$

Value m(A) may be interpreted as the degree of evidence or subjective belief that a specific element of X belongs to the set A but not to any special subset of A. Total evidence (or belief) is then calculated by Eq. (6). Sets A for which m(A)≠0 are called focal elements.

Measures E and C, given by Eqs. (2) and (3), respectively, emerged from the work of several researchers (Hohle, 1982; Yager, 1983; Dubois & Prade, 1987). It follows from this work that measure E characterizes dissonance (or conflict) in the representation of evidence, i.e., evidence supporting disjoint subsets of X; hence, it was given the appropriate name measure of dissonance. Measure C, on the other hand, characterizes the number of subsets of X that are supported by the given evidence and do not overlap or overlap only partially. The multitude of partially or totally conflicting representation of evidence is a source of confusion when applying a model. This measure of uncertainty is thus appropriately called a measure of confusion.

Measures E and C are usually called entropy-like measures since they collapse into the Shannon entropy when we restrict to probability theory. A fundamentally different type of uncertainty in evidence theory, which is connected with the size of subsets that are supported by the given evidence, is given by the function

$$V(m(A)|A\varepsilon P(X)) = \sum_{A\subset X} m(A)\log_2 |A|, \tag{7}$$

where $|A|$ denotes the cardinality of set A. The larger the sets A for which $m(A) \neq 0$, the less specific is the characterization given by m. This type of uncertainty has thus the meaning of nonspecificity in the representation of evidence.

The measure of nonspecificity was originally proposed for fuzzy set interpretation of possibility theory (Higashi & Klir, 1983) in the form

$$U(r(x)|x\varepsilon X) = \int_0^1 \log_2 |c(r,\alpha)| d\alpha, \tag{8}$$

where $|c(r,\alpha)|$ is the cardinality of the α-cut for the possibility distribution $(r(x)|x\varepsilon X)$ defined on a finite set X (Klir & Folger, 1988). When possibility theory is interpreted as a special case of evidence theory (based on the requirement that the focal elements are nested), the U-uncertainty can be expressed by the form

$$U(r) = \sum_{i=1}^{n} (\rho_i - \rho_{i+1})\log_2 i, \tag{9}$$

where $r = (\rho_1, \rho_2, \ldots, \rho_n)$ is an ordered possibility distribution $(\rho_j \geq \rho_k$ when $j<k)$ and $\rho_{n+1} = 0$ by definition.

The uniqueness of this possibilistic measure of nonspecificity
(usually called U-uncertainty) as well as the general measure of
nonspecificity (given by Eq. (7)) is now well established (Klir &
Mariano, 1987; Ramer, 1987; Klir & Folger, 1988). After careful
scrutiny, we can see that the U-uncertainty is a generalization of
the classical Hartley measure of information (Hartley, 1928; Rényi,
1970; Klir & Folger, 1988).

In addition to the entropy-like and nonspecificity measures in
evidence theory, another type of uncertainty exists within the framework of fuzzy set theory and fuzzy set interpretation of possibility
theory—uncertainty in the form of vagueness or fuzziness. A broad
class of possible measures of fuzziness was characterized by Higashi
and Klir (1982). They are based upon the idea proposed by Yager
(1979) that the degree of fuzziness of a fuzzy set should
characterize the lack of distinction between the set and its complement. A representative measure of fuzziness is given by the formula

$$f_c(A) = |X| - \sum_{x \in X} |\mu_A(x) - c(\mu_A(x))|, \qquad (10)$$

where A is a fuzzy subset of X defined by the membership function
μ_A and c is a fuzzy complement. In addition to fuzzy set theory and
possibility theory, measures of fuzziness are also applicable to
fuzzified evidence theory (Yager, 1986).

The applicability of the individual measures of uncertainty within
the mentioned mathematical theories is summarized in Figure 1. We
can see that some mathematical theories involve more than one type
of uncertainty. Fuzzy set theory as well as possibility theory
involve both nonspecificity and fuzziness; evidence theory involves
nonspecificity, dissonance, and confusion; when fuzzified (Yager,
1986), evidence theory involves also the measure of fuzziness.

Uncertainty and information are thus multidimensional entities in
all the considered mathematical theories except classical set theory
and probability theory. This means that the maximum and minimum
uncertainty principles, which are fundamental to inductive
modelling, lead to optimization problems with multiple objective
criteria. These principles have not been properly developed as yet.
Their development is currently a subject of active research.

Figure 1.   Applicability of the discussed measures of uncertainty in
            several mathematical theories.

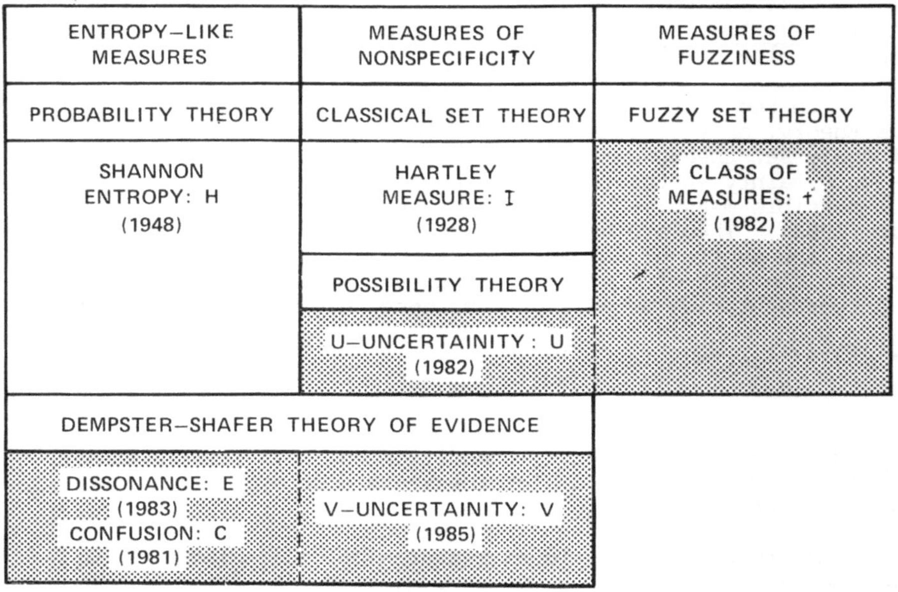

ENTROPY—LIKE MEASURES	MEASURES OF NONSPECIFICITY	MEASURES OF FUZZINESS
PROBABILITY THEORY	CLASSICAL SET THEORY	FUZZY SET THEORY
SHANNON ENTROPY: H (1948)	HARTLEY MEASURE: I (1928)	CLASS OF MEASURES: f (1982)
	POSSIBILITY THEORY	
	U—UNCERTAINITY : U (1982)	
DEMPSTER—SHAFER THEORY OF EVIDENCE		
DISSONANCE: E (1983) CONFUSION: C (1981)	V—UNCERTAINITY: V (1985)	

CONCLUSIONS

        The principles of maximum and minimum entropy are now
well established. When combined, these principles form a powerful
methodological tool for inductive (data—driven) systems modelling
based upon probability theory. Great skill in using this tool for
developing predictive models in many application areas was demonstrated by Christensen (1980–81, 1983, 1985, 1986).

In spite of the great significance and practical success of these
principles, it has increasingly been recognized that probability
theory is capable of representing only one type of uncertainty. It
seems that this is not sufficient for capturing the full scope of
uncertainty in systems that were characterized by Warren Weaver
(1948) as systems of organized complexity.

A turning point in our understanding of the concept of uncertainty
was reached when it became clear that more than one type of uncertainty must be recognized within evidence theory, and even within

the restricted domain of possibility theory. This was not
previously obvious since each of the two classical mathematical
theories (set theory and probability theory) represents only one
type of uncertainty.

The new insight into the concept of uncertainty was obtained by
examining the concept within a mathematical framework more general
than probability theory. A generalization of existing theories is
actually a current trend in mathematics, as exemplified by the
generalizations from classical geometries (Euclidean as well as non
Euclidean) into fractal geometry, from automata theory into dynamic
cellular automata, from two-valued logic into multiple-valued
logics, fuzzy logic, or logic of inconsistency, from classical set
theory into fuzzy set theory, or, as most relevant to the subject of
this paper, from probability measures to fuzzy measures. These
generalizations have enriched not only our insights but also our
capabilities for modelling the intricacies of the real world.

In this paper, I was only able to outline the new developments regarding the principles of uncertainty. A thorough and
self-contained treatment is included in a book I coauthored with
Tina Folger (Klir & Folger, 1988).

ACKNOWLEDGMENT

The work on this paper was partially supported by the
National Science Foundation under Grant No. IST 86-44676.

REFERENCES

Christensen, R. (1980-81). Entropy Minimax Sourcebook (4 volumes).
        Lincoln, Mass.: Entropy Limited.
Christensen, R. (1983). Multivariate Statistical Modeling. Lincoln,
        Mass.: Entropy Limited.
Christensen, R. (1985). Entropy minimax multivariate statistical
        modeling—I: Theory. Intern. J. of General Systems, 11,
        231-277.
Christensen, R. (1986). Entropy minimax multivariate statistical
        modeling—II: Applications. Intern. J. of General Sys-
        tems, 12, 193-271.
Dubois, D. & Prade, H. (1987). Properties of measures of information
        in evidence and possibility theories. Fuzzy Sets and
        Systems, 24, no. 2.
Gaines, B.R. (1976). On the complexity of causal models. IEEE Trans.
        on Systems, Man, and Cybernetics, SMC-6, 56-59.
Gaines, B.R. (1977). System identification, approximation and com-
        plexity. Intern. J. of General Systems, 3, 145-174.
Hartley, R.V.L. (1928). Transmission of information. The Bell Systems
        Technical J., 7, 535-563.

Higashi, M. & Klir, G.J. (1982). On measures of fuzziness and fuzzy complements. Intern. J. of General Systems, 8, 169–180.

Higashi, M. & Klir, G.J. (1983). Measures of uncertainty and information based on possibility distributions. Intern. J. of General Systems, 9, 43–58.

Hohle, U. (1982). Entropy with respect to plausibility measures. Proc. 12th IEEE Symp. on Multiple–Valued Logic, Paris, 167–169.

Klir, G.J. (1985). Architecture of Systems Problem Solving. New York: Plenum Press.

Klir, G.J. & Folger, T.A. (1988). Fuzzy Sets, Uncertainty, and Information. Englewood Cliffs, NJ: Prentice Hall.

Klir, G.J. and Mariano, M. (1987). On the uniqueness of possibilistic measure of uncertainty and information. Fuzzy Sets and Systems, 24.

Pearl, J. (1978). On the connection between the complexity and credibility of inferred models. Intern. J. of General Systems, 4, 255–264.

Ramer, A. (1987). Uniqueness of information measure in the theory of evidence. Fuzzy Sets and Systems, 24.

Renyi, A. (1970). Probability Theory. Amsterdam: North–Holland (Chapter IX, Introduction to Information Theory, 540–616).

Shafer, G. (1976). A Mathematical Theory of Evidence. Princeton, NJ: Princeton University Press.

Sugeno, M. (1977). Fuzzy measures and fuzzy integrals: a survey. In: Fuzzy Automata and Decision Processes, edited by M.M. Gupta, G.N. Saridis, and B.R. Gaines, Amsterdam and New York: North–Holland, 89–102.

Tsu, Lao (1972). Tao Te Ching. New York: Vintage Books (Sec. 71).

Yager, R.R. (1979). On the measure of fuzziness and negation. Part I: Membership in the unit interval. Intern. J. of General Systems, 5, 221–229.

Yager, R.R. (1983). Entropy and specificity in a mathematical theory of evidence. Intern. J. of General Systems, 9, 249–260.

Yager, R.R. (1986). Toward general theory of reasoning with uncertainty: nonspecificity and fuzziness. Intern. J. of Intelligent Systems, 1, 45–67.

Zadeh, L.A. (1978) Fuzzy sets as a basis for a theory of possibility. Fuzzy Sets and Systems, 1, 3–28.

COMPARISON OF MINIMUM CROSS-ENTROPY INFERENCE WITH MINIMALLY INFORMATIVE INFORMATION SYSTEMS

N. C. DALKEY*

Abstract

The Minimum Cross-Entropy (MXE) inference rule leads to information systems which are inconsistent, and which may have an expectation less than the prior. The Min-Score rule (a generalization of maximum entropy) applied to information systems generates consistent systems and has a guaranteed expectation at least as great as the prior. The guaranteed expectation for the Min-Score rule is always at least as great as that for MXE.

Introduction

In an earlier presentation to this workshop series I raised two caveats concerning minimum cross-entropy inference (MXE). [Dalkey, 1983] The first was that the framework of MXE is insufficient to permit determination of the consistency of inputs. Prior distributions and constraints on posteriors can be incompatible, but the MXE formalism will not identify the pathology. The second caveat was that MXE inference does not conform to the positive value of information principle (PVI). The conclusion of an inference can be less informative than the information determining the prior.

So far as I can tell, neither of these two caveats impressed the proponents of MXE, perhaps because I expressed them in too abstract a fashion. Since then it has been possible to obtain a sharper formulation of the difficulties with MXE, and to compare its performance with that of a more complete inference procedure based on the notion of a minimally informative information system.

Context

The first step was to find a context in which the role of the inference rule is quite clear -- that is, not obscured by issues such as whether the law of large numbers is involved. A suitable context turned out to be an inverse inference in which the prior probability is well known, but the likelihoods are incompletely known -- e.g., they are known only to the extent that they fulfill some partial constraints. Given an observation i, the prior probability combined with the incompletely known likelihoods, specify a class $K(i)$ of potential posteriors. MXE can then be applied to select a member of $K(i)$ that is

---

* Dept. of Computer Science, UCLA.
Research supported in part by National Science Foundation grant IST 84-05161.

G. J. Erickson and C. R. Smith (eds.),
Maximum-Entropy and Bayesian Methods in Science and Engineering (Vol. 1), 305–312.
© *1988 by Kluwer Academic Publishers.*

closest to the prior in the sense of cross-entropy. The derived estimate has all the favorable properties associated with MXE -- e.g., it is better than the prior as an estimate of the unknown posterior. [Shore & Johnson, 1981].

This context is quite different from one commonly employed in which the constraints on the posterior are furnished by incomplete statistics, e.g., knowing only the mean and standard deviation of some random variable for a set of observations. There are a number of conceptual difficulties with observational constraints that obscure the role of an inference procedure. This point is tangential to the present paper, so I will mention only one consideration. Using observational constraints involves ignoring the stochastic nature of the observations, i.e., the fact that a different sample might produce different statistics. [Dalkey, 1986]. In the context under consideration, it is assumed that the constraints on the likelihoods are furnished by a source other than direct observation; e.g., they are derived from physical laws or from the parameters of a model under investigation.

Within this context, if MXE is considered to be a generally applicable rule, then, in effect, it is a procedure for designing an information system. Before the fact, i.e., before any observations have been made, MXE prescribes an estimated posterior for each potential observation. Thus, the appropriate figure of merit for the rule is not how it performs given a specific observation, but how it performs over the entirety of the prescribed information system -- in short, its average performance over the ensemble of potential observations.

Information Systems

An information system is a joint probability distribution $P(E.I)$ on a set of events E (hypotheses) and a set of potential observations I, where the evaluation of the system is formulated in terms of the posterior distributions $P(E|I)$. An appropriate basis for the evaluation of an information system is the family of proper scoring rules. Let $S(P,e)$ be a reward function (payoff, rating, score, etc.) where P is a probability distribution and e is an instance of the event set E. $S(P,e)$ is called proper if it fulfills the condition:

$$\sum_E P(e) S(P,e) \geq \sum_E P(e) S(Q,e) \tag{1}$$

i.e., the expectation of the score is a maximum when the probability determining the expectation is the same as the probability determining the score. The logarithmic score, $S(P,e) = \log P(e)$, is widely used in communication theory; its expectation, $\sum P(e) \log P(e)$ is the negative of the entropy of P. However, there is an infinite family of functions which fulfill (1). Among these are decisional scores derived from the payoff function of an enterprise.

Abbreviate $\sum_E P(e) S(P,e)$ by $G(P)$ and $\sum_E P(e) S(Q,e)$ by $G(P,Q)$. $N(P,Q) - (G(P,Q)$ is the net score if P is the actual distribution and Q is the distribution that is implemented. From (1), $N(P,Q) \geq 0$. $N(P,Q)$ for the logarithmic score is the cross-entropy of P and Q.

The value of an information system is measured by

$$H(P) = \sum_I P(i) \, G(P(E|i))$$

where $P(i) = \sum_E P(e.i)$ is the initial probability of the observation i. $H(P)$ is the average expected score of the posterior distributions. A relative score $H(P, Q)$ can be defined for the case where an estimate $Q(E|i)$ of the posterior is posited.

$$H(P, Q) = \sum_I P(i) \, G(P(E|i), Q(E|i))$$

The net score $M(P, Q) = H(P) - H(P, Q)$ measures the loss if P is the actual information system, but Q is implemented. For the logarithmic score, $M(P, Q)$ is the generalized cross-entropy of information systems P and Q.

From (1),

$$H(P, Q) \leq H(P) \tag{2}$$

since the inequality holds for every i. Thus, $H(P)$ is a proper scoring rule for information systems. Let $P°(E) = \sum_I P(E.i)$ designate the prior probability distribution on E. It is straightforward to prove $H(P) \geq H(P°)$, where $P°$ is the degenerate information system that posits the prior as the posterior for every observation. [Dalkey, 1985] This is the positive value of information principle (PVI) -- overlooking costs, additional information is always advantageous (or at least, never harmful.)

Interpreting MXE as a design for an information system, the first salient consideration is that, in general, it will not produce a consistent information system. The posterior $P'(E|i)$ will not reproduce the prior via the summation rule $P°(e) = \sum P'(e|i) P(i)$. The reason is just that MXE selects a posterior separately for each observation i, ignoring the consistency condition.

As a result, it is possible to formulate problems in which the MXE designed information system will perform more poorly than a rule which simply ignores the information in the observations and chooses the prior distribution as the estimate of the posterior.

An elementary illustration of these two points can be constructed using a classic two-urn problem with balls of three different colors. Assume that the prior probability p of urn e is known, and the proportions r, s, t of balls in urn $\bar{e}$ are fully known, whereas for urn e the proportion w of one color is known, the other proportions u, v, fulfilling the condition $u + v = 1 - w$,

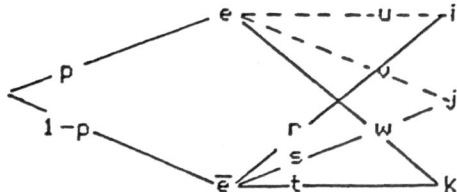

As a specific case, let $P = 3/8, r = .4, s = t = .3, w = .5$. The posterior $P(e \mid k)$ is fixed, and for the given inputs $= .5$. Considering the observations i and j separately $P'(e \mid i) = P'(e \mid j) = 3/8$ is allowed by the constraints, and thus these are the MXE estimates.

There is no average of $P'(e \mid i), P'(e \mid j), P'(e \mid k)$ that $= 3/8$ other than $P'(k) = 0$, but this is ruled out by the constraints, thus the MXE estimates are inconsistent. For the general case, $H(P')$ cannot be computed, since P' is inconsistent; however, for the present example, since $P'(e \mid i) = P'(e \mid j)$, and the initial probability for the disjunction $P'(i \lor j)$ is fixed, $H(P')$ can be computed, and for the logarithmic score is -.67341. The score of the prior $H(P^\circ) = -.66156$. The prior is more informative than the information system prescribed by MXE.

This result appears to be contrary to the Shore and Johnson finding alluded to in the second section that the MXE conclusion is a better estimate of the unknown posterior than the prior. Since this is true for each i, then it should also be true for an average overall i. The set of posteriors from which P' is selected is wider than the set of posteriors allowed by the constraints; but then, the better guess property should hold a-fortiori for the constrained set of posteriors. In a sense it does, but a precise analysis requires some results developed in the next sections (see Theorem 3 below).

For the present example, it is possible to compute the guaranteed expected score for P', i.e., $min\, H(P, P')$, where P ranges over the set of consistent information systems allowed by the constraints. This turns out to be -.64946, which is greater than $H(P')$ and also greater than $H(P^\circ)$. Colloquially, for this example, the MXE designed information system guarantees more than it promises.

Minimally Informative Information Systems

A more consistent way to deal with the case of incompletely known likelihoods is to define a class K of information systems, each of which is compatible with the given constraints; i.e., each P in K is a consistent information system. If the system P^* in K is selected, where $H(P^*) = \min_K H(P)$, it can be shown that P^* has two properties: (a) $H(P^*)$ is guaranteed, in the sense that whatever the actual system P, $H(P, P^*) \geq H(P^*)$, and (b) selecting P^* is the only rule that fulfills the PVI condition, $K \subset K'$ and $K \neq 0$ implies $H(P^*(K)) \geq H(P'(K'))$. The first property was originally derived using game-theoretic methods [Dalkey, 1985]; however, it can be more directly obtained using some results stated in the next section.

These two properties furnish a justification for using as an inference rule: given a set K of information systems, select P^* as an estimate. We can call P^* the minimally informative information system.

For the two-urn example, $P^*(e \mid i) = P^*(e \mid j) = .3$. $H(P^*) = -.64172$. Thus, for this example, we have

$$H(P^*) > H(P^\circ) > H(P')$$

A rough intuitive explanation of this result is that the constraints force the posteriors of a consistent information system to be more informative than the less constrained MXE estimates.

A Deeper Look

Additional insight into the role of MXE can be obtained by examining the consequences of applying MXE not to probability distributions, but to information systems. Thus, given a set K of information systems, with fixed prior, we can investigate the rule: select the information system P' in K which is closest to the prior system P° in the sense of $M(P, Q)$, i.e., select the P' in K with minimal $M(P, P^\circ)$. As noted earlier, $M(P, Q)$ for the log score is the extension to information systems of the notion of cross-entropy for probability distributions.

A basic result for net scores on information systems is the super-additive property.

Lemma 1. If $M(P', R) \geq M(Q, R)$, $P' + aP + (1 - a)Q, 0 \leq a \leq 1$ then $M(P, R) \geq M(P, Q) + M(Q, R)$.

> _Proof:_ By hypothesis, we have $H(P') - H(P', Q) \geq M(Q, R)$. Substituting $aP + (1 - a)Q$ for P' gives $aH(P, P') + (1 - a) H(Q, P') - aH(P, R) - (1 - a) H(Q, R) \geq M(Q, R)$. From (2) $(1 - a) H(Q) \geq (1 - a) H(Q, R)$. Substituting $(1 - a)$ H (Q)for$(1 - a) H(Q, R)$ and adding and subtracting $aH(P)$ and collecting terms gives $aM(P, R) \geq aM(P, P') + aM(Q, R)$. Thus, $M(P, R) \geq M(P, P') + M(Q, R)$. If $M(P, P')$ is continuous at $P' = Q$ -- i.e., at $a = 0$ -- then the result follows. If $M(P, P')$ is not continuous at $a = 0$, then the result holds for P' arbitrarily close to Q.

Theorem 1. If K is a convex and closed set of information systems, R is any information system, and Q is the system in K with minimal net score for R, i.e., $M(Q,R) \leq M(P,R)$, all P in K, then $M(P,Q) \leq M(P,R)$ all P in K.

Proof: The hypothesis of Theorem 1 fulfills the conditions for Lemma 1, hence $M(P,R) \geq M(Q,R) + M(P,Q)$. Thus, the result follows a fortiori.

Theorem 1 extends to information systems, and to all proper score rules, the "better guess" property demonstrated by Shore and Johnson for the log score and probability distributions.

Lemma 2. $H(P,P^\circ) = H(P^\circ)$

Proof: $H(P,P^\circ) = \sum_{E,I} P(i.e) S(P^\circ, e)$. But $S(P^\circ, e)$ is independent of i. Summing on I gives

$$H(P,P^\circ) = \sum_E P^\circ(e) S(P^\circ, e) = H(P^\circ).$$

Theorem 2. With the hypothesis of Theorem 1, and $R = P^\circ, H(P,Q) \geq H(Q)$.

Proof: From Theorem 1, $M(P,R) \geq M(P,Q) + M(Q,R)$. From Lemma 2, $H(P) - H(P^\circ) \geq H(P) - H(P,Q) + H(Q) - H(P^\circ)$. Cancelling the $H(P)$'s and $H(P^\circ)$'s gives the result.

Theorem 2 asserts the guaranteed score property of MXE. Whatever the unknown actual probability P, the relative score $H(P,Q)$ is at least as great as the "promised" score $H(Q)$. This property holds for any score rule.

From the definition of $Q, M(P,P^\circ) \geq M(Q,P^\circ)$. From Lemma 2, $H(P) - H(P^\circ) \geq H(Q) - H(P^\circ)$. Thus $H(P) \geq H(Q)$. Hence, MXE applied to information systems rather than probability distributions, is equivalent to the minimally informative information system inference.

The formalism now at hand allows a more careful analysis of the "better guess" property for MXE as originally conceived. Let $K(i)$ designate the set of posteriors $P(E \mid i)$ determined by ignoring the joint constraints on other members of I. The better guess property asserts that $G(P(E \mid i), P'(E \mid i)) \geq G(P(E \mid i), P^\circ)$ for all $P(E \mid i)$ in $K(i)$, where $P'(E \mid i)$ minimizes $N(P(E \mid i), P^\circ)$. [Dalkey, 1983]. As above, let K designate the set of (consistent) information systems compatible with the given constraints. Let P' designate the set of MXE estimates for all i's, i.e., the posterior matrix $P'(E \mid I)$.

Theorem 3. $H(P,P') \geq H(P^\circ)$ for all P in K.

Proof: By assumption, $P(E \mid i)$ is in $K(i)$, and thus
$\overline{G(P(E \mid i), P'(E \mid i))} \geq G(P(E \mid i), P^{\circ})$. Hence,

$$\sum_I P(i) \, G(P(E \mid i)) = H(P, P') \geq \sum_I P(i) \, G(P(E \mid i), P^{\circ}) = H(P, P^{\circ}),$$

which from Lemma 2 $= H(P^{\circ})$.

Thus, the better guess property can be extended to a form of PVI; whatever the unknown P, the expectation of P' is at least as good as that for P. This is a much stronger result than the original better guess property, since there is no guarantee that $G(P(E \mid i), P^{\circ}) \geq G(P^{\circ})$. The stronger result arises only from considering P' in the context of a complete information systems.

We can now formulate a precise comparison between the min-score conclusion P^* and the MXE conclusion P'. Since, as noted earlier, $H(P')$ is in general undefined because an MXE information system is inconsistent, a reasonable value for P' can be defined as the highest relative value it can guarantee, i.e., $\min_K H(P, P')$.

Theorem 4. $\quad \min_K H(P, P') \leq H(P^*)$

Proof: $\quad \min_K H(P, P') \leq H(P^*, P') \leq H(P^*)$.   The second inequality follows from (2).

Discussion

Informative systems are a much more powerful representation of knowledge than probability distributions. One major illustration of this point is the fact that information systems can exhibit dominance; i.e., one information system can guarantee a higher score than another for all proper scoring rules. A basic example is PVI, $H(P) \geq H(P^{\circ})$ for all score rules. In contrast, given any pair of probability distributions $P, Q, P \neq Q$, there is a score rule which gives $G(P) > G(Q)$ and another which gives $G(Q) > G(P)$.

Another advantage of taking information systems as the basic representation of knowledge is the fact that the min-score rule does not restrict the kind of uncertainty that can be treated. In practice, prior probabilities can be incompletely known as well as likelihoods. Thus K can represent the set of possible information systems allowing incomplete knowledge of both priors and likelihoods. The two basic properties of guaranteed score and PVI still hold for the min-score P in K. [Dalkey, 1986]

Thus, formulating inductive inference rules in terms of information systems, rather than probability distributions, is likely to generate stronger procedures.

On the other hand, from a practical point of view, computing a minimally informative information system is distinctly more difficult than computing a minimal cross-entropy probability distribution. As demonstrated in Theorem 3, MXE does generate an information system which guarantees PVI. One potentially fruitful research topic would be the identification of types of problems in which MXE, in its original form, is an acceptable approximation to its extension to information systems.

References

Dalkey, N. C. (1983). Updating Inductive Inference, presented at the third Workshop on Maximum Entropy and Bayesian Methods in Applied Statistics, University of Wyoming, August 1-4.

Dalkey, N. C. (1985). Inductive Inference and the Maximum Entropy Principle. In Maximum Entropy and Bayesian Methods in Inverse Problems, eds. C. Ray Smith and W. T. Grandy, pp. 351-64. Dordrect/Boston/Lancaster: D. Reidel.

Dalkey, N. C. (1986). Prior Probabilities Revisited. In Maximum Entropy and Bayesian Methods in Applied Statistics, ed. J. H. Justice, pp. 117-30. Cambridge: Cambridge University Press.

Shore, J. E. & Johnson, R. W. (1981) Properties of Cross-Entropy Minimization. IEEE Transactions on Information Theory, IT-27, 472-82.